Combinatorial Algorithms on Words

NATO ASI Series

Advanced Science Institutes Series

A series presenting the results of activities sponsored by the NATO Science Committee, which aims at the dissemination of advanced scientific and technological knowledge, with a view to strengthening links between scientific communities.

The Series is published by an international board of publishers in conjunction with the NATO Scientific Affairs Division

A Life Sciences **B Physics**	Plenum Publishing Corporation London and New York
C Mathematical and Physical Sciences	D. Reidel Publishing Company Dordrecht, Boston and Lancaster
D Behavioural and Social Sciences **E Applied Sciences**	Martinus Nijhoff Publishers Boston, The Hague, Dordrecht and Lancaster
F Computer and Systems Sciences **G Ecological Sciences**	Springer-Verlag Berlin Heidelberg New York Tokyo

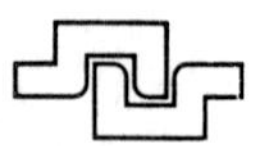

Series F: Computer and Systems Sciences Vol. 12

Combinatorial Algorithms on Words

Edited by

Alberto Apostolico

Department of Computer Sciences, Mathematical Sciences Building
West Lafayette, IN 47906/USA

Zvi Galil

Department of Computer Science, Columbia University
New York, NY 10027/USA

Springer-Verlag Berlin Heidelberg New York Tokyo

Published in cooperation with NATO Scientific Affairs Division

Proceedings of the NATO Advanced Research Workshop on Combinatorial Algorithms on Words held at Maratea, Italy, June 18–22, 1984

ISBN 3-540-15227-X Springer-Verlag Berlin Heidelberg New York Tokyo
ISBN 0-387-15227-X Springer-Verlag New York Heidelberg Berlin Tokyo

Library of Congress Cataloging in Publication Data.
NATO Advanced Research Workshop on Combinatorial Algorithms on Words (1984: Maratea, Italy) Combinatorial algorithms on words. (NATO ASI series. Series F, Computer and system sciences; vol. 12) "Proceedings of the NATO Advanced Research Workshop on Combinatorial Algorithms on Words held at Maratea, Italy, June 18–22, 1984"—T.p. verso. 1. Combinatorial analysis—Congresses. 2. Algorithms—Congresses. 2. Word problems (Mathematics)—Congresses. I. Apostolico, Alberto, 1948-. II. Galil, Zvi. III. Title. IV. Series: NATO ASI series. Series F, Computer and system sciences; no. 12. QA164.N35 1984 511'.6 85-8023
ISBN 0-387-15227-X (U.S.)

Printed in Germany

Printing: Beltz Offsetdruck, Hemsbach; Bookbinding: J. Schäffer OHG, Grünstadt
2145/3140-543210

Contents

PREFACE

Combinatorial Algorithms on Words refers to the collection of manipulations of strings of symbols (words) - not necessarily from a finite alphabet - that exploit the combinatorial properties of the logical/physical input arrangement to achieve efficient computational performances. The model of computation may be any of the established serial paradigms (e.g. RAM's, Turing Machines), or one of the emerging parallel models (e.g. PRAM , WRAM, Systolic Arrays, CCC).

This book focuses on some of the accomplishments of recent years in such disparate areas as pattern matching, data compression, free groups, coding theory, parallel and VLSI computation, and symbolic dynamics; these share a common flavor, yet have not been examined together in the past. In addition to being theoretically interesting, these studies have had significant applications. It happens that these works have all too frequently been carried out in isolation, with contributions addressing similar issues scattered throughout a rather diverse body of literature. We felt that it would be advantageous to both current and future researchers to collect this work in a single reference.

It should be clear that the book's emphasis is on aspects of combinatorics and complexity rather than logic, foundations, and decidability. In view of the large body of research and the degree of unity already achieved by studies in the theory of automata and formal languages, we have allocated very little space to them.

The material was divided, perhaps somewhat arbitrarily, into six sections. We encouraged several prominent scholars to provide overviews of their specific subfields. With its sizeable bibliography, the book seems well suited to serve as a reference text for a graduate course or seminar. Although there are no exercise sections, many open problems are proposed. Some of these may well serve as topics for term projects - a few may even blossom into theses.

Most of the papers contained in this volume originated as lectures delivered at the Workshop on Combinatorial Algorithms on Words, which was held in Maratea, Italy during the week of June 18-22, 1984. This workshop brought together researchers who had been active in the area so that a unified core of knowledge could be identified, and recommendations for future work could be outlined. The workshop was deliberately kept small and informal, and provided a congenial environment for lively discussions and valuable presentations.

The Maratea Workshop was sponsored by NATO under the Scientific Affair Division ARW Program, and it benefitted from the joint sponsorship of the University of Salerno and IBM Italia. Renato Capocelli, Mena De Santis, Dominique Perrin and Joel Seiferas joined us on the Program Committee for the Workshop; we thank them for their invaluable help. Mena De Santis did an excellent job of looking after the countless details of local arrangements. Finally, we would like to express our sincere gratitude to all the participants in the workshop.

New York City, December 1984 *A. Apostolico and Z. Galil*

Open Problems in Stringology

Zvi Galil*

Department of Computer Science
Columbia University
and
Tel-Aviv University

Abstract: Several open problems concerning combinatorial algorithms on strings are described.

0. Introduction

Every problem in theoretical computer science can be stated as a problem on strings (e.g. P = NP?). In this paper we restrict attention to combinatorial algorithms on strings. We list several open problems. We divide them into four groups: string matching, generalizations of string matching, index construction and miscellaneous. This list is far from being exhaustive.

In this paper Σ is a finite alphabet and all strings are in Σ^*. Sometimes we add special symbols \$ and # not in Σ. For a string x, x_i is the i-th symbol of x, $|x|$ is the length of x and x^R is the string x reversed. We say that x occurs in y if $x = y_{i+1} \cdots y_{i+|x|}$ for some i.

The string matching problem is the following: given two strings, the pattern and the text, find all the occurrences of the pattern in the text. About half of the open problems in this paper are about the string matching problem or its generalizations.

Our model of computation is the random access machine (RAM) with uniform cost (see [2]). Each register will typically store a symbol of Σ or an address.

1. Questions about String Matching

String matching is one of the most extensively studied problems in theoretical computer science. We briefly sketch the history of the problem. The problem was solved in linear time by the Knuth Morris and Pratt algorithm (KMP in short) [19], and then several versions of the problem were solved in real time, even by a Turing machine [12]. Another linear-time algorithm [6], (see also [11]) was designed that is sublinear in the average, assuming the text is already stored in memory [28]. Then attention was given to saving space while maintaining the optimality of the time complexity. This resulted in a linear-time (real-time) algorithm on a RAM

*This work was supported in part by National Foundation Grant MCS-8303139.

NATO ASI Series, Vol. F12
Combinatorial Algorithms on Words
Edited by A. Apostolico and Z. Galil

which uses only five (six) registers which store pointers to the input [15]. A simple **probabilistic** linear-time, constant-space algorithm was also designed [18].

A study of more theoretical nature followed [15]. String matching can be done by a Turing machine in linear time (or even real time) and logarithmic space. Moreover, a six head (eight head) two-way deterministic finite automaton (dfa) can perform string matching in linear time (real time). This study tries to identify the weakest computation model that can do string matching in optimal time and space.

Question 1: Can a one-way multihead dfa perform string matching?

More specifically, can such an automaton recognize the language $\{x\$uxv | u, x, v \in \Sigma^*\}$. Obviously a one head dfa cannot do the job. It was shown in [21] that a two-head dfa cannot do it either. (For the latter, there is a short alternative proof using Kolmogorov complexity.) It has been further claimed in [20] that a three-head one-way dfa cannot do string matching. Note that without loss of generality a one-way multihead dfa always halts, and it must do so in linear time. We believe that in fact a one-way multihead dfa **cannot** perform string matching.

Question 2: The number of states in a "Boyer-Moore dfa".

The original Boyer-Moore algorithm [6], BM in short, requires quadratic time in the worst case. The reason is that the BM "forgets" the part of the text it has seen when it slides the pattern after a mismatch. Consequently, many comparisons made by the BM are redundant, since the outcomes of these comparisons have already been established. Knuth [19] suggested using a dfa that will "remember" those parts of the pattern that need not be compared. The question is to find the exact number, or alternatively close upper and lower bounds on the number of states such a dfa must have.

An obvious upper bound is $2^{|x|}$. In [19], it is explained why a pattern x consisting of $|x|$ distinct symbols requires $\Omega(|x|^2)$ states. An $\Omega(|x|^3)$ lower bound for a pattern over a three letter alphabet is known [16]. The challenge is to narrow the still large gap.

This question is of theoretical interest only. If we use only **two** states and do not remember everything, linear time suffices for the BM [11]; if $|y| - |x|$ states are used, then the resulting algorithm requires at most $2|y| - |x|$ comparisons [3]. Recall that the KMP requires a similar number of comparisons, but with a smaller overhead.

Question 3: Parallel algorithms.

Recently, optimal parallel algorithms (those with $pt = O(n)$, where p = number of processors, t = time) were developed for string matching for a wide range of values of p: on the WRAM for $p \leq n/\log n$ and on the PRAM (for $p \leq n/\log^2 n$), first for fixed size alphabet [13], then for any alphabet [26]. (Recall that a WRAM [PRAM] is a collection of RAM's that share a common memory and are allowed simultaneous reads and writes [only simultaneous reads]. In the case of WRAM's, there is in addition a mechanism for resolving write conflicts.) Is it possible to design optimal parallel algorithms with a larger number of processors (e.g. $p \leq n$ on the WRAM, $p \leq n/\log n$ on the PRAM)? The algorithm in [26] needs concurrent writes only in preprocessing the pattern x but not during the search. Hence, the question for

PRAM is answered affirmatively if preprocessing is not included. Another question is whether it is possible to design optimal parallel algorithms on the more realistic model of a network of processors of constant degree.

2. Generalizations of String Matching

Question 4: A membership test for regular expressions.

Given a regular expression α over the alphabet Σ with operators $\bigcup$, $\cdot$, $*$ (union, concatenation and Kleene star) and a string x, test whether $x \in L(\alpha)$ (the language described by α).

The obvious algorithm converts α to a nondeterministic finite automaton (nfa) A in linear time; then inductively finds the set of states that A can be in after reading x_i for $i = 1, ..., |x|$. A similar algorithm works directly on α by inductively finding all places in α we can be in after reading x_i. (This construction is known as the dot construction.) The time bound of this algorithm is $O(|x||\alpha|)$. The open problem is whether this time bound can be improved. If $|\alpha|$ is fixed or very small the answer is positive. A time bound of $O(|x| + 2^{|\alpha|})$ can easily be obtained by converting A to a dfa first.

For two important special cases the problem can be solved in linear time ($O(|x| + |\alpha|)$).

(1) $\alpha = \Sigma^* u \Sigma^*$, where $u \in \Sigma^*$ and

(2) $\alpha = \Sigma^*(u^1 \bigcup u^2 \bigcup ... \bigcup u^k)\Sigma^*$, where $u^i \in \Sigma^*$.

These are the single pattern and multi-pattern string matching problems, respectively; in the first we ask whether u occurs in x and in the second whether one of the u^i's occurs in x. These linear-time algorithms [19,1] might seem encouraging. However, these special cases are solved efficiently because in both cases α can be converted in linear time to a dfa (of linear size) which is impossible in the general case.

Question 5: String matching with don't-cares.

We add to the alphabet Σ a new "don't-care" symbol ϕ which matches any single symbol. Given two strings x, y over $(\Sigma \bigcup \{\phi\})^*$ we want to find all occurrences of x in y. There is an occurrence of x at position $i + 1$ of y if, for all $j = 1, ..., |x|$ whenever x_j and y_{i+j} are not ϕ, they are equal. Observe that all known string matching algorithms simply do not work if we allow "mistakes" in the form of don't-cares or as in the next problem.

In [9] the problem was reduced to that of integer multiplication. If an algorithm multiplies two binary numbers of n and m bits in time $T(n, m)$, then the time bound for our problem is $O(T(|x|, |y|) \log |x| \log |\Sigma|)$. Using the multiplication algorithm that is currently asymptotically best, we obtain a bound of $O(|y| \log^2 |x| \log \log |x| \log |\Sigma|)$.

The open problem is whether we can do better. Perhaps we can do better in the case that the don't-care symbols do not appear in the text (y). This case is considered in [24]. Of course, in case of a small alphabet and a small (constant) number of don't-cares we can solve all the corresponding exact string matching problems. Note that the related problem of string matching with don't-care symbols in the pattern that can match an arbitrary string is easy: just treat each maximal pattern fragment separately.

Question 6: String matching with mistakes.

Given a pattern x and a text y and an integer k, find all occurrences of strings of length $|x|$ with distance at most k from x, where the distance between two equal size strings is defined as the number of positions in which they have different symbols.

Even for a small constant k we do not know how to solve the problem in time smaller that $O(|x||y|)$, which is an upper bound on the naive algorithm that computes the distance of x and each substring of y of length $|x|$.

The only known improvement seems to be in the case $k = 1$ [26]. A linear-time algorithm follows from a linear-time algorithm that finds for each position in the text the occurrence of the largest prefix of the pattern [22]. The latter is a modification of the KMP.

3. Index Construction

In efficient string matching we preprocess the pattern (or patterns) in linear time so we can search for it (or for them) in time linear in the length of the text. Another approach is to preprocess the text, to construct some kind of an index, and then use it to search for some pattern x (or answer some queries on x like finding the number of occurrences of x) in time linear in $|x|$. A number of linear-time algorithms are known, but all of them are closely related [7]. The first of these is the one discovered by Weiner [27].

Question 7: Can we construct an index in (almost) real time?

More specifically, can we read the text followed by one (or possibly more) pattern(s) one symbol at a time, spend a constant amount of time on each symbol, and identify occurrences immediately. This question can be rephrased as follows: Can we accept the language $L = \{uxv\$x\$|u, x, v \in \Sigma^*\}$ in real time. The word "almost" in Question 7 refers to the fact that we might still be constructing the index when we are already reading x.

One of the earliest results in computational complexity [17] implies that L cannot be accepted in real time by a multitape Turing machine, but Question 7 refers to RAM. It was shown in [17] that $L_1 = \{w_1\#w_2\#...\#w_k\$w^R|w_j \in \{0,1\}^*, k \geq 0, w = w_i$ for some $i\}$ cannot be accepted by a multitape Turing machine in real time, and the same proof implies that neither can $L_2 = \{w_1\#w_2\#...w_k\$w|w_j \in \{0,1\}^*, k \geq 0, w = w_i$ for some $i\}$. L_2 is a subset of L, and this is why the same is true for L. But L_1 and L_2 can be accepted easily in real time by a RAM [12]. Hence the results on Turing machines only show that the model is too weak.

A positive answer to Question 7 is claimed by Slisenko in [25] and in several of its earlier versions. In fact, the seventy-page solution claims to have characterized all periodicities of the text in real time. Unfortunately, so far I have not been able to understand the solution. So, the question may be interpreted as follows: Can we find a reasonably simple real-time construction?

Question 8: Parallel Algorithms.

There are three interpretations to this question:
(1) a parallel construction of indices for sequential searches;
(2) a parallel construction of indices for parallel searches; or even
(3) a sequential construction of indices for parallel searches.
In all cases we would like to design efficient parallel algorithms. In the second and third cases we first have to find a good way to define a specific index.

Question 9: Dependence on the alphabet size.

Most of the algorithms for string matching (with a single pattern) do not depend on the size of the alphabet Σ. In fact Σ can be infinite. The only assumption needed is that two symbols can be compared in one unit of time.

All the algorithms for index construction do depend on the size of Σ. Every index is either a tree or a dfa with up to $|\Sigma|$ successors for every node. If we use an array of size $|\Sigma|$ for every node, then the index construction takes time (and space) $O(|y||\Sigma|)$, while the search takes time $O(|x|)$. If we use lists of successors, then the former takes time $O(|y||\Sigma|)$ (but space $O(|y|)$) and the latter takes time $O(|x||\Sigma|)$. If we use a search tree for the set of successors of each node, the former takes time $O(|y| \log |\Sigma|)$ and the latter $O(|x| \log |\Sigma|)$. Alternatively, we can use hashing. The problem is to determine the exact dependence on the alphabet size.

A similar problem is to determine the exact dependence on the alphabet size in the problem of multi-pattern string matching.

Question 10: Generalized Indices.

Can we efficiently construct indices that will (efficiently) support harder queries. An example of such a query is finding the maximal number of **nonoverlapping** occurrences of a given string. Counting all the occurrences is a simple application of the known indices. If we insist on nonoverlapping occurrences, the best algorithm [5] is neither simple nor linear time.

4. Miscellaneous Problems.

Question 11: Testing unique decipherability.

Given a set Γ of n strings $c_1, ..., c_n$ over Σ of total length L, is there a string in Γ^* that can be parsed in two different ways, or is $\bigcup_{i \neq j} (c_i\Gamma^* \bigcap c_j\Gamma^*) = \Phi$?

There are several algorithms for solving the problem (see for example [4]). They are essentially the same, and they solve the problem in time $O(nL)$ by constructing a certain graph in a search for a "counter example". The time bound follows immediately from the fact that the graph may have up to L vertices of degree up to n. This time bound is quadratic in the worst case and is linear only for $n =$ constant. Can we do better?

Question 12: Solving string problems with two-way deterministic pushdown automata.

Two-way deterministic pushdown automata (2dpda's) have been closely related to string matching. The linearity of string matching (and palindrome recognition) can be easily established by first showing that a 2dpda accepts the corresponding language, since Cook showed that membership for a 2dpda language can be determined in linear time by a RAM [8] (see also [10]). There are some linear-time recognizable string languages that we do not know how to recognize with a 2dpda. Two such examples are the following.

$$\text{PREFIXSQUARE} = \{wwu|w, u \in \Sigma^*\}, \text{and PALSTAR} = (\text{PAL})^*,$$

where PAL $= \{w|w \in \Sigma^*, w = w^R, |w| > 1\}$ is the language of nontrivial palindromes. (The related language $\{ww^R u|w, u \in \Sigma^*\}$ can be recognized by a 2dpda.) For a linear-time recognition of PALSTAR and other similar open problems, see [14].

Question 13: String problems on DNA.

There are many interesting problems that arise in the study of DNA sequences. Here we mention one such problem. We assume that for some of the pairs of symbols (in $\Sigma \times \Sigma$) there is an associated positive real number. The meaning of this number is that if we fold a string so that these two symbols touch the associated number represents the amount of energy that is released. Given a string, we want to find an optimal way to fold it (a way that maximizes the energy released). The problem can be stated as finding a planar matching (graph matching, not string matching) of maximal weight. The problem can be solved easily in time $O(n^3)$ using dynamic programming [23]. Can we do better?

Acknowledgement: Alberto Apostolico, Stuart Haber and Joel Seiferas read an earlier version of the paper and gave me many helpful suggestions.

5. References

1. A.V. Aho and M.J. Corasick, Efficient string matching: An aid to bibliographic search, *Communications of the ACM 18* (1975), 333-340.

2. A. Aho, J. Hopcroft and J. Ullman, *The Design and Analysis of Computer Algorithms*, Addison-Wesley, Reading, MA 1974.

3. A. Apostolico and R. Giancarlo, The Boyer-Moore-Galil string searching strategies revisited, *SIAM J. on Computing*, to appear.

4. A. Apostolico and R. Giancarlo, Pattern matching machine implementation of a fast test for unique decipherability, *Information Processing Letters 18* (1984), pp. 155-158.

5. A. Apostolico and F. Preparata, A structure for the statistics of all substrings in a text string with or without overlap, *Proc. 2nd World Conf. on Math. at the Service of Man*, 1982, pp. 104-109.

6. R.S. Boyer and J.S. Moore, A fast string searching algorithm, *Communications of the ACM 20* (1977), 762-772.

7. M.T. Chen and J.I. Seiferas, Efficient and elegant subword tree construction, *Combinatorial Algorithms on Words*, A. Apostolico and Z. Galil eds., Springer Verlag Lecture Notes, 1985.

8. S. Cook, Linear time simulation of deterministic two-way pushdown automata, *Proc. IFIP Congress* (1971), pp. 172-179.

9. M.J. Fischer and M.S. Paterson, String-matching and other products, in: *Complexity of Computation* (*SIAM-AMS Proceedings 7*), R.M. Karp ed., American Mathematical Society, Providence, RI, 1974, pp. 113-125.

10. Z. Galil, Two fast simulations which imply some fast string matching and palindrome recognition algorithms, *Information Processing Letters 4* (1976), pp. 85-87.

11. Z. Galil, On improving the worst case running time of the Boyer-Moore string matching algorithm, *Communications of the ACM 22* (1979), 505-508.

12. Z. Galil, String matching in real time, *J. ACM 28* (1981), pp. 134-149.

13. Z. Galil, Optimal parallel algorithms for string matching, *Proc. 16th ACM Symposium on Theory of Computing*, 1984, pp. 240-248.

14. Z. Galil and J.I. Seiferas, A linear-time on-line recognition algorithm for "Palstar", *J. ACM 25* (1978), pp. 102-111.

15. Z. Galil and J.I. Seiferas, Time-space-optimal string matching, *J. Computer and System Sciences 26* (1983), pp. 280-294.

16. L. Guibas and A. Odlyzko, private communication.

17. J. Hartmanis and R.E. Stearns, On the computational complexity of algorithms, *Transactions of the American Mathematical Society 117* (1965), 285-306.

18. R.M. Karp and M.O. Rabin, Efficient randomized pattern-matching algorithms, manuscript.

19. D.E. Knuth, J.H. Morris, Jr., and V.R. Pratt, Fast pattern matching in strings, *SIAM Journal on Computing 6* (1977), 323-350.

20. M. Li, Lower bound on string-matching, Technical Report TR-84-636, Department of Computer Science, Cornell University, 1984.

21. M. Li and Y. Yesha, String matching cannot be done by a two-head one-way deterministic finite automaton, Technical Report TR-83-579, Cornell University, 1983.

22. M.G. Main and R.J. Lorentz, An $O(n \log n)$ algorithm for finding all repetitions in a string, *J. of Algorithms* (1984), pp. 422-432.

23. R. Nussinov, G. Pieczenik, J.R. Griggs, and D.J. Kleitman, Algorithms for loop matchings, *SIAM J. of Applied Math 35* (1978), pp. 68-82.

24. R. Pinter, Efficient string matching with don't-care patterns, *Combinatorial Algorithms on Words*, A. Apostolico and Z. Galil eds., Springer Verlag Lecture Notes, 1985.

25. A.O. Slisenko, Detection of periodicities and string-matching in real time, *J. of Soviet Mathematics 22*, Plenum Publishing Co. (1983), pp. 1316-1386.

26. U. Vishkin, Private communication.

27. P. Weiner, Linear pattern matching algorithms, *Proc. 14th IEEE Annual Symposium on Switching and Automata Theory*, 1973, pp. 1-11.

28. A.C.C. Yao, The complexity of pattern matching for a random string, *SIAM J. on Comput. 8* (1979), pp. 368-387.

1 - STRING MATCHING

EFFICIENT STRING MATCHING WITH DON'T-CARE PATTERNS*

Ron Y. Pinter
IBM Israel Scientific Center
Technion City
Haifa 32000, ISRAEL

ABSTRACT

The occurrences of a constant pattern in a given text string can be found in linear time using the famous algorithm of Knuth, Morris, and Pratt [KMP]. Aho and Corasick [AC] independently solved the problem for patterns consisting of a set of strings, where the occurrence of one member is considered a match. Both algorithms preprocess the pattern so that the text can be searched efficiently. This paper considers the extension of their methods to deal with patterns involving more expressive descriptions, such as don't-care (wild-card) symbols, complements, etc. Such extensions are useful in the context of clever text-editors and the analysis of chemical compounds.

The main result of this paper is an algorithm to deal efficiently with patterns containing a definite number of don't-care symbols. Our method is to collect "evidence" about the occurrences of the constant parts of the pattern in the text, using the algorithm of Aho and Corasick [AC]. We arrange the consequences of the intermediate results of the search in an array of small counters whose length is equal to that of the pattern. As soon as a match for the whole pattern is found, it is reported. If we assume that the counters can be incremented in parallel, the overall (time and space) complexity of the algorithm remains linear. Otherwise, the worst-case time complexity becomes quadratic, without changing the space requirements.

We include here a discussion of why alternative ways to solve the problem, especially those trying to preserve the purely automaton-driven constructions of [KMP] and [AC], do not work. Finally, we describe a minor extension to an algorithm of Fischer and Paterson [FP]. Originally, it could deal with don't-cares in the text; now it can also handle complements of single characters within the same computational complexity.

---

* This work was supported in part by the National Science Foundation, U.S.A., under grant No. MCS78-05849, and by a graduate fellowship from the Hebrew Technical Institute, New York, N.Y.

NATO ASI Series, Vol. F12
Combinatorial Algorithms on Words
Edited by A. Apostolico and Z. Galil

1. INTRODUCTION

String matching is a well known problem which has interesting practical and theoretical implications. An instance of the problem consists of a pattern and a text, and the objective is to find the first (or all) occurrence(s) of the string (or class of strings) specified by the pattern in the string (or strings) denoted by the text. Different variations of the problem are created by changing the ways in which the string(s) are specified by the pattern and the text. One can also find different attitudes towards the complexity measures applied to algorithms solving these problems; while the prevailing measure is counting the overall number of required comparisons (between individual characters), some consider also the overhead involved necessary to complete the tasks of the algorithm, besides the comparisons themselves, e.g. arithmetic operations, temporary storage accesses, output, etc.

The simplest, and hence easiest to solve, is the case in which both the pattern and the text are constants, i.e. both are words over a given alphabet, and all we want to know is whether, and if so — where, the pattern occurs as a substring in the text. A slight extension to this is the case in which the pattern consists of a finite set of constant strings (the text is still a single constant string), and we look for the first (all) occurrence(s) of any of the members in the pattern-set in the text; we call this the *set-pattern* version of the problem. These two cases have been shown to be solvable using comparisons whose number is linear in the length of the text (n) plus that of the pattern (m). In the case of the set-pattern problem, the length of the pattern is defined as the sum of the lengths of its members (i.e. the set is defined by the listing of its members, and hence the length of its description is as defined). In fact, the first problem (a single constant pattern) is solved in overall *time* linear in the length of the pattern and text combined $(m+n)$; this has been proven theoretically by Cook (cf. [AHU, Ch. 9]) using the result that each problem that can be solved by a 2DPDA can be solved in linear time by a RAM, and practically in [KMP], by supplying a specific algorithm which complies with this running time. The main idea in the KMP algorithm is to preprocess the pattern in such a way that enables us to deduce safe facts about the way it might match the text without having ever to rescan any part of the text itself; i.e. we can always move forward along the text. Moreover, the preprocessing phase is structurally similar to the algorithm when applied to the pattern itself, yielding the linearity in both the length of the pattern and that of the text. The set-pattern version is solved by a variation of the KMP algorithm, which has been discovered and described independently by [AC]: The main idea remains the same, preserving the property of the similarity between the preporcessing phase and the bulk of the algorithm. In this case, however, although the number of comparisons remains linear (in the same sense as above), the amount of bookkeeping needed to generate the result of the algorithm (where *all* occurrences are wanted) might involve a multiplicative factor, which is the maximum number of members in the pattern-set that are suffixes of other members (which, in turn, can be proportional to $\sqrt{m}$); but this factor is bounded by the order of the amount of data which has to be kept to record the result itself, so we cannot do any better anyway, and the optimality w.r.t. the number of comparisons is still achieved. By the way, this shows that we cannot solve the set-pattern version with a 2DPDA, by using the counterpositive of Cook's result on simulation of 2DPDA's by RAM's.

Another important version of the problem is the one in which the pattern is a full-blown regular expression. This might be very useful in practical situations, especially for text editing purposes, where we would like to match "generic" patterns which are not too complex but rich enough to justify themselves. The problem has been shown (cf. [AHU, Ch. 9]) to have time *and* number of comparisons complexities which are multiplicative in the lengths of the text and of the regular expression ($O(m \cdot n)$), where the regular expression is allowed to contain all (and only) the "conservative" (standard) regular operators (these are: concatenation, union and the reflexive-transitive closure, i.e. indefinite repetition, denoted by *), single character symbols and the empty string notation (λ). The problem is solved by constructing a NDFA which recognizes the language consisting of all words whose suffixes are defined by the original regular expression, and simulating the computation of the NDFA needed to find the pattern in an efficient manner; i.e. we do not construct the equivalent DFA (which would have caused an exponential blow-up) — we just simulate it for one given text at a time. Nevertheless, no non-trivial lower bound has been found for this problem, and it remains unknown whether more efficient techniques can be found to solve it.

Numerous variations can be made on this last version of the problem — both restrictions and extensions (or both at the same time). Some of them have been studied to some extent, while others seem to have skipped the attention of researchers. A partial list of them is given here:

(i) Allowing stronger regular operators in the pattern, i.e. making patterns which still denote regular sets, but are defined using operators whose expressive power is larger, e.g. squaring, explicit exponentiation (i.e. raising a subexpression to a definite power whose value is coded into the expression), upper limits on powers (i.e. a subexpression is allowed to occur repeatedly not more than a specified number of times), negation (complements of subexpressions), symmetric differences, abbreviations for subsets of the alphabet or the entire alphabet, etc. These extensions might make life much easier in a text-editing environment without changing the class of allowable patterns, but, of course, the complexity of the matching algorithms might change drastically in terms of the length of the pattern due to the possibility to obtain utterly succinct representations of patterns. In other words, the economy gained by the expressive power might transform into a substantial loss in the efficiency of the matching algorithm. A naive approach, according to which the pattern is preprocessed by translating it into a standard ("conservative") regular expression, will involve an exponential blow-up for most of the above mentioned operators, and thus is infeasible. In fact, for the limited extensions to be mentioned in what follows, none of the above questions has been studied in either way, i.e. neither efficient (sub-exponential) algorithms have been devised, nor non-trivial lower bounds have been proven. There are no known NP-completeness or P-space completeness results known for the matching problem (as opposed to related problems) either.

(ii) Another extension is "upgrading" the type of languages allowed as patterns in the Chomsky-hierarchy, e.g. let them be certain classes of context free languages. These problems have not been studied in this narrow context, primarily because their solutions fall out as special cases of general algorithms devised to solve more essential problems (such as parsing), but also because more elaborate constructs will be needed to describe the patterns (grammars, set notations with symbolic parameters, etc.), a fact which makes their usage cumbersome.

(iii) A common restriction is one in which only certain regular expressions are allowed, the most popular being the union of constant words (which has been discussed above), or — a mixture of this with extension (i) above — inclusion of the Σ symbol denoting the whole alphabet (sometimes called ϕ, for "don't care", as its presence in the pattern implies the possibility to match whatever text-symbol is present). The simpler version is the one allowing the usage of Σ only if it is closed, i.e. only the presence of Σ^* in the pattern is allowed. The meaning of this is that first we want to match that part of the pattern preceding Σ^*, and then start a match — search for the part following it exactly from where we left off. This technique can be applied wherever a Σ^* appears, and as long as the pieces between consecutive Σ^*'s are manageable by an efficient algorithm, the overall complexity does not change (as long as it is not sub-linear to start with!). As the union of constant words has been shown how to be dealt with by a simple extension to the KMP algorithm (i.e. its set-pattern version), and applying the above technique does not change the complexity of the matching algorithm, we can find — both in [KMP] and [AHU] — the conclusion that the regular language described by

$$(a_{1,1}+a_{1,2}+\ldots+a_{1,i_1})\Sigma^*(a_{2,1}+a_{2,2}+\ldots+a_{2,i_2})\Sigma^*\ldots$$
$$\Sigma^*(a_{p,1}+a_{p,2}+\ldots+a_{p,i_p})$$

can be matched using a linear number of comparisons using simple extensions of the KMP algorithm. A slightly more interesting case, is the one in which the Σ's are not "starred", i.e. each Σ denotes a *single* (arbitrary) character which has to be matched. The simple version of this case has one Σ for each don't care symbol we want to match, e.g. if we want to allow any sequence of s character between two constant parts of the pattern, we have to write down s Σ's at the appropriate place. A slightly more sophisticated version, which involves another extension from (i), allows us to encode the binary (or any non-unary) representation of s into the pattern using an external alphabet (i.e. with symbols not from Σ itself). An algorithm which solves both problems using a linear number of character-by-character comparisons and slightly more in overhead operations is presented in what follows.

(iv) A theoretically interesting extension was suggested by [FP], in which individual don't care (Σ) symbols (without *) are allowed *both* in the pattern and the text. Although this might have some interesting applications in the study of molecule structures in organic chemistry, it seems to have little — if any — relevance to string matching. Their theoretical result shows that pattern matching under such specifications is computationally equivalent (i.e. interreducible within a constant factor) to the problem of computing a boolean product (i.e. an inner product of two binary words with $\wedge$ and $\vee$), and, in fact, to a whole class of products. (It is interesting to note that their presentation is a formalization of an old APL programming trick.) Using this main result, a technique of embedding a boolean algebra in a ring (of integers) [FM], together with a simple alphabet-encoding-scheme, are used to show an upper bound on the number of comparisons and operations needed to find matches as described above. This upper bound depends on whatever best method we have for multiplication of integers; currently, this is the one due to Schoenhage and Strassen (cf. [AHU]), which yields an $O(n \cdot \log^2 m \log \log m \log q)$ (where q is the size of the alphabet, and n, m are as above) upper bound for the pattern matching algorithm, which is very impractical due to its indirect nature and the huge constant factor involved. Nevertheless, this is an interesting result, and a note concerning it, showing its applicability to a slightly broader version of the pattern matching problem, concludes this paper.

2. PATTERNS WITH DON'T-CARE SYMBOLS

The first question one asks when trying to extend the KMP algorithm (with all its virtues) to patterns with don't-care symbols (i.e. symbols each of which stands for the disjunction of all the alphabet characters), is whether we can use the schematic structure of the algorithm as it is, and just by changing a few minor details obtain a revised algorithm which is both accurate and as fast as the original one. To answer this question, we have to find out first what is wrong in running the algorithm without changing it at all. The problem is that comparing don't-care symbols (henceforth to be denoted by ϕ) makes the equality predicate non-transitive; i.e. although $a = \phi$ and $\phi = b$, this does not imply $a = b$! This very fact makes the idea of preprocessing the pattern (and deducing the *fail* and *next* functions) and then "sliding" it along the text infeasible if left unmodified; that is, we cannot use the algorithm as it is, because the presence of ϕ's in the pattern makes the processing "unfaithful" in the sense that it will allow slides that involve a wrong conclusion about the relation between earlier matched parts of the pattern and the text, and also will not enable the efficient computation of *fail* and *next*. Thus, if we want to retain the structure of the algorithm, we have to focus on the *meaning* of the atomic operations, and as the equality is the reason for the trouble, we have to start with it.

2.1 Outline of Solution Attempts

My attempts to adjust the KMP algorithm along these lines did not work out, and left the above question open. Nevertheless, it might be instructive to go shortly through the attempts I made and to see why they failed. The first idea is to modify the equality predicate and make it transitive — at the cost of losing its symmetry — for the preprocessing (of the pattern) phase, so the *fail* and *next* functions produced will provide us with safe "sliding" rules. The reasoning, therefore, stems from the will to avoid false conclusions about how previously matched portions of the text can be thought of as matching a smaller prefix of the pattern, thus enabling us to slide it. So, if we want to remain in the framework of the KMP algorithm on one hand, and retain faithfulness on the other, we are forced to avoid the occurrence of an unfaithful "transitive-link", i.e. to avoid an equality of two different characters deduced by equating each in turn to ϕ. Explicitly, if we had a text character matched by a ϕ in the pattern, at the time of preprocessing we do not know anything about the text which will be there, and thus we do not want to "slide" anything to that position, except for another ϕ; on the other hand, we do not mind sliding the pattern so as to putting a ϕ over any other character. Thus, while computing *fail* and *next*, two characters in the pattern are considered equal only if both are equal over the original alphabet, or if the one appearing *earlier* in the pattern (i.e. has a lower index in the pattern $y_1 \dots y_m$) is a ϕ. Formally, if we make the convention that the slidden position of the pattern always appears on the right-hand-side of the equality-test, we can define the modified predicate as:

$$(x \doteq y) = \begin{cases} \textit{true}, & \text{if } x,y \in \Sigma \text{ and } x = y; \\ \textit{true}, & \text{if } x = \phi \ (\text{thus } \phi \doteq \phi); \\ \textit{false}, & \text{otherwise.} \end{cases}$$

This way, $\doteq$ is transitive and enables us to produce efficient and faithful *fail* and *next* functions, and this is the least — as far as I could tell — we can do to fix these two problems. But, on the other hand, it introduces a new problem: it makes the pattern "slide" too fast, thus missing possible valid matches! i.e. it is too restrictive, and causes us to pass-by potential occurrences of the pattern and miss existing matches. This phenomenon is explained by the fact that ϕ's in the pattern might correspond, in special cases, to certain constants appearing earlier in the pattern, although we cannot make this conclusion in general; i.e. the possibility of "sliding" depends on the text rather than on the pattern alone, and preprocessing is not enough. The following example illustrates the above argument: the pattern $y = a\phi b\phi a$ over the alphabet $\Sigma = \{a,b\}$ will be preprocessed, using $\doteq$ for the computations of *fail* and *next*, to have the following values:

j	1	2	3	4	5
y_j	a	ϕ	b	ϕ	a
fail(j)	0	1	1	1	1
next(j)	0	1	1	1	0

It is interesting to note here, that the values of *next(j)* for those values of j at which $y_j = \phi$ (not to be confused with $y_j \dot{=} \phi$!) are immaterial as far as the part of the algorithm that matches the text itself is concerned, since these positions of the pattern will never cause a mismatch (ϕ matches everything). These values should be recorded, however, as they might be used by the preprocessing part which computes *next* in [KMP, p.328], or by going through the procedure defined later on that page to compute *next* without having to store *fail(j)* (denoted by *f(j)* in that paper); *note* that all equality tests with $=$ have to be replaced with $\dot{=}$ such that in each case the character occurring earlier in the pattern appears on the left-hand-side; in this particular case, all uses of $=$ have to be *reversed* when made into $\dot{=}$. This necessity is exemplified by the pattern $y' = a\phi\phi bba\phi b$, for which *next*(8) is computed using *next*(3) where $y'_3 = \phi$.

Now, coming back to the main example, if the pattern $y = a\phi b\phi a$ is going to be matched with the text $x = aabbba$ using the above values for *next*, the failure on matching $x_5 \overset{?}{\dot{=}} y_5$ will cause the "sliding" of y all the way to the right, trying to match $x_6 \overset{?}{\dot{=}} y_1$, thus missing the match $x_2 \ldots x_6 = y$. In other words, the attempt to avoid false deductions by defining $\dot{=}$, resulted in missing a valid match. Evidently, defining $\dot{=}$ in the opposite direction (i.e. make it evaluate to *true* whenever the right-hand-side is ϕ) does not make any sense, since it causes both problems (unfaithfulness and skipping matches) at the same time.

If a cure to the problem — which I might have overlooked — can be found, it will be of practical interest for another reason, which is the ease of a practical implementation of $\dot{=}$: it is even easier than the implementation of the usual (symmetric) extension of $=$ to handle $\Sigma \cup \{\phi\}$, because for $\dot{=}$ only one input line has to be distinguished and checked for holding ϕ on it.

The solution for the problem of matching patterns with definite occurrences of don't-care (ϕ) symbols, was finally found using another variant of the KMP algorithm, namely — the set-pattern version. This version's complexity in terms of the number of comparisons is the cumulative length of the members in the set (for preprocessing) *plus* the length of the text, thus we can still have linearity if we extend it carefully. The idea is simple: we define the sequences of characters out of Σ (i.e. words over Σ^+) between any two non-adjacent but consecutive occurrences of ϕ to be the *components* of the pattern (a formal definition follows); as we know the exact displacement between the beginning of any two consecutive components (which is the length of the first component + the number of ϕ's separating it from the next one), our goal is to find a series of occurrences of the components in the text whose distance from each other conforms exactly with these displacements. Matches of such series, which correspond to matches of the original pattern, can be found naively by gathering all the information about the components in terms of occurrences in the text, and then look for appropriate series; but doing this is not efficient in terms of space, as it involves storing information which can be easily checked for obsoleteness as the algorithm goes on; it also destroys the real-time nature of the KMP algorithm. Fortunately, there is a way to integrate the checking phase with the information accumulation about the matching of the components; in fact, two ways of implementing such integration were found, and are described in what follows.

2.2 Formal Discussion of the Proposed Solution

The problem of matching a pattern with definite occurrences of don't-care (ϕ) symbols is defined as follows:

Input: A text $x \in \Sigma^*$;

a pattern $y \in (\Sigma^+\phi^+)^*\Sigma^+$ where $\phi \notin \Sigma$.

We denote $n = |x|$, $m = |y|$, and index the characters in each $1,\ldots,n$ and $1,\ldots,m$, respectively.

Output: The first (or all) position(s) j in x s.t. $1 \leqslant j \leqslant n - m + 1$ and $x_j \ldots x_{j+m-1} = y_1 \ldots y_m$, where the equality is on a symbol-by-symbol basis, and two symbols are equal if they denote the same element of Σ or if one of them is ϕ (evidently, as only y contains the symbol ϕ, it is enough to say that either the text and pattern symbol match or the pattern symbol is ϕ)

Note that the problem in which we allow y to be in the more general class $\phi^*(\Sigma^+\phi^+)^*\Sigma^+\phi^*$ § can be easily reduced to the one stated above, since the inclusion of these ϕ's at both sides just implies the necessity of the

---

§ the use of * here does not imply an indefinite number of ϕ's; it just broadens the class for y.

presence of a certain number of characters on both sides of the "central-part" of the pattern (i.e. the part between the two extreme characters from Σ), and checking this condition is trivially achieved by matching the central part of the pattern on a text that is trimmed on both sides by the appropriate number of characters.

Definition: The *components* of a pattern y are the (maximal) words over Σ^+ which correspond to the Σ^+ portions in the syntactic characterization of y, *or:* The components of y are these substrings of it including only symbols from Σ and adjacent to a ϕ symbol on both sides (except for the first and the last one, which start with y_1 and end with y_m, respectively).

Definition: The ϕ-blocks (in short: blocks) of y are the longest consecutive runs of the symbol ϕ in y, *or:* These parts of y corresponding to the ϕ^+ in its syntactic characterization.

We number the components of y from 1 to p in the order of their appearance in y, and denote them by $\bar{y}_1,\ldots,\bar{y}_p$.

With each component $\bar{y}_r$, $r=1,\ldots,p-1$, we associate the block separating it from the next component, $\bar{y}_{r+1}$; thus, the blocks are numbered $1,\ldots,p-1$ and associate to the left. We also associate with each component $\bar{y}_r$, $r=1,\ldots,p-1$, its distance d_r, which is the distance between its first character and the first character of the next component, and is computed by adding the length of the component, ℓ_i (i.e. the number of characters in it), and the length of its corresponding block (i.e. the number of ϕ's in that block).

We recall now that the matching problem with don't-care symbols is solved using the set-pattern version of the general problem. So, we need to define the input for the set variant using the input to our problem. This is done simply by defining the components as the elements of the set. The blocks serve as element separators, and although they might consist of more than one (ϕ-) symbol, they are not redundant, as we need them in full for the other part of the input, which is – finding out the distances d_r for $r=1,\ldots,p-1$. The distances can be found easily during the preprocessing phase simply by counting the symbols encountered while reading the input and following the simple recognition scheme of the pattern's form (as a word in $(\Sigma^+\phi^+)^*\Sigma^+$). The distances found by this process will be stored along with the *output* sets associated with the trie (cf. [AC]): each word in the *output* set will have its distance associated with it. This burden on the storage requirements will be taken into consideration later on. Thus, the time-complexity of the preprocessing phase is the same as that of the set-pattern version.

At this point, we have to note that the set defined above using the components of the pattern naively might be a multiset; this will result in having more than one member of the set associated with those nodes of the trie corresponding to the set members with multiplicity of 2 or more. However, this is settled easily by associating with each word at the appropriate place all its distances, one by one; this does not result in any extra usage of space, as the distances had to be recorded anyway — one for each component, and the fact that some are put together does not change anything.

At this point I would like to digress from the description of the details of the algorithm, and give an outline of the general approach which led to the solution of the problem; although this approach is modified substantially, it is worth mentioning in order to enhance the intuition about the final version of the algorithm. The discussion of the preliminary approach will not be too formal to make it readily understood.

As part of the preprocessing phase, we set up an empty list for each component; this is the occurrence-list, which will hold all the matched occurrences of the corresponding component in terms of indices in the text. The lists will be created as we go along the text, and can be maintained easily so that they contain the occurrences in increasing order without spending extra time. Also, the first time a component is found (matched), we associate with its list its distance (copied from the trie). After the whole text is read and all occurrences of all components are recorded as sorted lists, we start a postprocessing phase, in which the occurrences found are checked to be in accordance with the original pattern's layout; the basic fact is that an occurrence of the original pattern (with the ϕ symbols) is found at a given position of the text *iff* we can find a sequence of occurrences of the components in their order, starting with the same position (i.e. the first component has to occur at that position), *and* such that the numerical differences between adjacent occurrences are equal to the corresponding distances associated with each component. This fact can be easily proven from the definitions of the components and their associated distances.

The technique used to check the above necessary and sufficient condition, using the occurrence-lists created by the scanning of the text, entails imposing a tree-structure on the lists as leaves to the effect of creating lists corresponding to runs of components — represented by internal nodes, until, eventually, an occurrence list for the whole pattern is found. At the first level, the p components are divided into $\lfloor p/2 \rfloor$ pairs (and an individual list, if p is odd), each corresponding to two adjacent components (i.e. $\bar{y}_1$ is paired with $\bar{y}_2$,$\bar{y}_3$ with $\bar{y}_4,\ldots,\bar{y}_{2r-1}$, with $\bar{y}_{2r}$, etc.). Then, we define the *merger* of two lists to be the list which contains those elements of the first list for which there exists an element in the second list whose value is exactly d units larger than it, where d is the distance associated with the first list; we also associate the sum of the distances of the two lists as the distance of the new list. Evidently, the newly created list, i.e. the merger, contains all and only those indices in the text which correspond to occurrences of the component associated with the first list followed by a run of characters whose number is the length of the associated ϕ-block followed by the component associated with the second list; also, the distance associated with the merger is the distance we would like to have between the first character of the pattern consisting of the two components put together (padded with their ϕ-blocks appropriately) and the next component (or its derivatives). After ending up with $\lceil p/2 \rceil$ lists describing occurrences of pairs of components (with the possibility of one last single component), we go on to the next level of finding quadruples (with the possibility of one residual component), and so on — forming occurrences of lists of super components until we are left with one list corresponding to the whole pattern. If any of the intermediate lists is found to be empty, we can abort the whole process because this is an evidence for the absence of a vital link in the "construction" of an occurrence of the original pattern. Note that the last component did not have a distance associated with it, because we did not need it as we do not care what follows it; this fact holds throughout the levels, i.e. we never need to know — and hence compute — the distance associated with the last (super-)component. On the other hand, the only distance we do not know how to compute due to lack of information about the distances of the underlying components is always the last one, because it is the only one using the last component of the previous level. Thus, everything works out well, and we end up with all the occurrences of the original pattern (without any associated distance). The depth of the computation-tree created is $\lceil \log_2 p \rceil$, but as all of it has to be traversed, the number of merger operations to be performed is $O(p)$ (actually, it is $p-1$ exactly). This same result could have been attained by just collapsing the lists one into its successor serially (i.e. merge the lists of $\bar{y}_1$ and $\bar{y}_2$, then merge the result into $\bar{y}_3$ and so on); this will yield p-1 list-merges, but does not exploit the potential parallelism as before, which is due to the associativity of the merging operation. It might also yield larger temporary results and postpone the "instant death" property mentioned above. However, I did not analyze this algorithm thoroughly as it is used only for presentation purposes.

The major drawbacks of the above algorithm are: (a) The fact that we have to wait until the whole text is scanned before we can get *any* answer (due to the postprocessing phase); i.e. no matter what time complexity it has, it cannot be used for real-time applications. (b) It might use up immense amounts of temporary storage, some of it might be sheer waste, because a most frequent occurrence of a certain component might be totally irrelevant due to the rareness (or even complete absence) of another, but it will still be recorded! However, one observation provides remedies for both problems: It is enough to read m consecutive symbols of the text (we recall that m is the length of the pattern) in order to decide whether they can be matched by the pattern or not; in other words, once we are past m positions from a given initial position, we should be able to make up our minds as to whether a match occurred at the original position or not, and whatever we are going to read afterwards will not make any difference. This can be phrased as follows: at each point along the scanning of the text, we are interested at most in the last m characters of it. Thus we use the following data structure (proposed by Ronald L. Rivest): An array of m counters, each capable of counting from 0 to p (i.e. having $\lceil \log_2(p+1) \rceil$ bits), that will "collect" evidence of the occurrence of a match of the original pattern by counting occurrences of the components; the array will hold information relevant only to the last m text symbols, and reuse its cells circularly (using a modular "wrap-around" scheme), and will be denoted by c.

Going back to the trie produced by the preprocessing phase, we recall that information concerning distances was associated with nodes that had non-empty output sets; this information was used in the naive version of the algorithm, and, as a matter of fact, was reorganized as soon as actual occurrences of components were found. In order to make the efficient version of the algoirthm work, we have to change this information slightly: now we associate with each component the distance from its last character to the first character of the original pattern; or, simply, the index of its last character in the pattern *minus 1* (i.e. if y_i is the last symbol of $\bar{y}_r$, then the node corresponding to the end of $\bar{y}_r$ will be associated with i-1). The reason for this is that whenever we succeed to match a component, we want to know the relative position of the character that will be aligned with the first character of the pattern if the component is really a part of an original match; or, in other words, we want to know which text-position (relatively) the match of the component serves as a "witness" for as part of an original match, thus knowing which counter to increment. It is easy to realize that producing this extra information is even easier than the former one, because it is really there (in the form of the index); we also make the initialization of the array c to all zeroes part of the preprocessing, and this takes only m steps. Thus, the time-complexity of the preprocessing phase remains unaltered.

The rest of the algorithm is now merely a form-retaining extension of the text-scanning phase of the set-pattern variant of the KMP algorithm with some additional overhead to be specified shortly; there is no postprocessing involved, and the occurrences of matches are "announced" as soon as they are found in [AC, p.334]: the variable i is used as an index into the text, going from 1 to n, and the trie nodes, numbered from 0 onwards, are visited using the pointer *state*; each node has its failure *(f)* and *output* functions associated with it, which are state and component-subset valued, respectively; the structure of the trie is maintained in the array g, which keeps for each valid combination of a node and an alphabet symbol (out of Σ) the corresponding descendant node of the trie, or holds the value *fail* otherwise. The only extensions to the algorithm are in two places:

(a) In the innermost level, that is only when an output set of the current trie-node (denoted by *state*) is non-empty; then, a component has been matched, and this fact has to be appropriately recorded (in c).

(b) Whenever the text index (i) is incremented; at this point (once we have reached the stage where the old $i \geq m$), a used cell of c has to be reused, and this is exactly the point where we have to check whether enough evidence for an original match has been accumulated in the counter to be reused. The reason for this is that the number of elements in c is exactly m, and due to the wrapping around scheme (mod m), a cell is about to be reused exactly after m text symbols have been scanned since it has been set-up; this will be made even clearer as we devise the details.

Before we go on, there is one technicality we have to introduce: As all indexing (into the text and the pattern) is done using 1-origin, we want to use the same convention for c. Since the index for c mimics the behaviour of that of the text except for the wrap-around phenomenon, we want to use a modular definition for it; but strict modular arithmetic does not conform with 1-origin, so we define a modification *mod'* over the domain of the positive integers, which has the same value as *mod* for arguments whose remainder is between 1 and m-1 (where m is the modulus), but m for exact multiples of m (rather than 0). If the argument is i (as it will be in the algorithm), i mod' m is simply defined as 1 + (i-1) mod m.

Now let us explain the counting mechanism: Each time a match of a component is found, the distance of its last character from the beginning of the pattern — as recorded at the trie node — is subtracted from the current text index (i); this yields the position in the text at which the original match can be found based on the fact that the above component has been found. We take this difference and find its value mod' m, and increment by 1 the cell of c having that index. As we go along, a counter that was set initially to 0, can reach p *iff* exactly m text characters have been read, and all the components of the pattern have been matched — thus an increment was performed for each of them; there is no way the components can interact to produce false evidence — each component has a unique distance associated with it and it can cause only *one* increment per occurrence and only at the right place. When do we check if the original pattern was fully matched? this happens exactly when we are about to reuse the corresponding cell of c, because on one hand we cannot be sure of a match until we read m consecutive symbols of the text, but once we read more — the text position starting more than m symbols behind becomes irrelevant. Thus, when we increment the text index i, before we reset the corresponding counter, we test if we have already read at least m symbols (i.e. new $i > m$?), and if so — we check if the old counter was already p; if this is found to be true, we can declare a match of the original pattern, and in any case — whether a match was found or not — we can reset the counter to 0. In fact, the check $i > m$ is redundant, since no counter can reach p before i reaches m and the mth text character is processed. Finally, the last point that has been checked this way corresponds to x_{n-m}, while the last iteration of the main loop may cause x_{n-m+1} to correspond to a match (aligned with the end of the text); thus one extra check is needed. The index of the text at which the match starts is then simply i-m. The full version of the text-scanning phase of the algorithm is, therefore (assuming g,f,c have been initialized, and that each element of *output(z)* — for any node z — has associated with it a value *cum_dist* which tells the distance between its last character and the first character of the pattern):

```
begin state ← 0
      for i ← 1 to n do
            begin if c(imod'm) = p then print (i-m);
                  c(imod'm) ← 0;
                  while g(state,x_i) = fail do state ← f(state);
                  state ← g(state,x_i);
                  if output(state) ≠ ∅ then
                        begin for z ∈ output(state) do
                              begin temp ← (i - cum_dist(z))mod'm;
                                    c(temp) ← c(temp) + 1;
                              end;
                        end;
            end;
     comment: Here is the extra test for the result of the last iteration;
     if c((n+1)mod'm) = p then print (n-m+1);
end;
```

The time-complexity of this algorithm is exactly the same as that of the algorithm given in [AC], since the innermost block in the algorithm had a print *output(state)* statement as part of it, which has to be written explicitly as for $z \in$ *output(state)* do print (z), i.e. it involves a traversal of all elements in *output(state)*, which is exactly the case here, with the only difference that here we have an arithmetic operation rather than an output statement. Notice that the total time it takes to count up to p (or less) in each register gets consumed by the overall linearity and does not cause a multiplicative factor of $\log_2 p$ to be created (cf., for example, [McI]).

The space overhead of this algorithm is the space for *cum_dist(z)* for elements of output sets, which all in all takes p words of length $\lceil \log_2 m \rceil$ bits each, and the array c which takes up m words of length $\lceil \log_2 (p+1) \rceil$ each.

It is important to note, that if a certain word over Σ^+ is equal to more than one component (i.e. some components are equal), this does not change the validity or the performance of the algorithm. This is certainly true because we do not need that the components be different for the reasoning to go through, but it is worthwhile to realize intuitively that a component appearing more than once in the pattern just results in having more than one *cum_dist* value associated with its corresponding trie node; what will happen then is that the occurrence of a substring equal to these components will serve as a witness for more than one starting position — once for each actual component, which is the right thing to happen! Such an occurrence can never cause a confusion in the counting, since it is used exactly once for each component, thus causing increments at different positions. In case increments of all registers can be done in parallel (this is not an unrealistic model of computation!), the time-complexity is still linear in the length of the pattern.

An extension to the domain of problems this algorithm can handle was mentioned shortly above, and will be made obvious now: Rather than specifying the number of ϕ's in each block using runs of ϕ's, i.e. using a *unary notation,* we can use an external alphabet, say $\Psi = \{\bar{0},\bar{1}\}$, to denote these numbers in a binary basis. The only change is in the preprocessing phase, and it even comes out to be a bit of saving: rather than having to count the ϕ's and manipulate the binary representation created by the counting process, the representation of the distances is readily there; the values of *cum_dist* are still found by subtracting 1 from the index, which is now incremented by more than 1 at a time — whenever we encounter a block. The class of patterns that can be handled is now defined by the language $\Sigma^+\{\phi\Psi^+\Sigma^+\}^*$ (the ϕ does not need to be there — it is just a "punctuation" put in for readability of the pattern).

3. AN EXTENSION TO THE FISCHER-PATERSON ALGORITHM

As a final remark, I would like to point out a possible minor extension to the domain of problems handled by the construction proposed by [FP] to solve the matching problem with don't-care (ϕ) symbols both in the pattern and the text. This extension entails a slight change to the construction, which maintains its original flavour and stays within its complexity bounds. The idea is to allow the usage of complemented alphabet symbols, in both the text and the pattern: for $\sigma \in \Sigma$, by saying $\bar{\sigma}$ at a given position in the pattern we mean that anything can be matched with this symbol *except* for the character σ itself; and, similarly, if $\bar{\sigma}$ appears in the text, it means that all we know is that this position certainly does not contain a σ in it. This kind of specification might be useful, since it gives us expressive power we did not have before.

The alphabet over which the patterns and texts can range now is $\Sigma \cup \bar{\Sigma} \cup \{\phi\}$, where $\bar{\Sigma} = \{\bar{\sigma} | \sigma \in \Sigma\}$ (naturally, $\bar{\phi}$ does not make any sense!).

The computational equivalence of the new problem with the problem of boolean products is proven using the *same* reduction as in [FP]: Obviously, the reduction of boolean products of ϕ-string products (in [FP]'s terminology) remains valid (the construction used comes up with a problem which certainly falls under the extension). As for the converse, we just have to define $q = |\Sigma| = |\bar{\Sigma}|$ more predicates, one for each symbol in $\bar{\Sigma}$:

$$H_{\bar{p}}(x) = \begin{cases} 1 & \text{if } x = \bar{p} ; \\ 0 & \text{if } x \neq \bar{p} \text{ (or } x = \phi). \end{cases}$$

and then

$$\underline{z} = \underline{x} \left(\overset{=}{\wedge}\right) \underline{y} = \neg\left(\left(\bigvee_{\substack{\sigma \neq \tau \\ \sigma,\tau \in \Sigma}} H_\sigma(\underline{x}) \left(\overset{\wedge}{\vee}\right) H_\tau(\underline{y})\right) \vee \left(\bigvee_{\sigma \in \Sigma} H_\sigma(\underline{x}) \left(\overset{\wedge}{\vee}\right) H_{\bar{\sigma}}(\underline{y})\right)\right)$$

The second disjunction of boolean products, involving $H_\sigma(x)$ and $H_{\bar{\sigma}}(y)$, is used to make sure that no character out of Σ is matched with its negation, and the rest of the proof follows exactly as in [FP]. The factor of q^2 involved with this reduction is reduced to $\lceil \log q \rceil$ in the same way as the one for the original construction, with the need for an extra bit (to tell elements of Σ from these in $\bar{\Sigma}$) which does not change anything.

Acknowledgements: I would like to thank Ron Rivest and Vaughan Pratt for many helpful discussions.

REFERENCES

[AC] Aho, A.V., and Corasick, M.J.: Efficient String Matching: An Aid to Bibliographic Search; CACM, Vol. 18, No.6 (June 1975), pp. 333-340.

[AHU] Aho, A.V., Hopcroft, J.E., and Ullman, J.D.: The Design and Analysis of Computer Algorithms; Addison Wesley, Reading, Mass., 1974.

[FM] Fischer, M.J., and Meyer, A.R.: Boolean Matrix Multiplication and Transitive Closure; 12th IEEE Symposium on Switching and Automata Theory (1971), pp. 129-131.

[FP] Fischer, M.J., and Paterson, M.S.: String-Matching and Other Products; Technical Report 41, Project MAC, M.I.T., 1974. *Also appeared* (under the same title) in: Karp, R.M. (ed.): Complexity of Computation, SIAM-AMS proceedings, Vol. 7 (1974), American Mathematical Society, Providence, R.I., pp. 113-125.

[KMP] Knuth, D.E., Morris, J.H., and Pratt, V.R.: Fast Pattern Matching in Strings; SIAM Journal on Computing, Vol. 6, No.2 (June 1977), pp. 323-350.

[McI] McIlory, M.D.: The Number of 1's in Binary Integers: Bounds and Extremal Properties; SIAM Journal on Computing, Vol. 3 (1974), pp. 255-261.

[T] Thompson, K.: Regular Expression Search Algorithm; CACM, Vol. 11, No. 6 (June 1968), pp. 419-422.

SUMMARY OF NOTATIONS

$x = x_1, \ldots, x_n$	the text; $n = \lvert x \rvert$.
$y = y_1, \ldots, y_m$	the pattern; $m = \lvert y \rvert$.
Σ	the input alphabet; $q = \lvert \Sigma \rvert$.
ϕ	the don't-care symbol.
p	the number of components in a definite don't-care pattern.
$\bar{y}_1, \ldots, \bar{y}_p$	the components themselves.
$\ell_1, \ldots, \ell_p$	the lengths of the components.
$d_1, \ldots, d_{p-1}$	the corresponding distances.

OPTIMAL FACTOR TRANSDUCERS

MAXIME CROCHEMORE

Laboratoire d'Informatique
*Universite de Haute-Normandie**
and L.I.T.P., Paris
FRANCE

0. Introduction.

Among string-matching algorithms the two most efficient methods of Knuth-Morris-Pratt [KMP 77] and Weiner [We 73] are usually considered as opposite of each others. In this paper we show that automata theory unifies the two approaches and gives a simple and efficient algorithm to construct the structure, a sequential transducer, underlying Weiner's method.

Consider a text x and a word u both on the same finite alphabet A. When u occurs inside x, a prefix of x belongs to the set of words A*u. Then, if a deterministic automaton for A*u is known, the search for u in x can be achieved in real time reading x from left to right [AHU 74]. Knuth-Morris-Pratt's algorithm does not explicitly built that automaton but rather a representation of it using default states (see for example [AU 79]). By the way, the difference between this algorithm and the original Morris and Pratt's one lies in the choice of default states. With this representation the space used by precomputed structures is independent of the size of the alphabet A, while the minimal deterministic automaton for A*u (which is complete) has |A||u| transitions. The time to compute these structures is linear in |u| the length of u.

Weiner's approach is to build a tree, the branch of which are labelled by suffixes of x. Nodes in the tree, called the suffix-tree of x, memorize positions of factors of x inside x itself. The tree needs to be compacted otherwise it could have a size quadratic in |x|. McCreight [Mc 76] gave an elegant and practical construction of the suffix-tree which is linear in |x|. Several variations or applications of suffix-trees are already known (see [AP 83], [Cr 83], [Ro 82]). The Boyer-Moore's string-matching algorithm itself [BM 77], even more efficient on average than Knuth-Morris-Pratt's algorithm, uses functions, precomputed on the pattern u from right to left, which are included in the suffix-tree of the mirror image of u.

Again inside Weiner's method the right tool is automaton. Recently Blumer, Blumer, Erhenfeucht, Haussler and McConnell [BBEHM 83] have given a linear construction of an automaton which recognizes F(x) the set of factors of x. They also showed that the size of their automaton is linear in |x| and independent of the size of the alphabet A.

---

* B.P.67 F-76130 Mont Saint Aignan

NATO ASI Series, Vol. F12
Combinatorial Algorithms on Words
Edited by A. Apostolico and Z. Galil

When positions of patterns inside x are needed the natural tool becomes the sequential transducer. Among different possible transducers, we consider here those which have exactly the same number of states as the minimal automaton for F(x). We give optimal bounds on the size (number of states and number of transitions) of the transducer together with words that reach the worst case. We also present a construction of the transducer closed to the construction of the minimal automaton for A*x. As a consequence the algorithm we get is particularly simple.

1. Minimal factor automata.

Words considered in this paper are elements of the free monoid A* generated by a finite alphabet A. the empty word of A* is denoted by 1 and A+ is A*-{1}. Letters of A are denoted by a,b,c,.., words of A* by x,y,z,u,v,w,... and |x| is the length of x. F(x) is the set of factors of x, that is

$$F(X) = \{y \in A^* / \exists u,v \in A^* \; x=uyv\}.$$

We note S(x) the set of suffixes of x. The occurrence of a factor y of x is its position inside the word x, and may be seen as (u,y,v) if x=uyv.

Let x be a word in A*. The definition of the minimal deterministic automaton $\mathfrak{F}(x)$ which recognizes F(x) is a straightforward application of Myhill and Nerode's theorem.

Let $\underset{x}{\sim}$, or simply ~, be the equivalence on A* defined by :

$$u \sim v \text{ iff } \forall w \in A^* \; uw \in F(x) \iff vw \in F(x).$$

Then, states of $\mathfrak{F}(x)$ are the equivalence classes, denoted by $[u]_x$ or [u], of words u in F(x). The initial state of $\mathfrak{F}(x)$ is [1], all states are terminal and the transition function of $\mathfrak{F}(x)$, for which we use a '.', is defined as follows :

$$\text{if } a \in A \text{ and } ua \in F(x), \; [u].a = [ua].$$

The right invariance of ~ makes this definition coherent.

EXAMPLE : Automaton of factors of aabbabb.

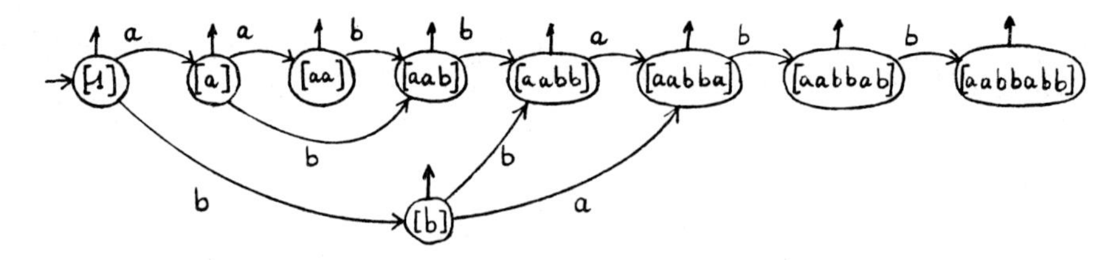

The minimal deterministic automaton $\mathcal{S}(x)$ which recognizes S(x) the set of suffixes of x could be defined in the same way. But, for algorithmic reason, we rather deduce S(x) from $\mathfrak{F}$(x\$) where \$ is a marker not in A : make terminal those states of $\mathfrak{F}$(x\$) on which a \$-transition is defined and delete \$-transitions together with state [x\$].

EXAMPLE : Automaton of suffixes of aabbabb.

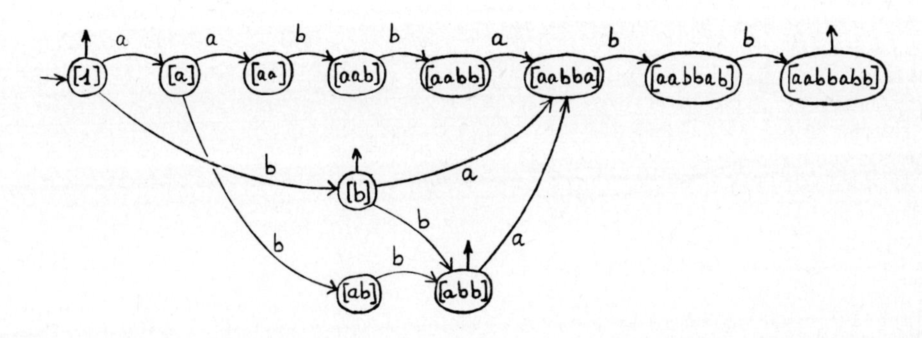

2. Minimal factor transducers.

In string-matching or related problems it is often not enough to know if a word is a factor of another given word, but an occurrence or a position of the searched word is also needed.

The natural way to deal with that problem is to use a (sequential) transducer instead of an automaton. The output of the transducer must give which occurrence of the input word has been found.

We are mainly interested here in those factor transducers for a given word x that can be built from $\mathcal{F}(x)$, without adding any new state.

The word x being fixed, our first example is the transducer associated to the left sequential function p defined as follows :

for u in F(x), p(u) is the shortest prefix of x which contains u.

The transducer corresponding to p is defined from $\mathcal{F}(x)$ by adding productions to transitions : If [u].a is a transition in $\mathcal{F}(x)$, the associated production is denoted by [u]*a and is equal to $p(u)^{-1}p(ua)$.

EXAMPLE : Transducer associated to function p on aabbabb.

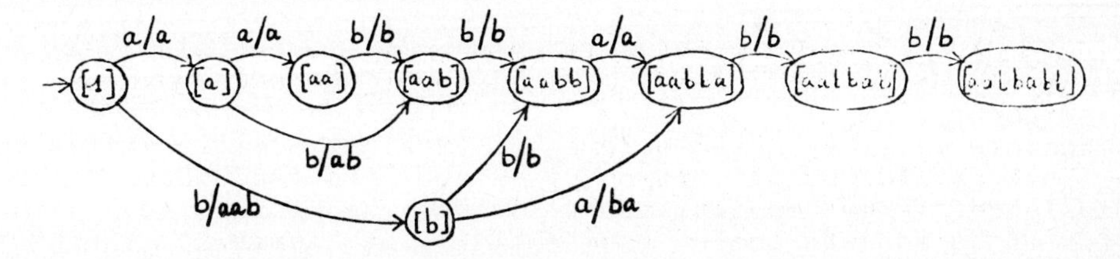

We will call T(x) the transducer associated to the function which gives for u in F(x) its first position in x : $|p(u)|-|u|$. It is worth noting that the non-commutative version of this function, which returns $p(u)u^{-1}$, leads to a transducer that may have more states than $\mathcal{F}(x)$.

EXAMPLE : Transducer T(aabbabb).

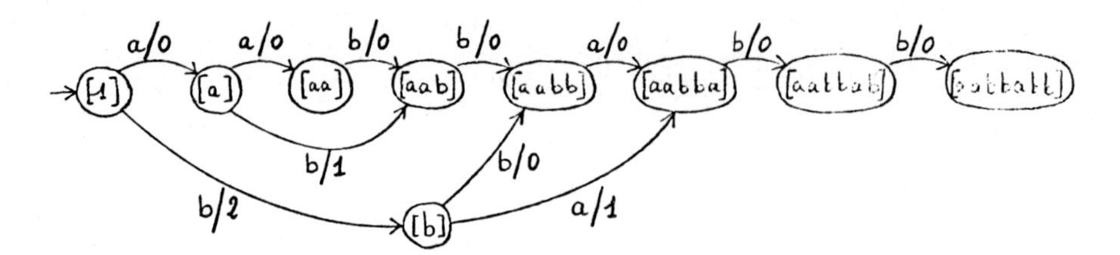

3. Size of T(x).

The fact that T(x) can be built in time linear in the length of the word x essentially comes from the linearity of its size. The situation is analogue to the construction of the minimal deterministic automaton which recognizes A^*x, and is given by the Knuth, Morris & Pratt's algorithm. This algorithm heavily uses what is called a failure function which helps determinizing the obvious non-deterministic automaton for A^*x. So we do in our algorithm with a function s called the suffix link. The definition of s we give also works for the failure function :

for u non empty in F(x), s(u) is the longest suffix of u such that $v \not\sim u$.

This definition is quite natural in term of automata and contrasts with the function used in McCreigth's suffix-tree algorithm (in this case s(u) is the longest proper suffix of u).

Lemma 1. Let $x \in A^*$ and $u, v, w, u' \in F(x)$. Then
 i. $u = u'vw$ and $w \neq 1$ imply $u \not\sim v$;
 ii. $u \sim v$ implies $p(u) = p(v)$;
 iii. w suffix of v, v suffix of u and $w \sim u$ imply $v \sim u$.

The proof is obvious. Note that ii implies that when two factors of x are equivalent one is a suffix of the other.

Lemma 2. Let $x \in A^*$ and $u, v \in F(x)$. Then
 i. $u \sim v$ implies $s(u) = s(v)$;
 ii. $u \not\sim p(u)$ implies $\exists w, w' \in F(x) - \{1\}$ $s(w) = s(w') = u$ and $w \not\sim w'$.

Proof :

i- Suppose $u \sim v$. As noted before v, for instance, is suffix of u. Then $s(v) \not\sim u$ by definition of s and the fact that $\sim$ is an equivalence. The maximality of $|s(v)|$ among the $|w|$ with w suffix of u and $w \not\sim u$ comes from part iii of lemma 1. Therefore $s(u) = s(v)$.

ii. Recall that p(u) is the shortest prefix of x which contains u, and then contains u as a suffix. If $u \not\sim p(u)$ there exists a shortest word v such that vu is suffix of p(u) and $vu \not\sim u$. Let $w = vu$; we have $s(w) = u$. Since $w \not\sim u$ there exists a prefix $x'u$ of x which does not end with w. As above we find a suffix w' of $x'u$ such that $s(w') = u$. Finally, $w \not\sim w'$ by part ii in lemma 1.

Property i in the preceeding lemma shows that s may be defined on states of T(x) (except on state [1]). This also gives a tree-structure to the set of states of T(x). A first upper bound on the number of states is a consequence of lemma 2.

Let e(x) be the number of states of T(x) (which is also the number of states of $\mathfrak{F}$(x)).

Proposition 3. For each x in A*, $|x|+1 \leq e(x) \leq 2|x|+1$.

Proof : Consider the tree which nodes are the states of T(x), s being the relation father. Apply on it the following transformation :

- replace each node without son by a leaf labelled with the corresponding prefix of x;
- to each node labelled by the equivalence class of a prefix of x and which has at least one son, add as a son a new leaf labelled by the prefix.

First rule is coherent by part ii of lemma 2 because a node without son must be labelled by the class of a prefix of x.

For the same reason and the rules applied, each internal node of the new tree has at least two sons.

The number of leaves being |x|+1 and by the fact that the tree is at least an extended binary tree we get the upper bound.

The lower bound is trivial.

We denote by t(x) the number of transitions in T(w) (or $\mathfrak{F}$(x)).

Proposition 4. For each x in A+, $|x| \leq t(x) \leq e(x)+|x|-2$.

Proof : As a labelled graph, T(x) is connex. This gives the lower bound. This also implies that we can find a spanning tree for T(x). Furthermore we may choose a spanning tree with a branch labelled by x. The number of edges in the tree is e(x)-1.

Consider a transition by a from q to q' which is not in the tree. Let y be the label of the unique path in the tree from [1] to q and z the longest suffix of x such that q'.z=[x]. To the transition is then associated a suffix yaz of x. This correspondence between extra edges and suffixes of x is one to one and neither 1 nor x are considered. Then the number of edges not in the spanning tree is bounded by |x|-1 which gives the upper bound.

EXAMPLE : Suffix function and modified tree for T(aabbabb).

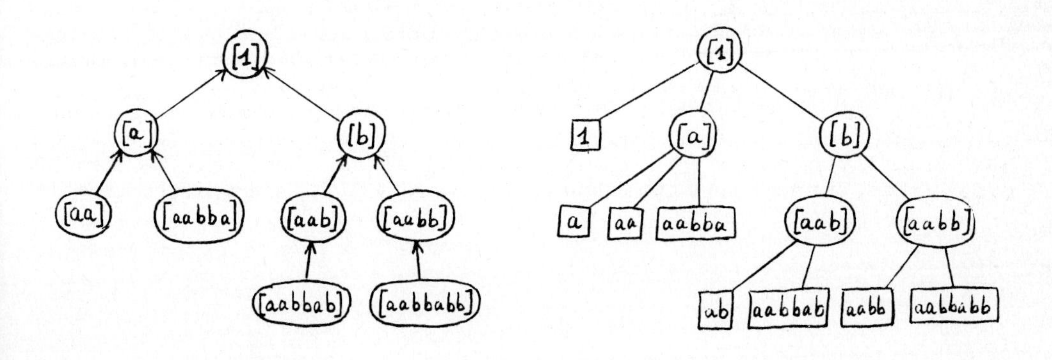

4. Optimal bounds.

The only bound which is not optimal in propositions 3 and 4 is the upper bound on $e(x)$ the number of states of $T(x)$. To refine it we look more precisely on how $T(xa)$ is built from $T(x)$. The next theorem gives also a base for the proof of our on-line algorithm to construct $T(x)$.

Theorem 5. Let $x \in A^+$. Define the words $u,v \in A^*$ as follows : u is $s(x)$ according to $\underset{x}{\sim}$; v is the longest prefix of u such that

(1) $|v| = \max\{|v'| \;/\; v' \in F(x) \text{ and } v' \underset{x}{\sim} v\}$.

Then, if $ua \in F(x)$, $e(xa) = e(x)+1$,

otherwise $e(xa) = e(xu^{-1}v)+2|v^{-1}u|+1$.

Proof : If $ua \in F(x)$ the only new states of $T(xa)$ are $[xa]_{xa}$ and $[vw]_{xa}$ such that $w \neq 1$ and vw is prefix of u.

Let w' be in $F(x)$ and w be a suffix of w' such that $w' \underset{x}{\sim} w$ and $w' \underset{xa}{\not\sim} w$. There exists z with $wza \in S(xa)$ and $w'za \notin F(xa)$. The word wz occurs at least twice in x and then must be a suffix of u. Furthermore $w'z \in F(x)-S(x)$.

If $ua \in F(x)$ we get a contradiction from $wza \in F(x)$, $w'za \notin F(xa)$ and $w' \underset{x}{\sim} w$.

Assume now $ua \notin F(x)$. Let $\bar{w}$ be the longest word such that $\bar{w} \in F(u)$ and $\bar{w} \underset{x}{\sim} w$. We have $\bar{w} \underset{xa}{\sim} w$ since otherwise u could not appear twice in x.

Consider a word vw with $w \neq 1$ and vw prefix of u. Let y be the longest word in $[vw]_x$; then $ua \notin F(x)$ leads to $y \underset{xa}{\not\sim} vw$. Together with the above properties, we may conclude that $[vw]_x$ are the only equivalence classes that are split in exactly two classes according to $\underset{xa}{\sim}$. This achieves the proof.

Corollary 6. Let $x \in A^*$, $e(x)$ be the number of states of $T(x)$.

If $|x| \leq 3$, $e(x) = |x|+1$.

If $|x| > 3$, $|x|+1 \leq e(x) \leq 2|x|-2$

and $e(x) = 2|x|-2$ iff $x \in ab^*c$, $a \neq b$, $b \neq c$.

Proof : It may be checked that if $|x| \leq 2$, $e(x)=|x|+1$ and if $|x|=3$, $e(x)=|x|+1=2|x|-2$.

Consider a word xc (c in A) with $|xc|>3$. Define u and v as in theorem 5. The result holds when $uc \in F(x)$ or $u=v$ by theorem 5 and induction hypothesis :

$$e(xc) = e(x)+1 \leq 2|x|-2+1 < 2|xc|-2.$$

Assume $uc \notin F(x)$ and $u \neq v$. By theorem 5 again :

$$e(xc) = e(xu^{-1}v)+2|v^{-1}u|+1.$$

When $|xu^{-1}v| \geq 3$, the induction hypothesis gives :

$$e(xc) \leq 2|xu^{-1}v|-2+2|v^{-1}u|+1 < 2|xc|-2.$$

If $|xu^{-1}v|<3$, the only possibility is $|xu^{-1}v|=2$, since otherwise x would be in a^* (for some a in A) and this would contradict $u \neq v$. We also deduce that $v=1$. Then

$$e(xc) = 3+2|u|+1 = 2|x| = 2|xc|-2.$$

In this situation the upper bound is reached. The word xu^{-1} may be written aa or ab (a and b in A, $a \neq b$). The case aa can be eliminated together with the case where u starts with a. It remains the case where x is in ab* which is the only possibility for xc to reach the upper bound. Hypothesis uc∉F(x) insures that c≠b.

<u>Corollary 7.</u> Let x∈A+, t(x) be the number of transitions of T(x).
If |x| ≤ 3 , |x| ≤ t(x) ≤ 2|x|-1.
If |x| > 3 , |x| ≤ t(x) ≤ 3|x|-4
and t(x) = 3|x|-4 iff x ab*c , a≠b , b≠c , a≠c.

Proof : Bounds comes from proposition 4 and corollary 6. For |x|>3, the upper bound can only be reached if x is in ab*c (a,b,c in A, a≠b, b≠c). If x is in ab*a, t(x)=3|x|-5. While, if x is in ab*c with a≠b, b≠c and a≠c, we have the maximum number of transitions in T(x), that is 3|x|-4.

EXAMPLE : Factor transducers for abbb and abbbc.

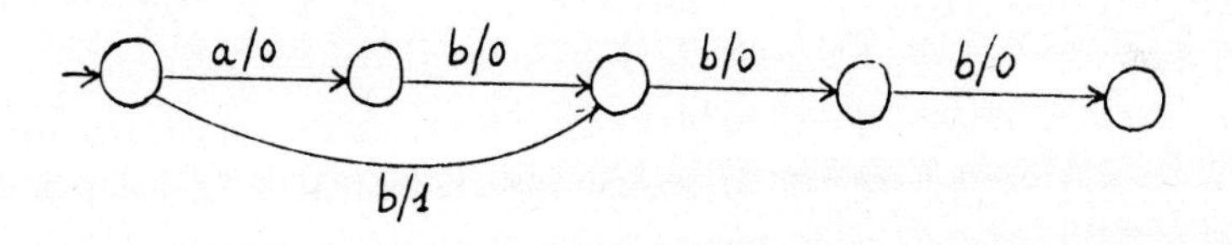

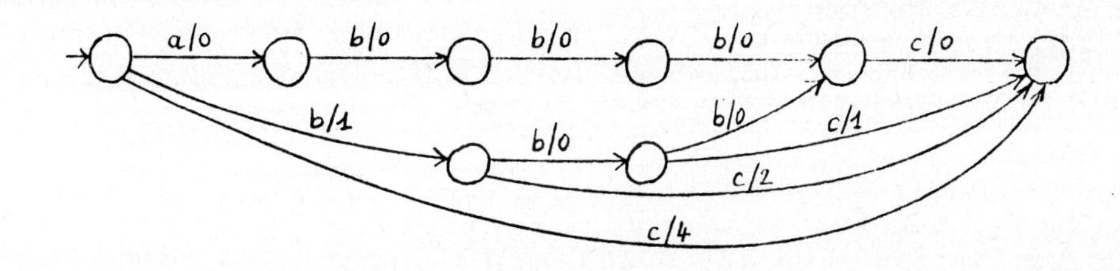

5. Construction of T(x).

Our algorithm to build T(x) for a given word x in A*, follows theorem 5 and its proof. It processes the word on-line. The structure of the algorithm is closed to that of Knuth, Morris and Pratt's algorithm, especially in the computation of function s through the procedure suffix which is written apart.

In the algorithm, init is the initial state of T(x) and last is init.y for the longest prefix of x already processed.

To each state q of T(x) are associated three attributes : s(q), p(q) and l(q). If w∈A* satisfies init.w=q then

s(q) is init.s(w) (see section 3),

p(q) is |p(w)| (see section 2),

and l(q) is the length of the longest such word w.

Assume that T(y) has just been built for a prefix y of x. Let u and v be defined on y as in theorem 5. Then the variable n has value |y|. When reading the next letter a of x some states of T(y) might be duplicated to get T(ya). Variables m and r help remember the path containing these states :

m is $|yu^{-1}v|$,

and r is init.v.

Are also used variables r', which is $init.yu^{-1}v$, and art which is an artificial state on which transitions by all the letters of the alphabet are defined and lead to init. It is as if a marker were at the beginning of the word x and behaved as any letter of x.

Each time a letter is read, a new state q is created and linked to last. According to the condition in theorem 5, new states are possibly created. The link s on q is then defined by a call to the procedure suffix (which is also called while states are duplicated). The aim of suffix is to return init.w for the longest suffix w of y (or u) such that wa is in F(y). it also creates transitions onto init.y if necessary. At the end of the main loop of the algorithm m,r,r' are possibly changed and last,n redefined.

There is (at least) one point that has not already been proved and which concerned the next value of variable r. This is done in the next lemma.

Lemma 8. Let x∈A* and u=s(x). If a∈A, ua∉F(x) and, for some w∈A*, wa is the longest suffix of xa (or ua) such that wa∈F(x) then

$$|w| = \max\{|w'| \;/\; w'\in[w]_{xa}\}.$$

Proof : Let w' be such that $w' \sim_{xa} w$ and |w'|>|w|. By lemma 1 it can be assumed that w'=bw for some letter b.

Since wa∈F(x), waza is suffix of xa for some z∈A*. Then bwaza∈F(xa) which proves that bwa-F(x). Therefore bw cannot be a suffix of x since this would contradict the definition of w.

From u=s(x), we know that u appears at least twice in x and this means that uya∈S(xa) for some y in A+. The word w being a suffix of u we have wya∈S(xa) and again from $bw \sim_{xa} w$ we get bwya∈F(xa). Since bw is not a suffix of u, bwya∉S(xa) and then bwya∈F(x). A contradiction arises since wy is a suffix of x with wy in F(x) and of longer length than w. This completes the proof.

While not necessary in the construction of T(x) in linear time, the above lemma gives all its simplicity to the algorithm. It allows to recompute the new words u and v as in theorem 5 (or more exactly init.v and $init.xu^{-1}v$) by only one test.

```
begin  create state art;  l(art) <- p(art) <- -1;
   create new state init;  l(init) <- p(init) <- 0;
   last <- init;           s(init) <- art;
   m <- n <- 0;  r <- art;  r' <- init;

   while  input not empty  do
      read next letter a;  art.a <- init;
      create new state q;  l(q) <- l(last)+1;  p(q) <- p(last)+1;
      last.a <- q;  last*a <- 0;

      if  s(last).a defined  then  s(q) <- s(last).a
      else  while  m<n  do
                  m <- m+1;  b <- x_m;  r' <- r'.b;
                  create copy r̄ of r.b
                               with same transitions and attributes;
                  l(r̄) <- l(r)+1;  s(r.b) <- r̄;  s(r') <- r̄;
                  r.b <- r̄;  r*b <- p(r̄)-p(r)-1;
                  s(r̄) <- suffix(r,b).b;
                  r <- r̄
               end while;
               r <- suffix(last,a);  s(q) <- r.a
      end if;

      if  m=n and l(r.a)=l(r)+1
      then  m <- n+1;  r <- r.a;  r' <- q  end if;
      last <- q;  n <- n+1;  x_n <- a
   end while
end.


function  suffix(r,b);
   if s(r).b undefined or l(s(r).b) ≥ l(r.b)
   then  s(r).b <- r.b;  s(r)*b <- p(s(r).b)-p(s(r))-1;
         return(suffix(s(r),b))
   else  return(s(r))  end if
end function.
```

Construction of factor transducers.

EXAMPLE : One step in the algorithm, from aabbabb to aabbabbb.

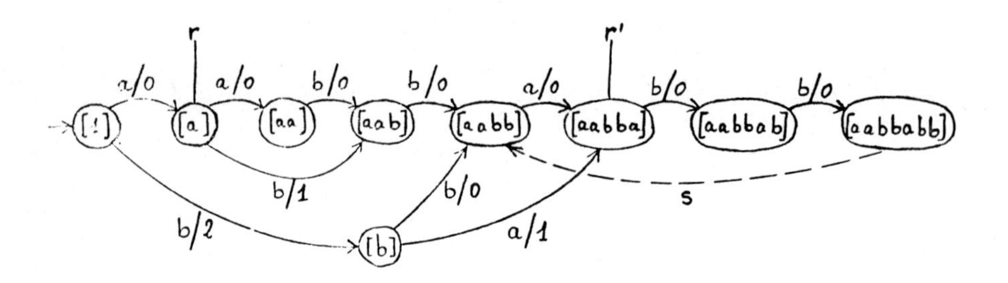

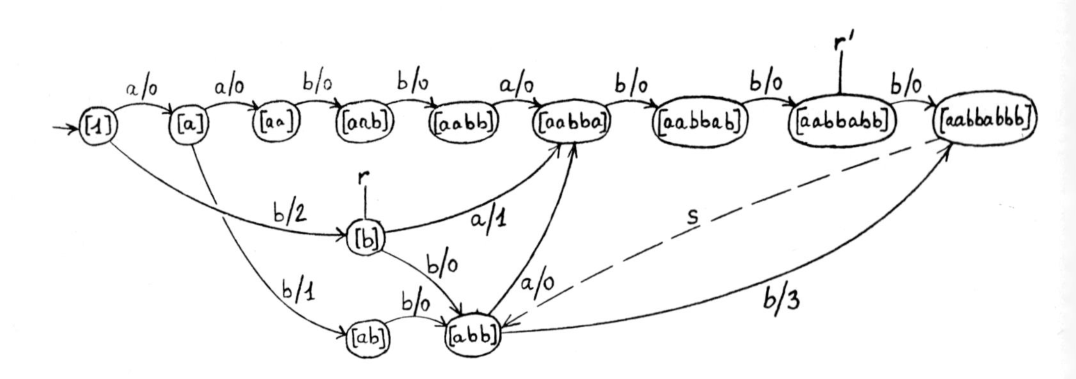

6. Implementation and complexity.

It is not hard to be convinced that the time to build T(x) is proportionnal to the sum of the three quantities :

- number of created states,
- number of created transitions,
- number of redefined transitions.

Instructions concerning transitions onto duplicated states together with calls to suffix contribute to the last number. The number of redefined transitions onto duplicated states is again proportionnal to the number of created states in the worst case. Each recursive call to suffix strictly shorten the word u in theorem 5 and then there are at most |x| such calls. Results of section 4 lead to an O(k.|x|) time complexity. The size of the alphabet A contributes to constant k when transitions are computed.

If an array indexed by letters is attached to states of T(x) and memorize transitions defined on the state, then an O(|A|.|x|) is obtained. However this seems to be a good way to implement T(x) when the alphabet is known before reading the word x and is relatively small.

In other cases, search trees or hashed tables can be used. The time complexity becomes O(|x|.log|A|) and the space complexity O(|x|).

Theorem 9. On a given alphabet A, the factor transducer T(x) of a word x in A* may be built in time linear in the length of x.

EXAMPLE : Two possible implementations of states of transducers, using arrays or AVL trees.

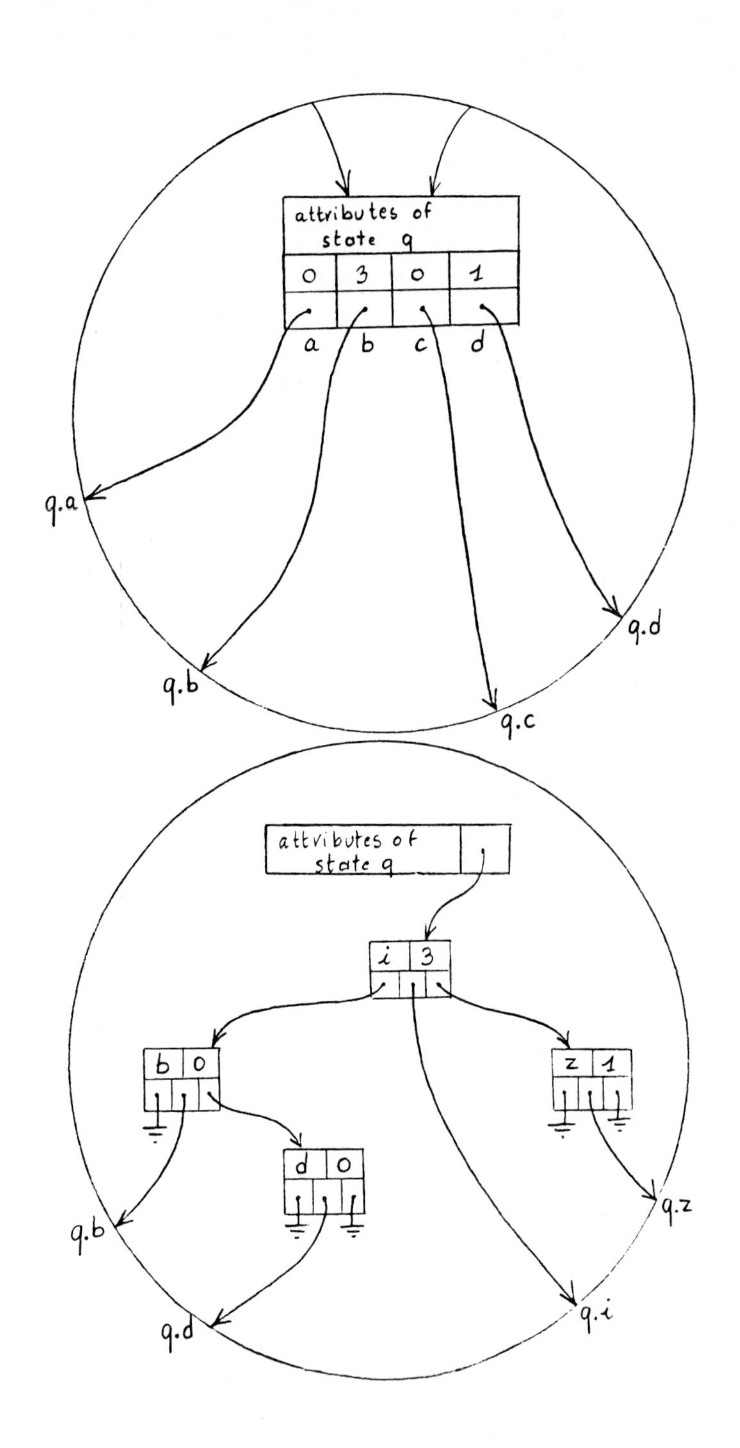

7. References.

[AHU 74] A.V.AHO, J.E.HOPCROFT, J.D.ULLMAN.
The design and analysis of computer algorithms,
(Addison-Wesley, Reading, Mass., 1974).

[AU 79] A.V.AHO, J.D.ULLMAN.
Principles of compiler design,
(Addison-Wesley, Reading, Mass., 1979).

[AP 83] A.APOSTOLICO, F.P.PREPARATA.
Optimal off-line detection of repetitions in a string,
Theor.Comput.Sci.,22(1983)297-315.

[BBEHM 83] A.BLUMER, J.BLUMER, A.EHRENFEUCHT, D.HAUSSLER, R.McCONNELL.
Linear size finite automata for the set of all subwords of a word,
Bulletin of the EATCS,21(1983)12-20.

[BM 77] R.BOYER, J.S.MOORE.
A fast string searching algorithm,
Commun.Assoc.Comput.Mach.,20,10(1977)762-772.

[Cr 83] M.CROCHEMORE.
Recherche lineaire d'un carre dans un mot,
C.R.Acad.Sci.Paris,t.296,serie I(1983)781-784.

[KMP 77] D.E.KNUTH, J.H.MORRIS, V.R.PRATT.
Fast pattern-matching in strings,
S.I.A.M.J.Comput.,6(1977)323-350.

[Mc 76] E.M.McCREIGHT.
A space-economical suffix tree construction algorithm,
J.A.C.M.,23,2(1976)262-272.

[Ro 82] M.RODEH.
A fast test for unique decipherability based on suffix tree,
I.E.E.E.Trans.Information Theory,28,4(1982)648-651.

[We 73] P.WEINER.
Linear pattern-matching algorithms,
in:(Proceedings of the 14th annual symposium on switching and automata theory, 1973)1-11.

RELATING THE AVERAGE-CASE COSTS OF THE BRUTE-FORCE AND KNUTH-MORRIS-PRATT STRING MATCHING ALGORITHM

Gerhard Barth
Fachbereich Informatik, Universität Kaiserslautern
D-6750 Kaiserslautern, West Germany

Key Words: String Matching, Analysis of Combinatorial Algorithms, Markov Chain Theory

1. Introduction

Among the algorithms which are able to check in $O(n+m)$ steps, if a string PATTERN = $p_1p_2...p_m$ occurs in a string TEXT = $t_1t_2...t_n$, the one developed by Knuth, Morris and Pratt (KMP, for short) plays kind of a fundamental role. Its linear worst-case time complexity contrasts sharply with the $O(n \cdot m)$ upper bound for the overhead of the brute-force strategy which naively probes each position i, for $1 \leq i \leq n-m+1$, in TEXT for a complete match

$$p_1p_2 \cdots p_m = t_it_{i+1} \cdots t_{i+m-1}$$

However, the average-case performance of the KMP algorithm has been suspected not to be drastically better than that of the naive method [9]:

> "The Knuth-Morris-Pratt algorithm is not likely to be significantly faster than the brute-force method in most actual applications, ..."

The main objective of this paper is to elaborate on this observation and to present a detailed and accurate average-case analysis of both the brute-force and the KMP algorithm. The analysis exploits results from Markov chain theory. This approach is believed to be practically sound, since string matching can be modeled conveniently by finite-state devices.

NATO ASI Series, Vol. F12
Combinatorial Algorithms on Words
Edited by A. Apostolico and Z. Galil

Specifically, the following results are derived for string matching with alphabets of c characters:

(1) The brute-force method performs an average number of

$$c^{m+1}/(c-1) - c/(c-1)$$

steps to locate the leftmost occurrence of PATTERN in TEXT.

(2) The KMP method performs an average number of

$$c^{m} + (1/(c-1))c^{m-1} + c - c/(c-1)$$

steps to locate the leftmost occurrence of PATTERN in TEXT.

(3) An accurate approximation for the ratio KMP/NAIVE, where KMP and NAIVE denote the average case complexities of the KMP and naive string matching algorithms, respectively, is given by the term

$$1 - (1/c) + (1/c^{2})$$

The last result clearly quantifies how close the costs of the naive and the KMP algorithm are to each other in most applications.

The method of analysis used in this paper can be conveniently applied to a variety of other combinatorial algorithms. Surprisingly enough, none of the standard textbooks on the design and analysis of computer algorithms contains (to the best of my knowledge) an analysis performed along the lines outlined in this paper.

2. Two Algorithms for Substring Searching

A problem frequently encountered in text processing is to search for occurrences of a string as a substring in an other one. To be more specific, let PATTERN = $p_1p_2...p_m$ and TEXT = $t_1t_2...t_n$ denote two strings. PATTERN is said to be a (contiguous) substring of TEXT, if

$$p_1p_2 \cdots p_m = t_k t_{k+1} \cdots t_{k+m-1} \qquad (1.1)$$

holds for some k, $1 \le k \le n-m+1$. Various forms of the substring searching problem aim at finding the smallest, the largest, any, or all indices k meeting the requirements of (1.1). Here we will concentrate on finding the smallest of these indices k. This amounts to locating the leftmost occurrence of PATTERN in TEXT. A straightforward solution to this problem aligns TEXT and PATTERN side by side and compares corresponding characters. As soon as a mismatch is detected, PATTERN is shifted one position towards the right end of TEXT and the search is resumed at the first character of PATTERN. Figure 1 pictorially describes this simple strategy.

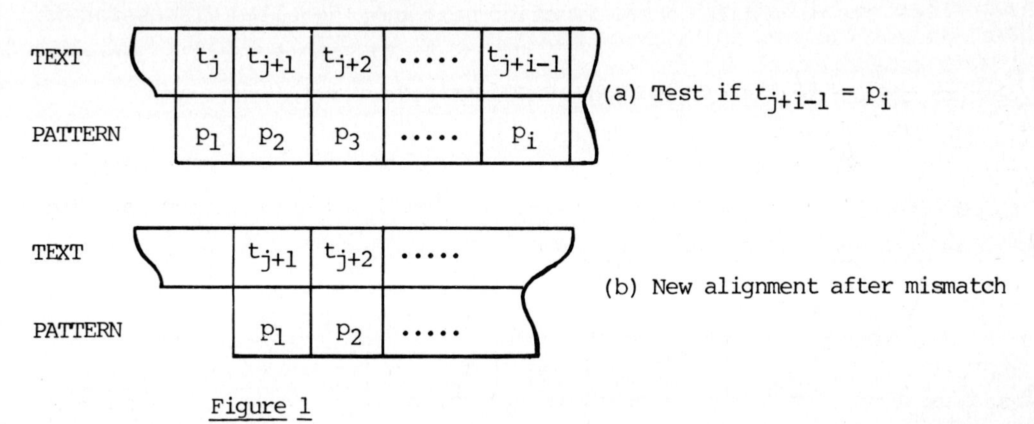

Figure 1

```
algorithm STRING_NAIVE
  pp := 1;                              {initialize pattern pointer}
  tp := 1;                              {initialize text pointer}
  while (pp ≤ m) and (tp ≤ n) do
    if  PATTERN[pp] = TEXT[tp]
        then pp := pp+1; tp := tp+1     {advance both pointers}
        else pp := 1; tp := tp-pp+2     {retract both pointers}
    endif
  endwhile;
  if pp > m then write ('leftmost occurrence at', tp-m)
         else write ('no occurrence found')
  endif
end STRING_NAIVE
```

Algorithm STRING_NAIVE given above implements this strategy. It is very simple to see that STRING_NAIVE may require as many as $O(n \cdot m)$ comparisons of characters in the worst possible case. A clever algorithm for substring searching with an $O(n+m)$ worst case complexity has been developed by Knuth, Morris and Pratt, see [7]. The basic idea of this method is never to retract pointer tp to the left. Instead, after a mismatch between PATTERN[pp] and TEXT[tp] has been detected, the search is resumed with comparing PATTERN[next(pp)] and TEXT[tp]. Thereby, next is a function defined as

$$\underline{\text{next}}(i) = \max\{k \mid 0 \le k < i,\ p_1 p_2 \dots p_{k-1} = p_{i-k+1} p_{i-k+2} \dots p_{i-1} \text{ and } p_k \neq p_i\}$$

for $1 \le i \le m$. The rationale behind function next and a detailed discussion of how to compute it is not given here, the reader is referred to [7]. Figure 2 depicts the critical step involved in this method.

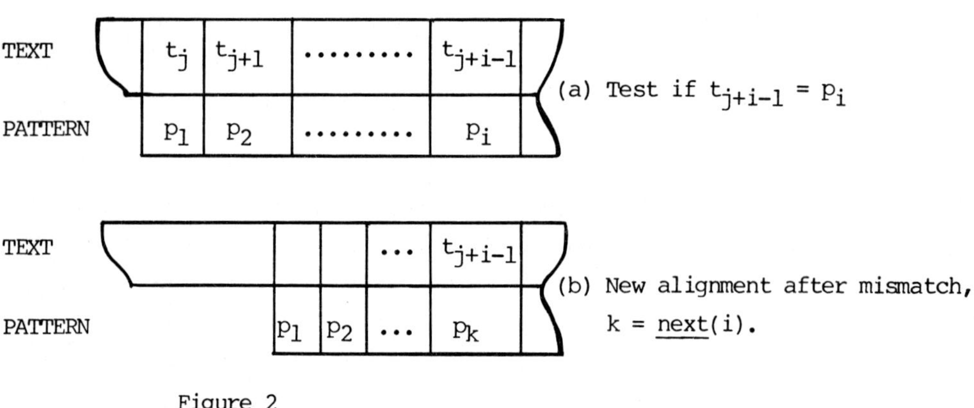

Figure 2

Algorithm STRING-KMP given below implements the strategy proposed by Knuth, Morris and Pratt.

```
algorithm STRING_KMP
  pp := 1; tp := 1;                          {initialize pointers}
  while (pp ≤ m) and (tp ≤ n) do
    while  (PATTERN[pp] ≠ TEXT[tp]) and (pp > 0) do
           pp := next(pp)
    endwhile;
    pp := pp+1; tp := tp+1                   {advance pointers}
  endwhile;
  if pp > m then write ('leftmost occurrence at', tp-m)
         else write ('no occurrence found')
  endif
end STRING_KMP
```

In the remainder of this paper the average case performances of both STRING_NAIVE and STRING_KMP will be investigated. The analyses center around Markov chain techniques. The required concepts will be briefly introduced in the subsequent section.

3. Markov Chains

An extensive treatment of Markov chains will not be given here, the reader is referred to the literature, like [10]. Instead, an intuitively appealing characterization shall suffice. A finite *Markov chain* is a stochastic process that at any given time is in one of a finite number of states $s_1, s_2, \ldots, s_r$, say. The probability that the process moves from s_i to s_j depends only on state s_i and not on the history of how s_i was reached. A transition probability p_{ij} is given for any pair of states. A state is *absorbing* if it is impossible to leave it. A Markov chain is called absorbing if

(i) it has at least one absorbing state, and
(ii) from every state it is possible to reach an absorbing state.

The transition probabilities p_{ij}, $1 \leq i,j \leq r$, can be stored in an $r \times r$ matrix P. For an absorbing Markov chain with absorbing states $s_{k+1}, \ldots, s_r$ and non-absorbing states $s_1, \ldots, s_k$ matrix P can be arranged as

$$P = \begin{bmatrix} Q_{k\times k} & A_{k\times(r-k)} \\ \\ 0_{(r-k)\times k} & I_{(r-k)\times(r-k)} \end{bmatrix}$$

where Q is the submatrix for transitions among non-absorbing states, A is the submatrix for transitions from non-absorbing into absorbing states, 0 and I are the null and identity submatrix, respectively. Matrix $F := (I-Q)^{-1}$ is called fundamental matrix. In essence, we need only one result from the theory of absorbing Markov chains to assist us in analyzing algorithms STRING_NAIVE and STRING_KMP.

THEOREM 3.1: f_{ij} of the fundamental matrix F equals the expected number of visits of state s_j, the process makes before absorption, provided it was started in state s_i. (For a proof see [10]).

COROLLARY 3.2: $\Sigma_{j=1}^{k} f_{ij}$ equals the expected number of steps the process makes from a start in state s_i until absorption.

Example (see next page): Figure 3a contains the description of an absorbing Markov chain with states 1,2,3,4 and 5. States 4 and 5 are absorbing, as follows from the transition probabilities $p_{4,4}$ and $p_{5,5}$ being 1. Figures 3b and 3c show the transition probability matrix P of this chain and its fundamental matrix F, respectively. The dashed lines in Figure 3b indicate the aforementioned partition of P into four submatrices. Matrix F tells us, for example, that when starting in state 2, the process may be expected to return to state 2 once, which results in a total number of two visits to state 2. Furthermore, we may expect that the process will be absorbed after 4 steps (either in state 4 or state 5), if being released in state 2.

(End of example)

Of course, the theory of absorbing Markov chains encompasses many more results than just Theorem 3.1 and Corollary 3.2, the interested reader may wish to consult [10]. Yet, the two facts cited above will suffice to analyse the average case performances of STRING_NAIVE and STRING_KMP.

4. Analysis of STRING NAIVE and STRING KMP

From Figure 4, it is straightforward to see that execution of algorithm STRING_NAIVE can conveniently be modeled as the activation of a deterministic finite automaton with m+1 states. At any moment, the current value of pp determines the state this automaton acts in. The characters $t_1, t_2, \ldots t_n$ of TEXT serve as successive input signals. When reading in state i character t_j, the automaton either advances to state i+1 or returns to state 1, depending on whether t_j matches p_i or not. As soon as state m+1 is entered the automaton terminates.

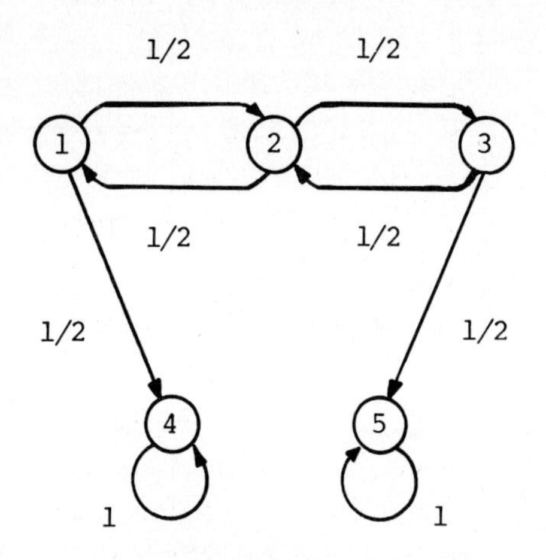

3a: A Markov Chain

$$
\left[\begin{array}{ccc|cc}
0 & 1/2 & 0 & 1/2 & 0 \\
1/2 & 0 & 1/2 & 0 & 0 \\
0 & 1/2 & 0 & 0 & 1/2 \\
\hline
0 & 0 & 0 & 1 & 0 \\
0 & 0 & 0 & 0 & 1
\end{array}\right]
$$

3b: Transition Probability Matrix P

$$
\begin{bmatrix}
3/2 & 1 & 1/2 \\
1 & 2 & 1 \\
1/2 & 1 & 3/2
\end{bmatrix}
$$

3c: Fundamental Matrix F

Figure 3

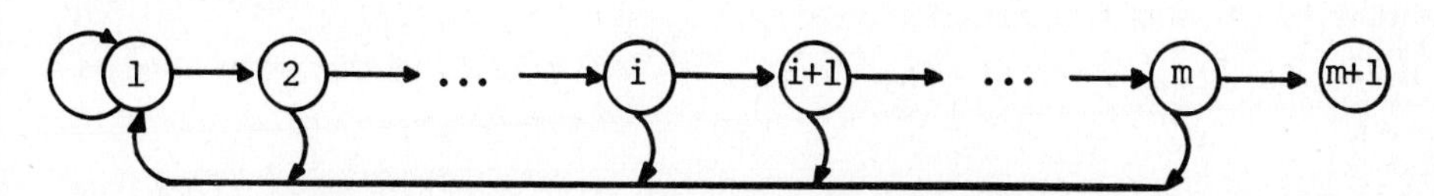

Figure 4: Automaton Modeling STRING_NAIVE

In order to apply Theorem 3.1 and Corollary 3.2, the automaton in Figure 4 is turned into an absorbing Markov chain by the following method:

1. Assign probabilities a_i (for "advance") to transitions from state i to state i+1, for $1 \le i \le m$.
2. Assign probabilities b_i (for "go back") to transitions from state i to state 1, for $1 \le i \le m$.
3. Add a transition from state m+1 to state m+1 and assign probability 1 to it.

Thus, we can construct an absorbing Markov chain with the single absorbing state m+1. Submatrix $Q_{m \times m}$ of the chain's transition probability matrix $P_{(m+1)\times(m+1)}$ contains the entries

$$q_{ij} = \begin{cases} b_i & \text{for } 1 \le i \le m,\ j = 1 \\ a_i & \text{for } 1 \le i < m,\ j = i+1 \\ 0 & \text{otherwise} \end{cases}$$

From there, the fundamental matrix $F_{m \times m} = (I_{m \times m} - Q_{m \times m})^{-1}$ can be computed, whose coefficients are

$$f_{ij} = \begin{cases} 1/(a_j a_{j+1} \cdots a_m) & \text{for } i = 1,\ 1 \le j \le m \\ (1 - a_i a_{i+1} \cdots a_m)/(a_j a_{j+1} \cdots a_m) & \text{for } i \ge 2,\ 1 \le j < i \\ 1/(a_j a_{j+1} \cdots a_m) & \text{for } i \ge 2,\ i \le j \le m \end{cases}$$

This can easily be verified by multiplying F with (I−Q) to obtain the identity matrix I. Let us elaborate on the results derived so far by substituting values for the parameters a_i and b_i. If both PATTERN and TEXT are random strings of characters drawn from an alphabet with c elements, the transition probabilities are $a_i = 1/c$ and $b_i = (c-1)/c$ for $1 \le i \le m$. Entries in the first row of the fundamental matrix F then become $f_{1j} = c^{m-j+1}$, for $1 \le j \le m$. Hence, we have the following result.

RESULT 4.1: For random strings over an alphabet with c characters, algorithm STRING_NAIVE performs an average number of

$$c^{m+1}/(c-1) - c/(c-1)$$

steps to locate the leftmost occurrence of PATTERN in TEXT.

Proof: From Corollary 3.2 and the fact that STRING_NAIVE always starts in state 1, it follows that we have to evaluate $\Sigma_{j=1}^{m} f_{1j}$.

$$\Sigma_{j=1}^{m} f_{1j} = \Sigma_{j=1}^{m} c^{m-j+1} = \Sigma_{j=1}^{m} c^{j} = c^{m+1}/(c-1) - c/(c-1).$$

As to non-random strings, it is not obvious at all which transition probabilities a_i und b_i should be substituted. Note that the technique employed in the above analysis is independent of any particular choice of values for the parameters a_i and b_i. The reader is invited to pick his favorite transition probabilities for non-random strings and to evaluate $\Sigma_{j=1}^{m} f_{1j}$ with coefficients $f_{1j} = 1/(a_j a_{j+1} \ldots a_m)$ as calculated above.

Let us now turn our attention to the analysis of STRING_KMP. First of all, an automaton has to be tailored to the given string PATTERN = $p_1 p_2 \ldots p_m$. Figure 5 presents the general picture.

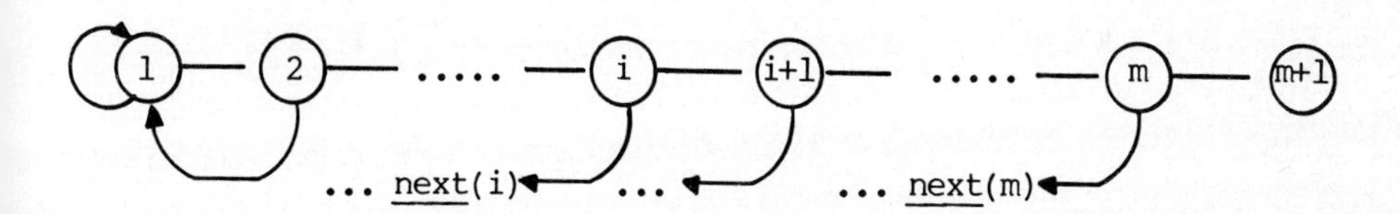

Figure 5: Automaton Modeling STRING_KMP

The specific values of next(i), $1 \le i \le m$, depend on the actual string PATTERN. For an average case analysis of STRING_KMP, the expected values $E[X_i]$ of the random variables X_i defined as $X_i = j$, iff control returns to state j after a mismatch in state i, have to be computed for $3 \le i \le m$. It holds that

$$E[X_i] = 1 \cdot (P\{\underline{next}(i) = 1\} + P\{\underline{next}(i) = 0\})$$

$$+ 2 \cdot P\{\underline{next}(i) = 2\}$$

$$+ \ldots\ldots\ldots$$

$$+ (i-1) \cdot P\{\underline{next}(i) = i-1\}$$

From the definition of function next, see Section 2, it follows that

$$P\{\underline{next}(i) = j\} = P\{p_1 \ldots p_{j-1} = p_{i-j+1} \ldots p_{i-1} \text{ and } p_j \neq p_i\}$$

$$\cdot P\{\text{no } k>j \text{ exists with } p_1 \ldots p_{k-1} = p_{i-k+1} \ldots p_{i-1}$$

$$\text{and } p_k \neq p_i\}$$

$$\leq P\{p_1 \ldots p_{j-1} = p_{i-j+1} \ldots p_{i-1} \text{ and } p_j \neq p_i\}$$

For random strings drawn from an alphabet with c symbols the probability of any two symbols to match is $a = 1/c$. Let b denote $1-a$, i.e. the probability of a mismatch between any two symbols. Hence we have $P\{\underline{next}(i) = j\} \leq a^{j-1}b$ and can continue as follows.

$$E[X_i] \leq P\{\underline{next}(i) = o\} + \Sigma_{j=1}^{i-1} ja^{j-1}b$$

$$\leq 1 + b \cdot \Sigma_{j=1}^{i-1} ja^{j-1}$$

$$< 1 + b \cdot \Sigma_{j=1}^{\infty} ja^{j-1}$$

$$= 1 + b/(1-a)^2 \qquad (\text{for } 0<a<1)$$

$$= 1 + 1/b$$

$$= 2 + 1/(c-1)$$

So we may conclude that for $c \geq 2$ (note that for unary alphabets the substring search problem is trivial!) the expected values $E[X_i]$ are bounded by $2 \leq E[X_i] \leq 3$. The sizes of actual computer codes (ASCII, EBCDIC, BCDIC, etc.) are large enough to warrant the usage of 2 as a very accurate approximation for every $E[X_i]$. It has to be admitted that this constitutes a heuristic approach, but since it gets rid of some complex conditioning it is believed to be practically sound. Consequently, the state diagram shown in Figure 6 may serve as a heuristic Markov chain model for the execution of STRING_KMP.

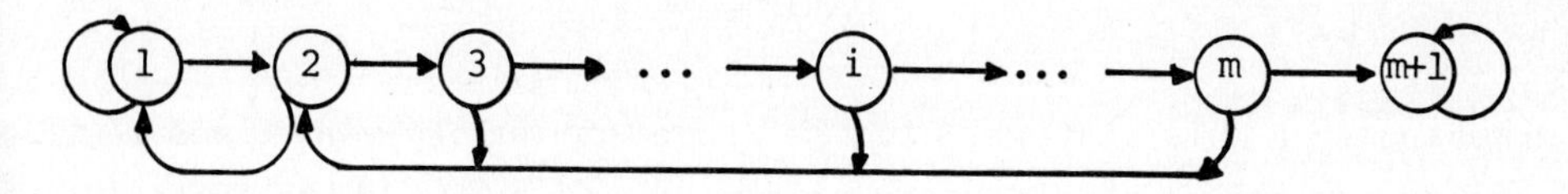

Figure 6: A Heuristic Markov Chain Model for STRING_KMP

The foregoing discussion has been based on the assumption that random strings are involved in a substring search implemented by STRING_KMP. Here, again, the identical remarks as made before about STRING_NAIVE apply, viz. that the applicability of Markov chain techniques to an average case analysis is not restricted to this situation. Each reader may choose a probability distribution he believes to be a good representation for the likelihood of matches and mismatches for a sample of non-random strings. Thereafter, he can persue the same way as described here to analyze STRING_KMP.
From Figure 6 follows that we have to compute the fundamental matrix $F_{m\times m} = (I_{m\times m} - Q_{m\times m})^{-1}$, where $Q_{m\times m}$ contains the following coefficients

$$q_{ij} = \begin{cases} b & \text{for } (j = 1;\ i = 1,2) \text{ or } (j = 2,\ 3 \leq i \leq m) \\ a & \text{for } j = i+1,\ 1 \leq i < m \\ 0 & \text{otherwise} \end{cases}$$

It turns out that $F_{m\times m}$ has the form

$$f_{ij} = \begin{cases} (a^{m-1}+b)/a^m & \text{for } i = 1,\ j = 1 \\ b/a^m & \text{for } i = 2,\ j = 1 \\ ((1-a^{m-i+1})b)/a^m & \text{for } 3\leq i\leq m,\ j = 1 \\ (1-a^{m-i+1})/a^{m-j+1} & \text{for } 2\leq j<i\leq m \\ 1/a^{m-j+1} & \text{otherwise} \end{cases}$$

The correctness of this statement may be proved by multiplying F and I-Q, which yields as product the identity matrix. Now we are ready to state our second result.

RESULT 4.2: For random strings over an alphabet with c characters, algorithm STRING_KMP performs an average number of

$$c^m + (1/(c-1))c^{m-1} + c - c/(c-1)$$

steps to locate the leftmost occurrence of PATTERN in TEXT.

Proof: By Corollary 3.2, the average number of steps performed by STRING_KMP is $\Sigma_{j=1}^{m} f_{1j}$ with coefficients f_{ij} as given above. Hence we get the sum

$$\begin{aligned} \Sigma_{j=1}^{m} f_{1j} &= (a^{m-1}+b)/a^m + \Sigma_{j=2}^{m} 1/a^{m-j+1} \\ &= 1/a + (1/a)^m - (1/a)^{m-1} + \Sigma_{j=1}^{m-1} (1/a)^j \\ &= c + c^m - c^{m-1} + (c^m-c)/(c-1) \\ &= c^m + c^m/(c-1) - c^{m-1} + c - c/(c-1) \\ &= c^m + c^{m-1}(c/(c-1)-1) + c - c/(c-1) \\ &= c^m + (1/(c-1))c^{m-1} + c - c/(c-1). \end{aligned}$$

Having derived quantities KMP and NAIVE, say, where

$$KMP = c^m + (1/(c-1))c^{m-1} + c - c/(c-1) \quad \text{and}$$

$$NAIVE = c^{m+1}/(c-1) - c/(c-1),$$

for the average case complexities of STRING_KMP and STRING_NAIVE, respectively, it is very informative to study the ratio KMP/NAIVE. We get

$$KMP/NAIVE \simeq [c^m + (1/(c-1)c^{m-1} + c]/[c^{m+1}/(c-1)]$$

$$= [c^{m+1} - c^m + c^{m-1} + c(c-1)]/c^{m+1}$$

$$= 1 - (1/c) + (1/c^2) + (c-1)/c^m$$

For $m = 2$ this ratio equals 1, which has to be the case, indeed, since then the Markov chains for STRING_NAIVE and STRING_KMP coincide (compare Figures 4 and 6). For m exceeding 2 the ratio KMP/NAIVE rapidly approaches $1 - (1/c) + (1/c^2)$, a quantity which is always smaller than 1, yet for computer alphabets of sizes $c = 64,128,\ldots$ by a negligible amount only.

We summarize the outcome of the foregoing analyses in stating that for random strings the difference in the run-times of both methods is on the average very close to zero. Taking into account the additional effort of computing function <u>next</u>, which takes $O(m)$ time, see [7], and requires m additional storage locations, we may even go a step further and expect that on the average STRING_NAIVE uses less resources than STRING_KMP.

5. Conclusion

Average case analyses of two substring searching algorithms have been conducted. In both cases methods from Markov chain theory have been employed. This has been expedient since both algorithms lend themselves to modeling their execution as the activation of finite automata. The latter statement is true for a large class of algorithms, whose average case analyses are therefore amenable to the application of Markov chain theory. Surprisingly enough, none of the standard textbooks on the design and analyses of computer algorithms, like [1],[5],[6],[8], contains (to the best of my knowledge) an analysis performed along the lines outlined in this paper. In [2], a more detailed discussion of the average case analyses of finite state algorithms by means of Markov chains is presented.

6. References

[1] A. Aho, J. Hopcroft, J. Ullman: The Design and Analysis of Computer Algorithms. Addison-Wesley, 1974

[2] G. Barth: Analyzing Algorithms by Markov Chains. Methods of Operations Research, Vol. 45, Athenäum Press, 1982, 405-418

[3] R. Boyer, J. Moore: A Fast String Searching Algorithm. CACM 20 (1977), 262-272

[4] L.J. Guibas, A.M. Odlyzko: String Overlaps, Pattern Matching and Nontransitive Games. Journal of Combinatorial Theory, Series A30 (1981), 183-208

[5] E. Horowitz, S. Sahni: Fundamentals of Computer Algorithms. Computer Science Press, 1978

[6] D. Knuth: The Art of Computer Programming, Vols. 1 and 3. Addison-Wesley, 1973

[7] D. Knuth, J. Morris, V. Pratt: Fast Pattern Matching in Strings. SIAM Journal on Computing 6 (1977), 323-350

[8] E. Reingold, J. Nievergelt, N. Deo: Combinatorial Algorithms. Prentice-Hall, 1977

[9] R. Sedgewick: Algorithms. Addison-Wesley, 1983

[10] J. Snell: Introduction to Probability Theory with Computing, Prentice-Hall, 1975

ALGORITHMS FOR FACTORIZING AND TESTING SUBSEMIGROUPS

Renato M. Capocelli
Dipartimento di Informatica ed Applicazioni
Universitá di Salerno
Salerno 84100, Italy

Christoph M. Hoffmann
Department of Computer Science
Purdue University
West Lafayette, Ind. 47907

ABSTRACT

Given a finite subset Γ of a fixed, finite alphabet Σ, we construct the basis B of the minimum subsemigroup of Σ^+ containing Γ, such that B has various properties. The properties we consider are that B be a uniquely decipherable, a finitely decipherable, a synchronizable, or a prefix code. The algorithm for constructing the uniquely decipherable and the finitely decipherable code B requires $O(n^2 L + L^2)$ steps, the algorithm for constructing the synchronizable code B requires $O(n L^2)$ steps, and the algorithm for constructing the prefix code B requires $O(L^2)$ steps. Here n is the cardinality of Γ and L is the sum of the lengths of the words in Γ. Finally, given a synchronizable or finitely decipherable code Γ, we also show how to determine its synchronizability or decipherability delay, in $O(n L)$ steps.

1. Introduction

Consider the problem of transmitting messages written in a source alphabet over a channel which has a different, smaller alphabet. In such a situation, the source message has to be encoded, and a particularly convenient method is to substitute, for each letter of the source alphabet, a nonempty word over the channel alphabet. Given a fixed channel alphabet Σ, a code is then a subset of Σ^+, the set of all nonempty words over Σ, and if we are presented with such a subset as proposed code, it is natural to investigate certain properties of this set.

The properties of unique decipherability, of finite decipherability, and of synchronizability are central to Coding Theory. They formalize whether a coded message can be unambiguously decoded, whether this decoding is possible without first receiving the entire message, and whether an infinite message can be decoded without knowing its beginning and its end.

Since every encoded message is a concatenation of words, these properties can also be understood as concepts from Semigroup Theory. Specifically, for finite sets $\Gamma \subset \Sigma^+$, the set Γ of code words generates the subsemigroup Γ^+ of Σ^+, where Σ^+ is the free semigroup with $|\Sigma|$ generators. Viewed in this way, Γ is uniquely decipherable precisely when Γ^+ is a free subsemigroup of Σ^+, Γ is finitely decipherable whenever Γ^+ is a weakly prefix subsemigroup [Cap 80], and Γ is synchronizable iff Γ^+ is a very pure subsemigroup [Res 74]. In each case, Γ must be the minimum generating set of Γ^+ as well.

In particular, Γ has a decipherability delay of zero iff no code word is prefix of another code word. Such a code is called a *prefix* code, and Γ^+ is a left unitary subsemigroup. Prefix codes play an important role in Coding Theory both for their intrinsic properties, as well as

NATO ASI Series, Vol. F12
Combinatorial Algorithms on Words
Edited by A. Apostolico and Z. Galil

because they are easily constructed.

Sardinas and Patterson [SP 53] and Levenshtein [Lev 62, Lev 64] provided criteria for testing whether a code is uniquely decipherable, finitely decipherable, or synchronizable. Variations of these criteria can be found in [Mar 62, Eve 63, Eve 64]. A unified treatment of the above properties can be found in [Ril 67, Spe 75, Cap 79].

In case a set Γ does not possess some or all of these properties, one might want to determine the "most convenient" codes Δ possessing them and such that Γ^+ is a subset of Δ^+. It turns out that such codes are the unique minimum generating sets for the minimum subsemigroup of Σ^+ containing Γ and possessing the desired properties. This question has been investigated in [Spe 75, Cap 77, Cap 80]. The purpose of this paper is to give efficient algorithms for finding such codes with the desired properties, given a set Γ not enjoying them. Such algorithms find application in, e.g., data compression [CR 78]. Past work on such algorithms has been done in [Spe 75, Cap 77, DR 79, Ber 79, Cap 80], but specific complexity bounds on their performance have not been established except in the most general sense. We rectify this situation by deriving efficient algorithms for the above mentioned problems and analyzing in detail their complexity. Moreover, we give algorithms for determining the decipherability delay and the synchronization delay for codes.

Most of our algorithms depend on a new test for unique decipherability whose performance puts it on equal footing with the most efficient unique decipherability tests previously reported [Rod 82, AG 84]. Unlike these other algorithms, the one given here has special properties which make it uniquely suitable to our purposes.

2. Terminology and Basic Algorithms

In the following, Σ is a fixed finite alphabet, Σ^+ the free semigroup generated by Σ, and $\Sigma^* = \Sigma^+ \cup \{\lambda\}$ is the free monoid over Σ. Let Γ be a finite subset of Σ^+. We call elements of Γ *code words*. Γ^+ is the set of all words in Σ^+ which are concatenations of code words. The elements of Γ^+ are usually called *messages*. When considering Σ^+ as the free semigroup generated by Σ, Γ^+ may be viewed as a subsemigroup of Σ^+.

Definition
A subsemigroup P of Σ^+ is *left unitary* if, for all $p \in P$ and all $m \in \Sigma^+$, $m \in P$ whenever $pm \in P$.

Definition
A set $\Gamma \subset \Sigma^+$ is a *prefix code* if no code word is prefix of another code word.

It is easily verified that a subsemigroup of a free semigroup is left unitary iff its unique minimum generating set, $P - P^2$, is a prefix code. Here P^2 is the set of all elements of P obtained as p_1p_2, where $p_1, p_2 \in P$. In the following, we refer to this unique minimum generating set as the *basis* of P.

Definition
A set $\Gamma \subset \Sigma^+$ is *uniquely decipherable* if, for all $c, c' \in \Gamma$ and $x, y \in \Gamma^*$ the equation $cx = c'y$ has no solution with $c \neq c'$.

If Γ is not uniquely decipherable then the solution to $cx = c'y$ yields the ambiguous message cx, which can be deciphered in at least two different ways. Clearly Γ is uniquely decipherable iff Γ^+ is a free subsemigroup of Σ^+ with basis Γ (see [Lal 79]).

Definition
A subsemigroup S of Σ^+ is ***very pure*** if, for all $x, y \in \Sigma^+$, $xy \in S$ and $yx \in S$ imply that $x, y \in S$.

Definition
A set $\Gamma \subset \Sigma^+$ is a ***synchronizable code*** if there exists a number $s > 0$ such that, for all $f \in \Gamma^s$, there exist $f_1, f_2 \in \Sigma^*$ such that $f = f_1 f_2$ and, for all $u, v \in \Sigma^*$, $ufv \in \Gamma^*$ implies that both $uf_1 \in \Gamma^*$ and $f_2 v \in \Gamma^*$.

The smallest integer s for which Γ satisfies this definition is the ***synchronization delay*** of Γ.

Intuitively, Γ is synchronizable if it is possible to decipher message fragments of sufficient length. Synchronizability delay s here means that whenever a fragment contains s consecutive code words, then a correct code word boundary can be determined. Note that our definition of synchronizability delay differs from the one made in terms of synchronizing pairs (e.g., [Lal 79]). However, the two definitions of *synchronizable code* are clearly equivalent.

Obviously, a very pure subsemigroup is a free subsemigroup. A finitely generated subsemigroup is very pure iff its basis is a synchronizable code [Res 74].

Definition
A subsemigroup D of Σ^+ is said to be ***weakly prefix*** if, for all $a \in D$ and $x, y \in \Sigma^+$, the relations $ax \in D$, $xy \in D$, and $yx \in D$, together imply $x, y \in D$.

Obviously a weakly prefix subsemigroup of Σ^+ is also a free subsemigroup. Moreover, a left unitary (or very pure) subsemigroup is also weakly prefix.

Definition
A subset $\Gamma \subset \Sigma^+$ is ***finitely decipherable*** if there is an integer $s \geq 0$ such that, for all $u \in \Sigma^*$, $x \in \Gamma^+$, and $y \in \Gamma^s$, $xyu \in \Gamma^*$ implies that $yu \in \Gamma^*$.

The smallest integer for which Γ satisfies the definition is called the ***decipherability delay*** of Γ. Intuitively, Γ has decipherability delay s if the next code word can be correctly deciphered only after s subsequent code words have been received (unless, of course, the message has ended). Conversely, a uniquely decipherable code with infinite decipherability delay does not permit a correct decoding until the entire message has been received. It can be shown that if D is a finitely generated, weakly prefix subsemigroup, then its basis is a code with finite decipherability delay [Cap 80].

We note that the notions of synchronizable codes and finitely decipherable codes are the natural extensions of comma-free codes [GGW 58] and prefix codes, respectively.

It is clear that the intersection of a family of left unitary (free, very pure, weakly prefix) subsemigroups of Σ^+ is again a left unitary (free, very pure, weakly prefix) subsemigroup of Σ^+. Therefore, given $\Gamma \subset \Sigma^+$, there always exists the ***minimum*** left unitary (free, very pure, weakly prefix) subsemigroup of Σ^+ containing Γ^+. The proofs can be found for free subsemigroups in

[Til 72], for left unitary subsemigroups in [Sch 73, Cap 77, Ber 79], for very pure subsemigroups in [DR 79, Cap 80], and for weakly prefix subsemigroups in [Cap 80].

The respective bases have been constructed for the minimum left unitary subsemigroup in [Cap 77, Ber 79], for the minimum very pure subsemigroup in [DR 79, Cap 80], for the minimum free subsemigroup in [Spe 75, Cap 77, Ber 79], and for the minimum weakly prefix subsemigroup in [Cap 80].

The Defect Theorem implies that if Γ is not uniquely decipherable or not finitely decipherable, then the cardinality of the basis of the minimum subsemigroup containing Γ must be less than the cardinality of Γ, [Ber 79, Cap 80b]. For the basis of the minimum very pure subsemigroup [DR 79] and the minimum left unitary subsemigroup [Ber 79] only a weaker statement holds, namely, that the cardinality of the basis cannot exceed the cardinality of Γ.

Let $\Gamma \subset \Sigma^+$, and assume that P, F, S, and D are, respectively, the minimum left unitary, free, very pure, and weakly prefix subsemigroup of Σ^+ containing Γ^+. Then the following inclusions hold:

$$\Sigma^+ \supset S \supseteq D \supseteq F \supseteq \Gamma^+$$

$$\Sigma^+ \supset P \supseteq D \supseteq F \supseteq \Gamma^+$$

The construction of bases for these minimum subsemigroups is considered next.

2.1 Algorithms for Minimum Subsemigroups

We now present the conceptual algorithms of Capocelli for obtaining the bases for minimum subsemigroups containing Γ^+ which are left unitary, free, very pure, or weakly prefix. The treatment given is, we believe, a good deal simpler than the alternatives put forth in the literature, as it is based explicitly on certain fundamental combinatorial properties of code words and their sequences.

Intuitively, the algorithms work by eliminating those code words which deprive Γ of the desired properties, substituting appropriate shorter words.

Definition
Let $\Gamma \subset \Sigma^+$. An *L-sequence* of prefix s_0 and suffix s_n is a finite sequence $\sigma = \langle s_0, c_1, ..., c_n, s_n \rangle$ where the c_i are in Γ such that

$$\begin{aligned} c_1 &= s_0 s_1, & & \\ c_2 &= s_1 s_2, & \text{or } s_1 &= c_2 s_2, \\ c_3 &= s_2 s_3, & \text{or } s_2 &= c_3 s_3, \\ & \cdots & & \\ c_n &= s_{n-1} s_n, & \text{or } s_{n-1} &= c_n s_n, \end{aligned}$$

and where the s_i, $0 \leq i \leq n$, are strings in Σ^+. The L-sequence is *minimal* if, for $0 \leq i, j < n$ and $i \neq j$, we have $s_i \neq s_j$

Definition
An L-sequence σ is a *dividing sequence* if s_0 and s_n are code words. If c_1 is the second code word in σ, then we also say that σ *divides* c_1.

Definition
An L-sequence σ is a *loop sequence* if $s_0 \in \Gamma$ and $s_n = s_i$ for some $i < n$.

Definition
An L-sequence σ is *cyclic* if $s_0 = s_n$

Intuitively, an L-sequence specifies a pair of messages x and y such that $x = s_0 u$ and $y = u s_n$, for some string u, or $x = s_0 y s_n$. The code words of the L-sequence specify how x and y may be deciphered, and the strings s_i show how prefixes of x and y which are in Γ^* are related.

Example
Consider the L-sequence $\langle s_0, c_1, ..., c_5, s_5 \rangle$, where $s_1 = c_2 s_2$, $s_2 = c_3 s_3$, $c_4 = s_3 s_4$, and $c_5 = s_4 s_5$. Then the message x is $c_1 c_5 = s_0 y s_5$, where $y = c_2 c_3 c_4$. ∎

Theorem (Capocelli)
The set $\Gamma \subset \Sigma^+$ is not uniquely decipherable (not finitely decipherable, not synchronizable) iff there exists a dividing sequence (loop sequence, cyclic sequence).

The algorithm for finding the basis of the minimum subsemigroup containing Γ which is free (weakly prefix, or synchronizable) proceeds by finding a dividing sequence (a dividing sequence or a loop sequence; a dividing sequence, a loop sequence, or a cyclic sequence) and replacing the leading code word $c = s_0 s_1$ of the sequence by the shorter words s_0 and s_1. Note that in a dividing sequence and in a loop sequence s_0 is in Γ.

Algorithm A

1. Initialize C to $\Gamma \subset \Sigma^+$.
2. Construct a dividing sequence (dividing sequence or loop sequence; dividing sequence, loop sequence, or cyclic sequence) σ. If none exists, then go to Step 6.
3. Set $U := C$.
4. Let $\sigma = \langle s_0, s_1, ... \rangle$ be the sequence, $c = s_0 s_1 \in U$. Set $U := U - \{c\} \cup \{s_0, s_1\}$.
5. Set $C := U$ and return to Step 2.
6. Stop. C is now the basis for the minimum subsemigroup containing Γ which is free (weakly prefix, very pure).

The algorithm for finding the basis of the smallest left unitary subsemigroup containing Γ is the following

Algorithm B

1. Initialize C to $\Gamma \subset \Sigma^+$.
2. If no element of C is prefix of another element of C then go to Step 6.
3. Set $U := C$.
4. If an element p of U is prefix of another element q of U, say $q = pr$, then set $U := U - \{q\} \cup \{r\}$.
5. Set $C := U$ and return to Step 2.
6. Stop. C is now the basis for the minimum subsemigroup containing Γ which is left unitary.

Correctness of Algorithm A has been shown in [Cap 77, Cap 80], and correctness of Algorithm B in [Cap 77].

3. An Algorithm for Unique Decipherability

We wish to sketch an algorithm for testing unique decipherability on which we will base the implementation of Algorithm A. The unique decipherability algorithm is due to Hoffmann [Hof 84], and its asymptotic performance, $O(n\,L)$, is equal to that of the other two fast algorithms in the literature [Rod 82, AG 84].

There is a considerable literature dealing with unique decipherability testing, e.g., [SP 53, Mar 62, Lev 62, Eve 63, Blu 65, Spe 75, Cap 79, Rod 82, AG 84]. All proposed algorithms are similar in that they attempt to construct a counter example to unique decipherability. However, most algorithms do not construct an explicit representation for all deciphering ambiguities.

There are three efficient unique decipherability tests, namely [Rod 82, AG 84], and the algorithm to be given. Conceptually, our algorithm is related to Spehner's, [Spe 75], whereas the other two are implementations of the algorithm given in [SP 53]. Moreover, the others do not explicitly construct a representation for all deciphering ambiguities, and it is precisely this explicit representation which is needed later.

Spehner's algorithm could be modified to construct a graph whose vertices are the set of suffixes of the words in Γ. There is an edge (u,v) if $u = cv$ or $c = uv$ for some $c \in \Gamma$. Finding a dividing sequence now means finding a path from a certain set of initial suffixes to the empty suffix.

Instead of working with suffixes as vertex set, our algorithm takes the set of prefixes as vertex set of an equivalent graph. Each prefix simultaneously encodes all suffixes whose concatenation results in a code word. This graph is explored breadth-first to determine the existence of a path from a distinguished vertex set to a vertex representing a code word. Any such path is shown to yield a deciphering ambiguity, and such a solution can be constructed from it in linear time. We make use of the pattern matching technique of [AC 75].

3.1 The Graph $R(\Gamma)$

Let Γ be a finite set of nonempty words over a fixed alphabet Σ, and assume we wish to test whether Γ is uniquely decipherable. The heart of the algorithm to be presented is a graph $R(\Gamma)$, which makes explicit a number of properties of Γ. We define the set

$$Prefix(\Gamma) = \{u \in \Sigma^* \mid u < c,\ c \in \Gamma\},$$

the set of all prefixes of words in Γ. Of course $\Gamma \subset Prefix(\Gamma)$. We will consider a directed graph $Tree(\Gamma)$, defined by

(1) The vertex set of $Tree(\Gamma)$ is $Prefix(\Gamma)$.
(2) For $u, v \in Prefix(\Gamma)$, $a \in \Sigma$, if $ua = v$, then (u,v) is an edge of $Tree(\Gamma)$.
(3) Nothing else is an edge of $Tree(\Gamma)$.

It is clear that $Tree(\Gamma)$ is a tree whose edges are directed from the root λ towards the leaves. In the context of pattern matching [AC 75], $Tree(\Gamma)$ has been called the *goto function*.

We now define the graph $R(\Gamma)$. The vertex set of $R(\Gamma)$ is $Prefix(\Gamma)$. There are two kinds of edges, a set

$$E_{\mathrm{Reach}} = \{ (u, v) \mid uc = v,\ u \neq \lambda,\ c \in \Gamma \}$$

whose members we will call *reach edges*, and the set

$$E_{\mathrm{Divisor}} = \{ (u, v) \mid c = uv,\ u \neq \lambda,\ v \neq \lambda,\ c \in \Gamma \}$$

whose members we will call *divisor edges*. We will show how to construct this graph using the Aho-Corasick pattern matching algorithm [AC 75].

In $R(\Gamma)$, we also distinguish the subset Γ of the vertices and the subset

$$S(\Gamma) = \{ c \in \Gamma \mid c < c',\ c' \in \Gamma \}$$

consisting of those words in Γ which are prefix of some other word(s) in Γ. In Section 3.2 below we will demonstrate that a set Γ is not uniquely decipherable if, and only if, in $R(\Gamma)$ there is a nontrivial path from a vertex in $S(\Gamma)$ to a vertex in Γ. Furthermore, we will show that properties such as finite decipherability and finite synchronizability are manifest as other properties of this graph.

3.2 Testing Unique Decipherability

In the following, $\Gamma = \{c_1, ..., c_n\}$, and $L = |c_1| + ... + |c_n|$. We assume that the reader is familiar with the pattern matching algorithm of [AC 75]. We give an algorithm for testing whether Γ is uniquely decipherable, but in this paper we are primarily interested in the first three steps which construct the graph $R(\Gamma)$.

Algorithm UD

1. Construct a pattern matching machine M with Γ the set of patterns.
2. Construct the edge set E_{Reach}.
3. Construct the edge set E_{Divisor}.
4. Search $R(\Gamma)$ breadth-first to locate a path from $S(\Gamma)$ to Γ. If such a path exists, then Γ is not uniquely decipherable; if no such path exists, then Γ is uniquely decipherable.

Step (1) is implemented using the algorithm of [AC 75]. A routine modification enables it to locate all matches in a subject, rather than the first match only. As shown in [AC 75], Step (1) can be implemented in $O(L)$ steps. Note that M contains *Tree*(Γ) in its description.

Steps (2) and (3) will have to perform the following operations on *Tree*(Γ):

(a) Given a vertex v of *Tree*(Γ), find the length $|v|$ of v.
(b) Given a vertex v of *Tree*(Γ), find the vertex u of a prefix u of v of prescribed length.

It is clear that both operations can be implemented in constant time assuming a preprocessing step creating index structures requiring $O(L)$ steps. Briefly, with every leaf v of *Tree*(Γ) is associated a vector V such that $V[i]$ points to the prefix of length i of v. Each interior vertex u uses the vector associated with an arbitrarily selected leaf in the subtree rooted at u.

Consider the determination of E_{Reach} in Step (2). Here we run M on *Tree*(Γ). Each time we have a match of the set $\{d_1,\ ...,\ d_k\} \subset \Gamma$ at vertex u, we add the arcs (u_i, u), $1 \leq i \leq k$, where u_i is the prefix of length $|u| - |d_i|$ of u.

Note that a traversal of *Tree*(Γ) for matching purposes requires at most $O(L)$ steps, since the sum of the lengths of all root to leaf paths is bounded by L. Therefore, Step (2) is $O(L+q)$, where q is the number of edges in E_{Reach}. It is easy to see that $|E_{\text{Reach}}| \leq n\,L$, hence Step (2) requires $O(n\,L)$ steps.

Recall from [AC 75] the notation $f(u) = v$ which means that v is a maximal proper suffix of u which is in *Prefix*(Γ), and recall that M contains the graph of f for all strings in *Prefix*(Γ). Consider $c \in \Gamma$, and assume that $f(c) = y_1$, $f(y_1) = y_2$, ..., $f(y_k) = \lambda$, where $y_k \neq \lambda$. Note that $k < |c|$. We add the arcs (u_i, y_i) to E_{Divisor}, $1 \leq i \leq k$, where u_i is the prefix of c of length $|c| - |y_i|$. Tracing f from c, this can be implemented in $O(|c|)$ steps, hence Step (3) requires $O(L)$ steps in all.

Step (4) is a standard breadth-first search of $R(\Gamma)$. We have a "current" vertex set and all arcs originating in the set are explored. Newly reached vertices become the next current vertex set. Exploration ends when a vertex in Γ is reached or all edges have been examined. If a vertex in Γ is reached, then Γ is not a uniquely decipherable, otherwise it is. The initial current vertex set is $S(\Gamma)$ which is easily identified in Step (1). Clearly the time required for Step (4) is $O(|E_{\text{Reach}}| + |E_{\text{Divisor}}|)$ which is $O(n\,L)$. Correctness of the algorithm will be shown in Section 4. In summary, we have:

Theorem 3.1
Let $\Gamma = \{c_1, \ldots, c_n\}$ be a set of words in Σ^+, Σ a fixed alphabet, and let $L = |c_1| + \ldots + |c_n|$. Then in $O(n\,L)$ steps we can construct $R(\Gamma)$ and test whether Γ is uniquely decipherable.

4. Code Properties and $R(\Gamma)$

A number of properties of Γ correspond to structural properties of $R(\Gamma)$. We explain this correspondence now. In particular, we show how L-sequences correspond to paths in $R(\Gamma)$, and from this fact will follow the correctness of Algorithm UD of Section 3.

Proposition 4.1
There is a path in $R(\Gamma)$ from a vertex $s_0 \neq \lambda$ to a vertex q with $qs_n \in \Gamma$ iff $\sigma = \langle s_0, c_1, \ldots, c_n, s_n \rangle$ is an L-sequence for Γ.

Proof Recall the definition of L-sequence from Section 2. We will show how to find a path in $R(\Gamma)$ from the L-sequence σ. A complete proof of the Proposition will be apparent from this construction.

Since $c_1 = s_0 s_1$, s_0 is in *Prefix*(Γ), and s_1 is a code word suffix. We let $q_0 = s_0$ and observe that $q_0 s_1 \in \Gamma$.

If $c_2 = s_1 s_2$, then s_1 is also in *Prefix*(Γ), and there is an edge (s_0, s_1) in E_{Divisor}. In this case, let $q_1 = s_1$. If $s_1 = c_2 s_2$, then $s_0 c_2$ is in *Prefix*(Γ), and there is an edge $(s_0, s_0 c_2)$ in E_{Reach}. In this case, let $q_1 = s_0 c_2$. In either case, observe that $q_1 s_2 \in \Gamma$. By induction, therefore, there is a path in $R(\Gamma)$ corresponding to σ, from $q_0 = s_0$ to a vertex q_{n-1} and $q_{n-1} s_n \in \Gamma$. ∎

Corollary 4.2
A set $\Gamma \subset \Sigma^+$ is not uniquely decipherable iff there is a nontrivial path in $R(\Gamma)$ from a vertex in $S(\Gamma)$ to a vertex in Γ.

Proof Clearly Γ is not uniquely decipherable iff there is a dividing sequence. Let $\sigma = \langle s_0,\ c_1,\ \ldots,\ c_n,\ s_n\rangle$ be a dividing sequence. By Proposition 4.1, there is a path in $R(\Gamma)$ beginning at s_0 which is in $S(\Gamma)$, and ending at a vertex q_{n-1} where $q_{n-1}s_n \in \Gamma$. Since $s_n \in \Gamma$, there is a divisor edge from q_{n-1} to s_n in $R(\Gamma)$. Hence the path ends at a vertex in Γ and is nontrivial. ∎

The properties of finite decipherability delay as well as of synchronizability of Γ can be expressed in terms of L-sequences. Specifically, Γ is not finitely decipherable (synchronizable) iff there is a loop sequence (cyclic sequence) [Cap 79]. Because of the strong correspondence of L-sequences with paths in $R(\Gamma)$, we readily obtain the following

Corollary 4.3
A uniquely decipherable code $\Gamma \subset \Sigma^+$ is not finitely decipherable iff there is a cycle in $R(\Gamma)$ which is reachable from a vertex in $S(\Gamma)$. Moreover, Γ is synchronizable iff $R(\Gamma)$ is acyclic.

In Section 8 we explore in detail how to determine the decipherability and synchronization delays of a code given the graph $R(\Gamma)$.

5. The Minimum Free Subsemigroup

We give an $O(n^2L+L^2)$ algorithm which finds the basis of the minimum free subsemigroup of Σ^+ containing a finite set $\Gamma \subset \Sigma^+$. Here n is the cardinality of Γ and L is the sum of the lengths of the code words in Γ. Of course, if Γ is a uniquely decipherable code, then the basis of the minimum free subsemigroup is Γ. The algorithm is developed from a simpler, $O(n\,L^2)$ implementation of Algorithm A, in which Algorithm UD is used to find the required dividing sequences. The faster version is obtained from the simpler algorithm by grouping several iterations of Steps 2–5 in Algorithm A, such that the new set Γ', derived from Γ, has cardinality at most $|\Gamma| - 1$.

We wish to exploit the relationship between L-sequences and paths in $R(\Gamma)$ which we touched upon above in Section 4. Specifically, from the proof of Proposition 4.1 we readily obtain the following:

Proposition 5.1
Let $\sigma = \langle s_0, c_1, c_2, \ldots\rangle$ be an L-sequence for Γ. If $c_2 = s_1 s_2$, then there is a divisor edge (s_0, s_1) in $R(\Gamma)$. If $s_1 = c_2 s_2$, then there is a reach edge $(s_0, s_0 c_2)$ in $R(\Gamma)$.

The crucial operation of Algorithm A is splitting the code word c_1 of a dividing sequence. We now state how this operation is done using the graph $R(\Gamma)$, referring to the version of Algorithm A which determines the basis of the minimum free subsemigroup as Algorithm A^{FS}. Note that this is the simple, $O(n\,L^2)$ implementation.

Corollary 5.2
Let Γ be a finite subset of Σ^+, and assume there is a path in $R(\Gamma)$ from $c \in S(\Gamma)$ to $c' \in \Gamma$ containing at least one divisor edge, corresponding to the dividing sequence σ. Let (u, v) be the first divisor edge of the path, and let w be defined by $cw = uv$ Then the set

$$\Gamma' := \Gamma - \{uv\} \cup \{w\}$$

is the set U of Algorithm A^{FS} after executing Step 4 for the first time with σ the chosen dividing sequence.

Proof Let $\sigma = \langle c, c_1, \ldots \rangle$ be a dividing sequence. Then there is a corresponding path from the vertex $c \in S(\Gamma)$ to a vertex $c' \in \Gamma$. Without loss of generality, this path terminates with a divisor edge. Let $c_1 = c s_1$. Then $c_1 = uv$ and $w = s_1$. ■

Consider the following implementation of Algorithm A:

1. Construct $R(\Gamma)$ and locate a path from $S(\Gamma)$ to Γ. If none exists, stop: Γ is now the desired set of generators for the minimum free subsemigroup.
2. Locate the first divisor edge (u, v) on the path. ¿From it, determine uv and w. Remove uv from Γ and add w to Γ. Return to Step 1.

Clearly, this is a correct implementation of Algorithm A. As for the timing, note that the new set derived in Step 2 has cardinality not greater than the original set. Moreover, the longer word $uv = cw$ is replaced with the shorter word w, hence Step 2 cannot be executed more than L times, where L is the sum of the lengths of the words in the original set Γ. Since Step 1 requires $O(n\,L)$ steps, whereas Step 2 requires only $O(L)$ steps, it follows that the algorithm requires at most $O(n\,L^2)$ steps.

If $w \neq v$, then it is clear that w is split eventually into $w = c_2 \ldots c_k v$, where the c_i are in Γ. Hence one may, equivalently, replace uv with v in Step 2. The advantage here is that v and its prefixes already exist in the graph $R(\Gamma)$, hence no new vertices need to be added to the underlying *Tree*(Γ). While this simplifies the algorithm somewhat, it does not alter its time complexity.

Another possible modification is to split *all* words c occurring in *some* dividing sequence of Γ. This involves locating all divisor edges (u, v) in $R(\Gamma)$ which are on some path from $S(\Gamma)$ to Γ. Then the words uv are removed from Γ, and the words v are added. By processing these edges by decreasing length uv one can assure that a word v which already is in Γ and is split through some other divisor edge (u', v'), $v = u'v'$, is removed from Γ right away. The resulting algorithm constructs the generating sets produced by the algorithm proposed by Berstel et al. in [Ber 79].

Curiously, we can only give an $O(L^3)$ time bound for this second version of the algorithm, since the intermediate sets Γ' produced may be of greater cardinality than the original set Γ, and also the sum of the word lengths may temporarily increase. Of course, an $O(L^3)$ bound is contrary to intuition. Yet this size problem, stemming from splitting the same code word in more than one way, would also impair a similar attempt at speeding up the $O(n^2 L + L^2)$ algorithm which we describe next.

In the faster $O(n^2 L + L^2)$ algorithm we exploit the following idea: After splitting the code word c_1 in the dividing sequence $\sigma = \langle c, c_1, c_2, \ldots \rangle$, there is a new dividing sequence σ' for the set Γ' which is obtained by substituting $c_0 s_1$ for every occurrence of c_1 in σ, followed by dropping the leading occurrences of c_0. The order of words in the resulting sequence might have to be changed somewhat to account for the way in which the shorter words c_0 and s_1 overlap with each other and with the other words in the sequence. The corresponding path may be constructed without reconstructing the graph $R(\Gamma')$; rather, we modify $R(\Gamma)$ accordingly. After this has been repeated for these "dividing sequence tails", we eventually obtain a new set $\tilde{\Gamma}$ whose cardinality is smaller than that of Γ. We illustrate the idea by an example:

Example

Let $\sigma = \langle c_0, c_1, c_2, c_1, c_3, c_4 \rangle$ be a dividing sequence as shown in Figure 5.1. Here $c_1 = c_0 s_1$, $s_1 = c_2 s_2$, $c_1 = s_2 s_3$, and $s_3 = c_3 c_4$. We split $c_1 = c_0 s_1$ and replace c_1 with s_1 in Γ, thus obtaining Γ'.

Substituting $c_0 s_1$ for c_1 in the sequence and dropping the leading pair of c_0, we obtain the new dividing sequence $\langle c_2, s_1, c_0, c_3, s_1, c_4 \rangle$ shown in Figure 5.2. Note that the order of code words has been slightly altered. Here s_1 can be split $s_1 = c_2 s_2$, and so on.

After several of these split and modify steps, we eventually obtain the dividing sequence of Figure 5.3. Here c_4 is split, but since both u and s_2 are in the current set Γ', the next set $\tilde{\Gamma}$ obtained from this split has cardinality at most $|\Gamma| - 1$. ∎

After each such group of split and modify steps, we construct the graph $R(\tilde{\Gamma})$ and repeat this operation for a new dividing sequence until we obtain a uniquely decipherable code, i.e., until no more dividing sequences exist. Note that in each split and modify step a longer word is replaced with a shorter one.

We now explain how the split and modify step is expressed as operations on the graph $R(\Gamma)$. Assume that we have fixed a path from $S(\Gamma)$ to Γ such that the path does not have cycles and the last path edge is a divisor edge.

Step A (Splitting c_1)
Let (u, v) be the first divisor edge on the path, and assume that $c_1 = uv$. Define s_1 by $c_0 s_1 = c_1$. Delete c_1 from Γ and add s_1 to Γ, thereby obtaining Γ'. This may involve creating a new path from the root of *Tree*(Γ) for s_1. Adjust the path in $R(\Gamma)$ as specified by Operation B.

Step B1 (Adjusting the first occurrence of c_1)
Delete the edge (u, v) from the path. If the edge was the first one, then $v = s_1$ and the path now begins at s_1. Otherwise, let $(p_1, p_2), (p_2, p_3), ..., (p_{k-1}, p_k)$ be the leading reach edges of the path, i.e., $p_1 = c_0$ and $p_k = u$. Define c_i by $p_i = p_{i-1} c_i$, $2 \le i \le k$. Note that $c_i \in \Gamma$ and that $c_2 c_3 ... c_k < s_1$. Remove the edges (p_{i-1}, p_i) from the path and add in their place the reach edges $(c_2, c_2 c_3), ..., (c_2 ... c_{k-1}, c_2 ... c_k)$. Add the divisor edge $(c_2 ... c_k, v)$. In this case, the new path begins at c_2. Note that $c_2 \in S(\Gamma')$. If any of the new edges terminate at a vertex in Γ, then delete all subsequent path edges.

Step B2 (Adjusting a subsequent reach edge belonging to c_1)
Let (p, q) be a reach edge on the path with $q = pc_1$. Replace the edge with the two reach edges (p, pc_0) and (pc_0, q). Note that $q = pc_0 s_1$. If $pc_0 \in \Gamma$, then remove the edge (pc_0, q) and all subsequent path edges.

Step B3 (Adjusting a subsequent divisor edge belonging to c_1)
Let (p, q) be a divisor edge on the path where $c_1 = pq$. Since the path was without cycles, we have $p < c_0$ or $c_0 < p$.
If $p < c_0$, let r be the prefix of q of length $|q| - |s_1|$, i.e., $q = rs_1$. Replace the edge (p, q) with the divisor edge (p, r), belonging to c_0, and the reach edge (r, q). The change is shown in Figure 5.4.
If $c_0 < p$, there are two cases:

(1) The edge (p, q) is preceded by the reach edges $(p_k, p_{k-1}), ..., (p_2, p_1)$, where $p_1 = p$, $p_k < c_0$, and $c_0 < p_{k-1}$. See Figure 5.5.

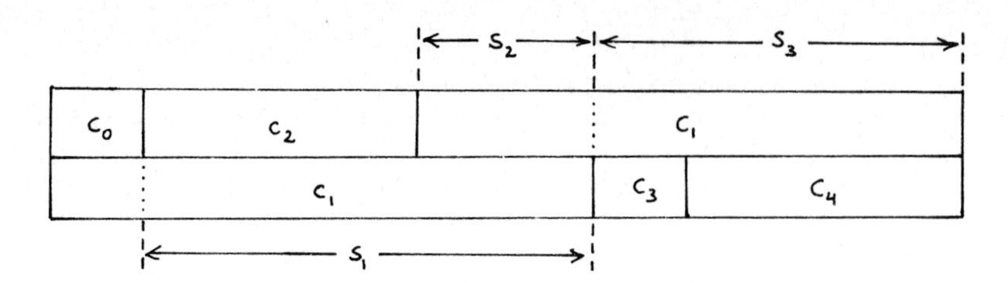

Figure 5.1

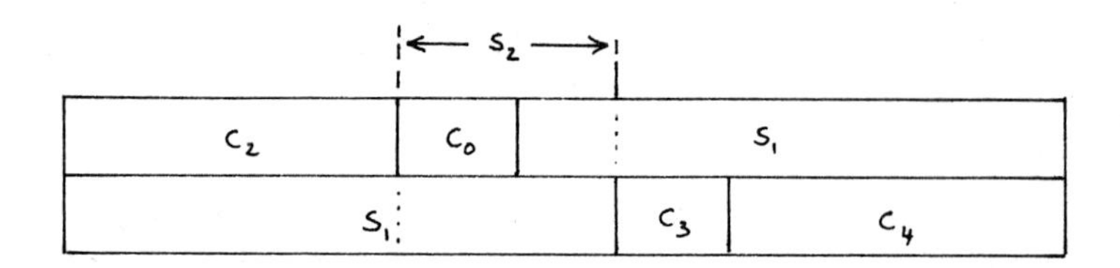

Figure 5.2

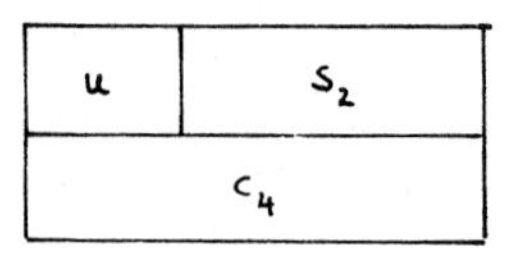

Figure 5.3

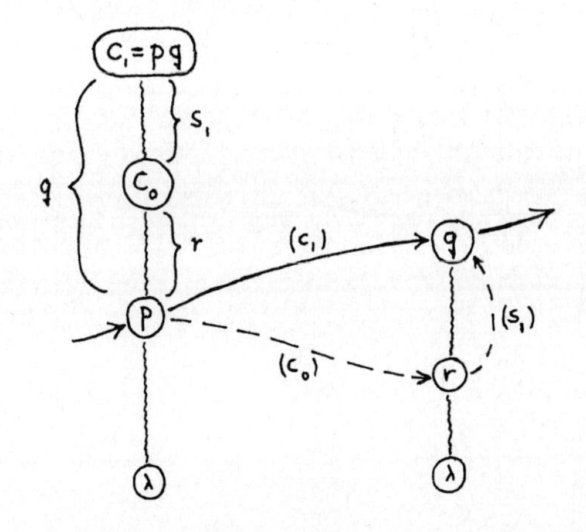

Figure 5.4
Pathmodification, Step B3, $p < c_0$

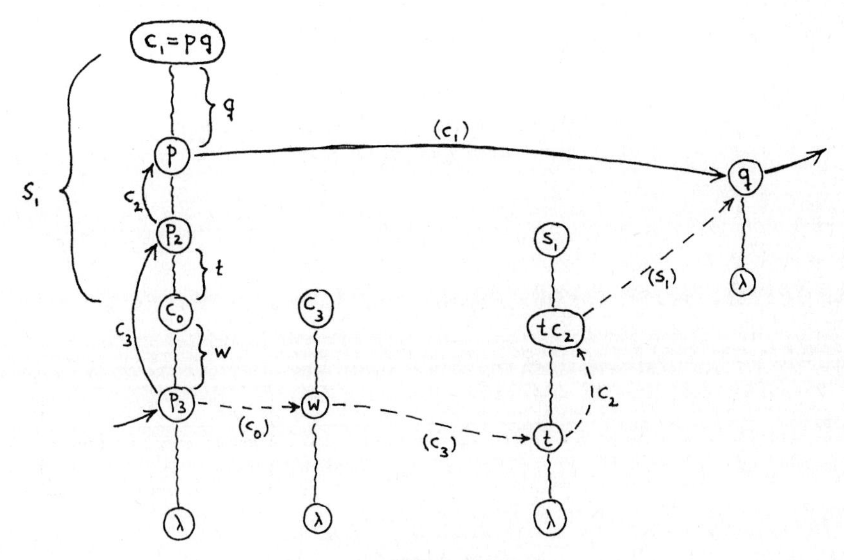

Figure 5.5
Pathmodification, Step B3, $c_0 < p$
Case (1), $k = 3$

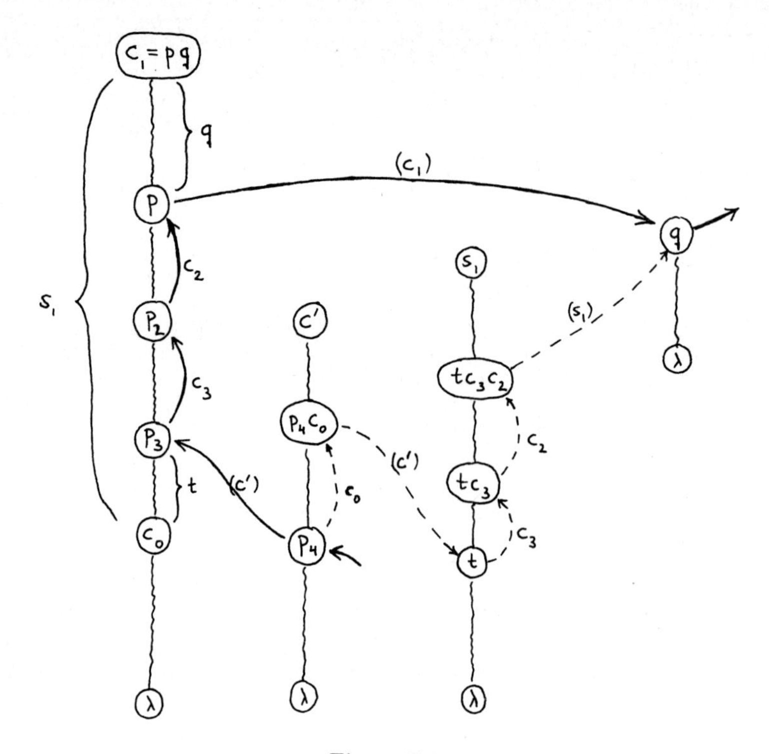

Figure 5.6
Pathmodification, Step B3, $c_o < p$
Case (2), $k = 4$

(2) The edge (p,q) is preceded by zero or more reach edges $(p_{k-1},p_{k-2}),\ldots,(p_2,p_1)$, where $p_1 = p$, and by the divisor edge (p_k,p_{k-1}). Moreover, $c_0 < p_{k-1}$. See also Figure 5.6.

In Case (1), define c_i by $p_{i-1} = p_i c_i$, $2 \le i \le k$. Replace the reach edges (p_i,p_{i-1}) with the divisor edges (p_k,w) and (w,t), where $p_k w = c_0$ and $wt = c_k$, followed by the reach edges $(t,tc_{k-1}),(tc_{k-1},tc_{k-1}c_{k-2}),\ldots,(tc_{k-1}\ldots c_3,tc_{k-1}\ldots c_3c_2)$. Also, replace the divisor edge (p,q) with the divisor edge $(tc_{k-1}\ldots c_2,q)$. Note that $tc_{k-1}\ldots c_2q = s_1$. The transformation is shown in Figure 5.5.
In Case (2), define t from $p_{k-1} = c_0t$ and let $c' = p_kp_{k-1}$. If $k > 2$, define c_i by $p_{i-1} = p_ic_i$, $2 \le i < k$. Replace the divisor edge (p_k,p_{k-1}) with the reach edge (p_k,p_kc_0) followed by the divisor edge (p_kc_0,t). Replace the reach edges (p_i,p_{i-1}), $2 \le i < k$, with the reach edges $(t,tc_{k-1}),\ldots,(tc_{k-1}\ldots c_3,tc_{k-1}\ldots c_2)$, and replace the divisor edge (p,q) with the divisor edge $(tc_{k-1}\ldots c_2,q)$. Note that $tc_{k-1}\ldots c_2q = s_1$. The transformation is shown in Figure 5.6.

In either case, if one of the new vertices on the path is in Γ, then delete all subsequent path edges.

Note that initially the path fixed in $R(\Gamma)$ does not contain any cycles. In the path transformation, however, cycles may be created. It is therefore necessary to make a pass over the transformed path deleting cycles, to insure that the maximum path length is bounded by L, the sum of the lengths of words in Γ. This results in the following

Algorithm FS

1. Initialize C to $\Gamma \subset \Sigma^+$.
2. Construct the graph $R(C)$ using Algorithm UD.
3. By breadth-first search, determine a path π in $R(C)$ from $S(C)$ to C, such that π is acyclic and ends with a divisor edge.
4. If no such path π exists, then stop; C is the basis of the minimum free subsemigroup of Σ^+ which contains Γ.
5. While π is not empty, do Steps 6–8. Thereafter, return to Step 2.
6. Perform Step A on the path π as described above.
7. Transform the path by performing the adjustments of Steps B1–B3 above.
8. Eliminate all loops from the transformed path.

Theorem 5.3
Algorithm FS correctly determines the basis of the minimum free subsemigroup containing Γ in $O(n^2L + L^2)$ steps.

Proof Correctness follows from Corollary 5.2 and the fact that the transformed path is a path in $R(C_1)$, where $C_1 = C - \{c_1\} \cup \{s_1\}$.

For the timing, observe first that Steps 2–4 are executed at most n times, since the loop of Steps 5–8 decreases the cardinality of C. Hence the total amount of work done by Steps 2–4 is at most $O(n^2L)$ steps.

Now consider Step 6. Here c_1 is split into c_0 and s_1, and the sum L of the lengths of words in C is decreased by $|c_0|$. It follows that Steps 6–8 can be executed at most L times ever. It remains to show that a single execution of Steps 6–8 requires no more than $O(L)$ steps.

In executing Steps 6–8, Step 7 dominates. Now it is easy to see that if a dividing sequence contains m words, then the corresponding path has $m - 2$ edges. Since path transformation corresponds to the substitution of two words for each occurrence of c_1, it follows that the transformed path contains at most twice as many edges as the original path. As loops are eliminated from the transformed path, each path considered in Step 7 has $O(L)$ edges maximum. Moreover, given the index structures of Algorithm UD, it is clear that the number of steps required to transform the path is proportional to the number of edges on the path. Consequently, a single execution of Steps 6–8 requires $O(L)$ steps, hence the total work done in Steps 5–8 is at most $O(L^2)$. ∎

Instead of eliminating all loops on the path, one may eliminate all edges following the first vertex u which repeats on the path. This is correct since u eventually becomes a word in C, when encountered for the first time by the splitting operation, hence it anticipates, in essence, the elimination of subsequent path edges due to an edge terminating at a vertex in C. Strictly speaking, one may lose the property that the path ends with a divisor edge, which we have assumed throughout in order to simplify the presentation. It is not difficult, however, to augment the transformations either eliminating this requirement, or replacing a terminal reach edge with the corresponding divisor edge.

6. The Minimum Weakly Prefix and Very Pure Subsemigroups

Algorithm A can be specialized to find the basis of the minimum subsemigroups containing Γ which are weakly prefix or very pure. The corresponding implementations differ from Algorithm FS in that different types of paths in $R(\Gamma)$ are considered, but are otherwise identical.

Recall Steps A and B used in Algorithm FS. Clearly these operations can be applied to paths corresponding to loop sequences and cyclic sequences. Here a path corresponding to a loop sequence is a path in $R(\Gamma)$ which begins at a vertex $c_0 \in S(\Gamma)$ and contains some vertex u both as an intermediate as well as the final path vertex. We will call such a path a *loop path*, and call u the *lead vertex*. Without loss of generality, we may assume that u is the only vertex occurring twice.

Consider the algorithm for finding the basis for the minimum weakly prefix subsemigroup containing Γ. Here we must consider all loop and dividing sequences. Since in either case the path begins at a vertex in $S(\Gamma)$, it follows that all intermediate sets Γ' have cardinality no greater than Γ. Moreover, when the path is completely split, the resulting set $\tilde{\Gamma}$ has cardinality at most $|\Gamma|-1$. We give the algorithm below as Algorithm WP. Note the similarity to Algorithm FS.

Algorithm WP

1. Initialize C to $\Gamma \subset \Sigma^+$.
2. Construct the graph $R(C)$ using Algorithm UD.
3. Determine a path π in $R(C)$ from $S(C)$ to C, or from $S(C)$ to a cycle of $R(C)$. The path should not contain unneccesary cycles.
4. If no such path π exists, then stop; C is the basis of the minimum weakly prefix subsemigroup of Σ^+ which contains Γ.
5. While π is not empty, do Steps 6–8. Thereafter, return to Step 2.
6. Perform Step A on the path π as described above.

7. Transform the path by performing the adjustments of Steps B1–B3 above.
8. Eliminate all edges following the first repeated vertex u from the transformed path.

Theorem 5.3
Algorithm WP correctly determines the basis of the minimum weakly prefix subsemigroup of Σ^+ containing Γ and requires at most $O(n^2 L + L^2)$ steps.

The theorem is proved like Theorem 5.3.

Now consider the algorithm for determining the basis of the minimum very pure subsemigroup containing Γ. Here we must consider all loop sequences, cyclic sequences, and dividing sequences. The algorithm to be given requires $O(n\,L^2)$ steps. This inferior time bound is due to the fact that splitting a cyclic sequence need not result in a smaller set $\tilde{\Gamma}$, since in the cyclic sequence $\langle s_0, c_1, ..., c_m, s_0\rangle$ the word s_0 need not be in Γ. Hence splitting c_1 may result in a new set Γ' of cardinality $|\Gamma| + 1$. Here splitting the sequence tail completely eventually reduces the cardinality to $|\Gamma|$ or less. Note that the initial splitting step generates a set which is not a uniquely decipherable code.

Because of the initial increase in cardinality, it is crucial to split the tail sequence completely before considering other cyclic sequences. Otherwise we cannot guarantee that all intermediate sets have cardinality $O(|\Gamma|)$. However, as s_0 is added to Γ', we may equivalently split a suitable dividing sequence instead. The resulting algorithm is Algorithm VP:

Algorithm VP

1. Initialize C to $\Gamma \subset \Sigma^+$.
2. Construct the graph $R(C)$ using Algorithm UD.
3. If $R(C)$ contains no path from $S(C)$ to C and is acyclic, then stop: C is the basis of the minimum very pure subsemigroup of Σ^+ which contains Γ.
4. If there is a cycle π in $R(C)$, let (u, v) be a divisor edge on the cycle. Set $C' := C - \{uv\} \cup \{u, v\}$.
5. Using Algorithm WP, find the basis U of the minimum weakly prefix subsemigroup containing C'.
6. Set $C := U$ and goto Step 2.

Theorem 6.2
Algorithm VP is correct and requires at most $O(n\,L^2)$ steps.

Proof Correctness is evident. For the timing, recall that in Step 4 we have $|C'| \leq |C|+1$ and $L' \leq L$, where L and L' are the sum of the lengths of words in C and C', respectively. Moreover, C' is not uniquely decipherable, hence $|U| \leq |C|$, and $L'' < L$, where L'' is the sum of the lengths of words in U. Hence Step 5 is executed at most L times. By the proof of Theorems 5.3 and 6.1, therefore, all invokations of Algorithm WP require a total of $O(n\,L^2)$ steps. The time bound now follows. ∎

7. The Minimum Left Unitary Subsemigroup

Algorithm B determines the minimum left unitary subsemigroup of Σ^+ which contains Γ.

It is easily implemented by repeatedly constructing *Tree*(C), followed by splitting each root to leaf path $p = u_1u_2 \cdots u_k$ in accordance with which prefixes are in C. That is, p is split into the set u_1, $u_2, \ldots, u_k$, where u_1, $u_1u_2, \ldots$, $u_1u_2 \cdots u_k \in C$. The set is added to the next version of C. The algorithm terminates as soon as *Tree*(Γ) has no interior vertices in C. Evidently, the time bound for this implementation is $O(L^2)$, since at each stage the length of the words in C is decreased by at least one. Of course, if Γ is already a prefix code, then Γ is returned.

8. Determination of Synchronizability and Decipherability Delays

Corollary 4.3 implies an $O(n\,L)$ test whether a uniquely decipherable set Γ is synchronizable and whether it has finite decipherability delay. We now develop an algorithm for determining the decipherability delay of a finitely decipherable code, and the synchronization delay of a synchronizable code. It turns out, that both problems can be solved by essentially the same algorithm. So, we develop the algorithm for determining the synchronization delay first, followed by explaining how to modify it to determine the decipherability delay of a given code.

Recall that a synchronizable code Γ has the synchronization delay s if s contiguous code words in a message fragment suffice to detect a correct code word boundary in the fragment.

Example
Let $\Gamma = \{a, baa\}$. Clearly Γ is a synchronizable code. In the fragment aa of a possibly longer message we can correctly announce a word boundary following the second a, but in the fragment a we could not. Therefore, Γ has the synchronization delay 2.

We want to define a valuation of L-sequences which relates the synchronization delay of a code to path properties in the graph $R(\Gamma)$. First, we associate with each L-sequence a pair of strings in Σ^+, as already sketched in Section 2.1:

Definition
Let σ, be an L-sequence for Γ. The strings $x_\sigma, y_\sigma \in \Gamma^*$ *associated* with σ are defined by the following rules:

(1) If $\sigma = \langle s_0, c_1, s_1 \rangle$, then $x_\sigma = c_1$ and $y_\sigma = \lambda$.

(2) Let $\sigma = \langle s_0, c_1, \ldots, c_m, s_m \rangle$, for $m > 1$, where $\mu = \langle s_0, c_1, \ldots, c_{m-1}, s_{m-1} \rangle$. If $x_\mu s_{m-1} = s_0 y_\mu$, then $x_\sigma = x_\mu c_m$ and $y_\sigma = y_\mu$. Moreover, if $x_\mu = s_0 y_\mu s_{m-1}$, then $x_\sigma = x_\mu$ and $y_\sigma = y_\mu c_m$.

Lemma 8.1
Let x_σ, y_σ be the strings associated with the L-sequence σ. Then $x_\sigma s_m = s_0 y_\sigma$ or $x_\sigma = s_0 y_\sigma s_m$.

The lemma is easily proved by induction on the length of σ. Note that the code words $c_1, \ldots, c_m$ specify how to decipher the strings x_σ and y_σ.

Next, we associate with the L-sequence σ a pair of nonnegative integers $L(\sigma) = (i, j)$, where i and j are the number of code words in the strings x_σ and y_σ, respectively. The definition of $L(\sigma)$ closely parallels the definition of the strings x_σ and y_σ:

Definition
Let σ, be an L-sequence for Γ. The ***word length pair*** of σ, denoted by $L(\sigma)$, is defined as follows:

(1) If $\sigma = \langle s_0, c_1, s_1 \rangle$, then $L(\sigma) = (1, 0)$.
(2) Let $\sigma = \langle s_0, c_1, ..., c_m, s_m \rangle$, for $m > 1$, where $\mu = \langle s_0, c_1, ..., c_{m-1}, s_{m-1} \rangle$ and $L(\mu) = (r, s)$. Then if $x_\mu s_{m-1} = s_0 y_\mu$, then $L(\sigma) = (r+1, s)$. Moreover, if $x_\mu = s_0 y_\mu s_{m-1}$, then $L(\sigma) = (r, s+1)$.

Finally, we define a valuation $\|\sigma\|$ of σ, which is directly related to the synchronization delay.

Definition
Let σ be an L-sequence with $L(\sigma) = (p, q)$. If $x_\sigma = s_0 y_\sigma s_m$, then $\|\sigma\| = \max(p-1, q+1)$. If $x_\sigma s_m = s_0 y_\sigma$, then $\|\sigma\| = \max(p, q)$.

Lemma 8.2
Let Γ be a synchronizable code with synchronization delay s. Then, for all L-sequences σ, $s \geq \|\sigma\|$.

Proof Let σ be an L-sequence with $L(\sigma) = (p, q)$. If $x_\sigma s_m = s_0 y_\sigma$, define u by $u s_m = y_\sigma$. Then u has a prefix of $q-1$ consecutive code words and a suffix of $p-1$ consecutive code words which are incompatible decipherings of u. Hence $s > \max(p-1, q-1)$.

If $x_\sigma = s_0 y_\sigma s_m$, then y_σ consists of q contiguous code words and contains a factor of $p-2$ contiguous code words leading to an incompatible deciphering of y_σ. Hence $s > \max(p-2, q)$. ∎

Theorem 8.3
Let $s = \max(\{ \|\sigma\| \mid \sigma \text{ an L-sequence for } \Gamma \})$, where Γ is a synchronizable code. Then Γ has the synchronization delay s. If there are no L-sequences, then Γ has delay one.

Proof By Lemma 8.2, the synchronization delay of Γ can be no smaller than s. Hence it suffices to show that in any message fragment containing some $f \in \Gamma^s$ we may correctly determine a code word boundary.

Assume that $f \in \Gamma^s$ cannot be correctly partitioned on a code word boundary. Then there must be an L-sequence σ such that either $L(\sigma) = (p, s)$ and $x_\sigma = s_0 y_\sigma s_m$, or $L(\sigma) = (s+1, q)$ and $x_\sigma s_m = s_0 y_\sigma$. All other cases are equivalent to one of these two. In either case we have $\|\sigma\| > s$, contradicting the definition of s. Hence all $f \in \Gamma^s$ can be correctly partitioned. ∎

We relate the valuation of σ to a metric defined on the path associated with σ in $R(\Gamma)$. The connection is established by determining $L(\sigma)$ from the edges in the path, but here we need to remember whether y_σ is a proper substring of x_σ:

Definition
Let p be a path in $R(\Gamma)$. The *modified length pair*, $L'(p)$ of p, is defined as follows:
(1) $L'(\langle p_0 \rangle) = (0, 1; 1)$.
(2) If $(p_{m-1}, p_m) \in E_{\text{Reach}}$, $m > 1$, and $L'(\langle p_0, ..., p_{m-1} \rangle) = (i, j; k)$, then $L'(p) = (i+1, j; k)$.
(3) If $(p_{m-1}, p_m) \in E_{\text{Divisor}}$, $m > 1$, and $L'(\langle p_0, ..., p_{m-1} \rangle) = (i, j; k)$, then $L'(p) = (j, i+1; \bar{k})$, where $\bar{1} = 0$ and $\bar{0} = 1$

Lemma 8.4
Let $\sigma = \langle s_0, ..., s_m \rangle$ be an L-sequence for Γ, p the path associated with σ as in Proposition 4.1. If $L'(p) = (i, j; k)$, then $\|\sigma\| = \max(i+k, j-k)$.

Proof By induction, it is easy to see that $L(\sigma) = (j, i)$ if $k = 1$, and $L(\sigma) = (i, j)$ if $k = 0$. Moreover, if $k = 0$, then $x_\sigma s_m = s_0 y_\sigma$, and if $k = 1$ then $x_\sigma = s_0 y_\sigma s_m$. The lemma now follows from the definition of $\|\sigma\|$. ∎

It is clear that for each path p in $R(\Gamma)$ we can compute $L'(p)$ in time proportional to the number of edges in p. By Theorem 8.3 and Lemma 8.4, we may determine the synchronization delay of Γ by evaluating $L'(p)$ for all paths in $R(\Gamma)$ and computing $\max(i+k, j-k)$, retaining the largest value obtained. This is of course impractical, as the number of paths in $R(\Gamma)$ may be exponential. We therefore seek to evaluate all paths simultaneously, retaining only those values $L'(p)$ which may contribute to the maximum value sought.

Specifically, consider a vertex u of $R(\Gamma)$ at which m paths p_t terminate, where $L'(p_t) = (i_t, j_t; k_t)$, $1 \le t \le m$. If $m \ge 2$, we select two values (not necessarily from distinct paths), $(i_1, j_1; k_1)$, and $(i_2, j_2; k_2)$, such that, for any value $(x, y; k)$ among the $L'(p_t)$, we have $x + k \le i_1 + k_1$ and $y - k \le j_2 - k_2$. These two values suffice to determine $\max(i+k, j-k)$, as is clear from the following, straightforward:

Lemma 8.5
Let p_1, p_2, p_3 be three paths in $R(\Gamma)$ ending at the vertex u, (u, v) an edge in $R(\Gamma)$, and let p_{t+3} be the path p_t extended by (u, v), for $t = 1, 2, 3$. For $1 \le t \le 6$, let $L'(p_t) = (i_t, j_t; k_t)$, and assume that $i_3 + k_3 \le i_1 + k_1$ and $j_3 - k_3 \le j_2 - k_2$. Then either $i_6 + k_6 \le i_4 + k_4$ and $j_6 - k_6 \le j_5 - k_5$, or $i_6 + k_6 \le i_5 + k_5$ and $j_6 - k_6 \le j_4 - k_4$.

The above derivation suggests the following approach to determining the synchronization delay of a code Γ:

Algorithm SD

1. Construct $R(\Gamma)$ and test whether Γ is a uniquely decipherable code, using Algorithm UD.
2. By topologically sorting $R(\Gamma)$ determine whether Γ is a synchronizable code. If Γ is not synchronizable, stop.
3. If $R(\Gamma)$ has no edges and $S(\Gamma)$ is empty, then output one and stop; otherwise, initialize $s := 1$.
4. Assign $(0, 1; 1)$ to each root of $R(\Gamma)$.
5. For each edge (u, v) in $R(\Gamma)$, in sorted order, perform Steps 6 and 7. Thereafter, output s and stop.
6. For each triple $(i, j; k)$ at u, if (u, v) is a reach edge, then place $(i+1, j; k)$ at vertex v, and if (u, v) is a divisor edge, then place $(j, i+1; \bar{k})$ at v.
7. If the number of triples at v is greater than 1, select up to two triples, $(i_1, j_1; k_1)$ and $(i_2, j_2; k_2)$, such that $i_1 + k_1$ and $j_2 - k_2$ are maximum, and discard the others. Update $s := \max(s, i_1 + k_1, j_2 - k_2)$.

Theorem 8.6
Algorithm SD is correct and requires at most $O(n\,L)$ steps, where n is the cardinality of Γ, and L is the sum of the lengths of the code words in Γ.

Proof Correctness is evident from Theorem 8.3 and Lemmas 8.4 and 8.5. For the timing, note that $R(\Gamma)$ has $O(L)$ vertices and $O(n\,L)$ edges. Steps 1–4, therefore, take $O(n\,L)$ steps. Since the edges are processed in sorted order, at most two triples $(i, j; k)$ are propagated along each edge, hence the work of Steps 5–7 is proportional to $O(n\,L)$. ∎

Now consider determining the decipherability delay of a finitely decipherable code Γ. Here we need only consider those L-sequences $\sigma = \langle s_0, \ldots, s_m \rangle$ for which $s_0 \in \Gamma$. In analogy to Lemma 8.2, we have the obvious

Lemma 8.7
Let Γ be a code with finite decipherability delay d. If $\sigma = \langle s_0, ..., s_m \rangle$ is an L-sequence with $s_0 \in \Gamma$, then $d \geq \|\sigma\|$.

Theorem 8.8
Let $d = \max(\{ \|\sigma\| \mid \sigma = \langle s_0, ..., s_m \rangle$ an L-sequence for Γ with $s_0 \in \Gamma \})$, where Γ is a finitely decipherable code. Then Γ has decipherability delay d, unless the maximum is taken over the empty set, in which case the delay is zero.

Proof By Lemma 8.2, the decipherability delay of Γ can be no smaller than d. Hence it suffices to show that in any message prefix containing some $f \in \Gamma^{d+1}$ we may correctly determine the first code word.

Assume there exist $c \in \Gamma$, $u \in \Gamma^d$, and $w \in \Sigma^*$ such that $cuw \in \Gamma^*$ and $uw \notin \Gamma^*$. Then there is $c' \in \Gamma$ and $v \in \Gamma^*$ such that $c'v = cuw$ and $c \neq c'$. Since Γ is uniquely decipherable, we have $w \neq \lambda$.

If $c' < c$, then there is an L-sequence $\sigma = \langle c', c, ..., s_m \rangle$, where $x_\sigma s_m = c' y_\sigma$ and $x_\sigma = cu \in \Gamma^{d+1}$. If $c < c'$, then there is an L-sequence $\sigma = \langle c, c', ..., s_m \rangle$, where $x_\sigma = c y_\sigma s_m$ and $y_\sigma = u \in \Gamma^d$. In either case, $\|\sigma\| > d$, contradicting the definition of d. By Lemma 8.7, therefore, Γ has decipherability delay d. If there are no L-sequences for Γ, then Γ is a prefix code, hence has decipherability delay zero. ■

In view of Theorem 8.8, a simple modification of Algorithm SD enables us to determine the decipherability delay of Γ. Instead of processing the entire graph $R(\Gamma)$, we process the subgraph of all edges reachable from a vertex in $S(\Gamma)$. Clearly paths in this subgraph correspond to L-sequences with $s_0 \in \Gamma$, and vice versa. Of course, if $S(\Gamma)$ is empty, then Γ is a prefix code and has decipherability delay zero. We thus obtain

Theorem 8.9
Let Γ be a finite subset of Σ^+. Then in $O(n\,L)$ steps we can decide whether Γ is finitely decipherable and determine its decipherability delay.

Note that according to our definitions the synchronization delay s of a code is no smaller than its decipherability delay d. This fact can be recovered from the algorithms by observing that for the determination of d only a subgraph of $R(\Gamma)$ is considered, whereas the entire graph is processed for the determination of s.

References

[AC 75] A. V. Aho and M. J. Corasick
"Efficient string matching: an aid to bibliographic search", *CACM 18:6* (1975) 333–343

[AG 84] A. Apostolico and R. Giancarlo
"Pattern matching machine implementation of a fast test for unique decipherability" *Inf. Proc. Letters*, 18 (1984), 155–158

[Ber 79] J. Berstel, D. Perrin, J. F. Perrot, A. Restivo
"Sur le theoreme du defaut", *J. of Algebra 60* (1979) 169–180

[Blu 65] E. K. Blum
"Free subsemigroups of a free semigroup", *Mich. Math. J. 12* (1965) 179–182

[Cap 77] R. M. Capocelli
"On some free subsemigroups of a free semigroup", *Semigroup Forum 15* (1977) 125–135

[Cap 79] R. M. Capocelli
"A note on uniquely decipherable codes", *IEEE Trans. on Inf. Thy. IT-25* (1979) 90–94

[Cap 80] R. M. Capocelli
"On optimal factorization of free semigroups into free subsemigroups", *Semigroup Forum 19* (1980) 199–212

[Cap 80b] R. M. Capocelli
"On weakly prefix subsemigroups of a free semigroup", *Advances in Communications*, D. G. Lainiotis and N. S. Tzannes, eds., Reidel Publishing, Dordrecht 1980, 123–129

[CR 78] R. M. Capocelli and L. M. Ricciardi
"A heuristic approach to feature extraction for written languages", *Proc. 4th Intl. Conf. on Pattern Recognition*, Kyoto, Japan, 1978, 364–368

[DR 79] A. DeLuca and A. Restivo
"Synchronization and maximality for very pure subsemigroups of a free semigroup", *Springer Lect. Notes in Comp. Sci. 74* (1979) 363–371

[Eve 63] S. Even
"Tests for unique decipherability" *IEEE Trans. on Inf. Thy. IT-9* (1963) 109–112

[Eve 64] S. Even
"Tests for synchronizability of finite automata and variable length codes", *IEEE Trans. on Inf. Thy. IT-10* (1964) 185–189

[GGW 58] S. W. Golomb, B. Gordon, L. R. Welch
"Comma-free codes", *Can. J. of Math. 10* (1958) 202-209

[Hof 84] C. M. Hoffmann
"A note on unique decipherability", 11th *Intl. Symp. Math. Found. of Comp. Sci.*, (1984), Springer Lecture Notes in Comp. Sci.

[KMP 77] D. Knuth, J. Morris, V. Pratt
"Fast pattern matching in strings", *SIAM J. on Comp. 6:2* (1977) 323–350

[Lal 79] G. Lallement
"Semigroups and Combinatorial Applications", J. Wiley & Sons, New York (1979)

[Lev 62] V. I. Levenshtein
"Certain properties of code systems", *Sov. Phys. Doklady 6* (1962) 858–860

[Lev 64] V. I. Levenshtein

"Some properties of coding and self-adjusting automata for decoding messages", *Problemy Kyberneticki 11* (1964) 63–121

[Mar 62] A. A. Markov
"Non-recurrent coding", *Problemy Kyberneticki 8* (1962) 169–189

[McC 76] E. M. McCreight
"A space-economical suffix tree construction algorithm", *J. ACM 23:2* (1976) 262–272

[Res 74] A. Restivo
"On a question of McNaughton and Papert", *Inf. and Control 25* (1974) 93–101

[Ril 67] J. A. Riley
"The Sardinas-Patterson and Levenshtein theorems", *Inf. and Control 10* (1967) 120–136

[Rod 82] M. Rodeh
"A fast test for unique decipherability based on suffix trees", *IEEE Trans. on Inf. Thy. IT-28:4* (1982) 648–651;

[SP 53] A. A. Sardinas and C. W. Patterson
"A necessary and sufficient condition for the unique decomposition of coded messages", *IRE Intl. Conv. Rec. 8* (1953) 104–108

[Sch 73] B. M. Schein
"Semigroups whose every transitive representation by function is a representation by invertible functions", *Izv. Vyss. Ucebn. Zaved, Matem.* 7 (1973) 112–121; in Russian.

[Spe 75] J. C. Spehner
"Quelques constructions et algorithmes relatifs aux sous-monoides d'un monoide libre", *Semigroup Forum 9* (1975) 334–353

[Til 72] B. Tilson
"The intersection of free submonoids of free monoids is free", *Semigroup Forum 4* (1972) 345–350

2 - SUBWORD TREES

THE MYRIAD VIRTUES OF SUBWORD TREES

ALBERTO APOSTOLICO
Department of Computer Science
Purdue University
West Lafayette, Indiana 47907

ABSTRACT

Several nontrivial applications of subword trees have been developed since their first appearance. Some such applications depart considerably from the original motivations. A brief account of them is attempted here.

INTRODUCTION

Subword trees fit in the general subject of digital search indexes [KN]. In fact their earliest conception is somewhat implicit in Morrison's 'PATRICIA' tries [MO]. Several linear time and space subword tree constructions are available today [MC, PR, SL] (see also [AH]), following the pioneering work by Weiner [WE]. More compact alternate versions have been introduced recently in [BL, BE, CS2]. The data structures developed in this endeavor are variously referred to as B-trees, position trees, suffix (or prefix) trees, subword trees, repetition finders, directed acyclic word graphs, etc. A concise account of the similarities and discrepancies among the various approaches is presented in [SE1, CS1]. On line (though not linear time) constructions are discussed in [MR]. In this paper, we choose to refer mostly to the version in [MC], to which we also conform as much as possible as for basic definitions and notations. However, the properties presented here are to a large extent independent of the particular incarnation of a subword tree, and, from the conceptual standpoint, so are indeed the associated criteria and constructions. This paper addresses itself to a reader with scarce previous exposure to the subject, but it does assume some familiarity with elementary facts and concepts in combinatorics on words. The paper is also self-contained in the description of the various applications presented. However, some proofs are only sketched; the reader is also pointed to the referenced literature when it comes to constructions too elaborate to be given here in full details. Finally, the list given here is not meant to be exhaustive. In particular, it reflects some recent involvements of this author, and his personal perspective.

The paper is organized as follows. Basic properties and applications of subword trees are outlined in the next section. In Section 2, such trees are treated as a unifying framework for the description of a class of linear time sequential data compression techniques that is becoming increasingly popular. In Section 3, we take steps from one such data compression paradigm and use subword trees to decide whether a word contains a square subword, in linear time. We show next how subword trees can be used also to spot all such squares, as well as to establish bounds on the number of cube subwords in a string. Augmented subword trees are suited to allocate the statistics without overlap of all subwords of a textstring, as highlighted in Section 4. In Section 5, we mention two applications in which subword trees are outperformed by other approaches.

NATO ASI Series, Vol. F12
Combinatorial Algorithms on Words
Edited by A. Apostolico and Z. Galil

1. PRELIMINARIES

We shall deal with *strings (words)* of *symbols* from a finite alphabet I. If x is a word, $|x|$ will denote the *length* (i.e., the number of symbols) of x. Sometimes we will implicitly assume $|x|=n$. The set of all distinct nonempty substrings of x (*subwords*) is called the *vocabulary* of x, denoted V_x. We say that $x_i x_{i+1} \cdots x_{i+|w|-1}$ is an *occurrence* of $w \in V_x$ in x if $x_{i+k}=w_k$ $(k=0,1,...,|w|-1)$. Let $\$ \notin$ I be a special end-marker. For each i in the set $P=\{1,2,...,n+1\}$ of *positions* of $x\$$, suf_i denotes the ith *suffix* of $x\$$. Since $\$ \notin$ I, it is always possible to write $\mathrm{suf}_i=\mathrm{head}_i \cdot \mathrm{tail}_i$, with tail_i nonempty and head_i the longest prefix of suf_i which is also a prefix of suf_j for some $j<i$. The *subword tree* T_x associated with x is defined here as the digital search tree with $n+1$ leaves and at most n interior vertices such that: each edge is labeled with an occurrence of a subword of x via a pair of pointers to a common, randomly accessible, copy of x; each leaf is labeled with a position in P; the labels on the path from the root to leaf labeled i describe suf_i. This labeling policy enables to maintain an $O(n)$ space allocation for any subword tree. Figure 1 displays a portion (i.e., all suffixes starting with a) of T_x for $x=abaababaabaababaababa$.

Any vertex α of T_x distinct from the root describes a subword $w(\alpha)$ of x in a natural way: vertex α is called the *proper locus* of $w(\alpha)$. In general, the *locus* of $w \in V_x$ in T_x is the unique vertex of T_x such that w is a prefix of $w(\alpha)$ and $w(\mathrm{FATHER}(\alpha))$ is a proper prefix of w.

The obvious approach to the construction of T_x is to start with the empty tree T_0 and inserts suffixes in succession into an increasingly updated version of the tree, as follows.

$$\textbf{for } i:=1 \textbf{ to } n+1 \textbf{ do } T_i \leftarrow \text{insert}(T_{i-1}, \mathrm{suf}_i)$$

A brute force implementation of *insert* would lead to an algorithm taking $O(n^2)$ time in the worst case. The time consuming subtask of *insert* is that of finding the locus of head_i $(i=1,2,...n+1)$ in T_{i-1} (head_i might not have a proper locus in T_{i-1}, but it certainly will in T_i). McCreight's construction [MC] exploits auxiliary "suffix links" to retrieve the locus of head_i $(i=1,2,...n+1)$ in overall linear time. Basically, this is made possible by the simple fact that if $\mathrm{head}_i=aw$ $(i=1,2,...n)$ with $a \in I$, then w is a prefix of head_{i+1}. All clever variations of subword trees are built in linear time by resorting to similar properties.

The original motivation behind Wiener's construction of the first subword tree [WE] was that of transmitting and/or storing a message with excerpts from a main string in minimum time or space. It became soon apparent that the structure of such indexes is ideally suited to several other, almost straightforward, applications.

- By treating T_x as the state transition diagram of a finite automaton it is possible to decide whether or not $w \in V_x$, for an arbitrary w, in $O(|w|)$ time. This is of use in multiple searches for different patterns in a fixed set. The particular role played by $ makes it possible to tell also whether w is a suffix of x, for the same cost.
- Assume that each vertex of T_x bears the label of the smallest leaf label in its subtree (this is not difficult to maintain during the construction of T_x or it can be achieved in one appropriate walk of T_x). Then it is possible to find in $O(|w|)$ steps and for arbitrary w what is the first occurrence of w in x (whence also whether w is a prefix of x). Notice that to find the last occurrence of w in $O(|w|)$ time for any w requires a walk through T_x, after its construction: similar

asymmetries are inherent to other variations of the tree as well.

- Let $w \in V_x$ and α the locus of w in T_x. By inspecting the leaves in the subtree of T_x rooted at α we can pinpoint all the occurrences of w in x in $O(|w|+\text{output})$ time.
- Consider the *weighted vocabulary* (V_x, C), where the weighting functions C associates, with each $w \in V_x$, the number of occurrences of w in x. To allocate (V_x, C) it is sufficient to traverse T_x bottom up weighting each vertex with the sum of the weights of its offsprings (leaves have weight 1). Then for each $w \in V_x$, $C(w)$ is retrieved in $O(|w|)$ time by accessing the (not necessarily proper) locus of w in T_x.
- Let head_i^* be the longest prefix of suf_i which has a non-leaf locus in T_x; let $\text{suf}_i = \text{head}_i^* \text{tail}_i^*$ and assume that a is the first symbol of tail_i^*. The string $\text{head}_i^* \cdot a$ is the shortest subword of x that occurs only at position i. This is the *substring identifier* for i [AH]: it tells how much of a pattern is necessary to identify a position in the text x completely, which can spare time during searches.
- The head_i of maximum length is the longest repeated subword of x. The tree associated with the string $x\#y\$$ ($\# \notin$ I) makes it possible to find the longest common substring of x and y in $O(n+m)$ time, where $m = |y|$. It is remarkable that this problem has such a straightforward solution once T_x is given. A previous algorithm [KMR] could solve it only in $O((n+m)\log(n+m))$, and, as is reported in [KMP], Knuth had conjectured in 1970 that linear time performance was impossible to achieve.
- The longest subword common to k out of m strings of total length n can be also found in $O(n)$ time, although by more elaborate constructions [PR]. This is not trivial, since the straightforward extension of the case $m=2$ produces an algorithm taking $O(n \cdot m)$ time.

2. A FRAMEWORK FOR LINEAR TIME SEQUENTIAL DATA COMPRESSION

Subword trees T_Q for the set of suffixes suf_j where $j \in Q \equiv \{i_1, i_2, ... i_m\}$ and Q is an ordered subset of P are the natural habitat for a class of sequential data compression techniques based on textual substitution. As pointed out elsewhere in this book [ST], this class embodies the few optimization problems in the realm of textual substitution that can be solved in polynomial (actually linear) time. In fact the techniques in this class also feature asymptotic optimality in the information theoretic sense [ZI, ZL, ZL1, LZ, SZ1, SZ2].

The idea is in general that of interleaving the construction of a (possibly partial) subword tree with a *parse* of the textstring into *phrases*. Compression is achieved whenever phrases are susceptible of a more compact representation.

The set Q is retrieved from P by means of a *generative process*, which is actuated by following a set of rules to identify, for each suffix of x with starting position $i_j \in Q$, the associated jth *reproduction* rep_j of x and its strictly related *production* prod_j. The exact nature of rep_j depends on the particular generative process chosen. In all cases, however, rep_j will coincide with a suitable prefix of a suffix suf_{i_f}, with $i_f \in Q$ and $f < j$; prod_j is always:

$$\text{prod}_j = \text{rep}_j \cdot x[i_j + |\text{rep}_j|]$$

Thus prod_j is fully individuated by setting some suitable pointer(s) to the previous suffix and by providing the (possibly new) terminal symbol. This information is the *identifier* for prod_j, denoted by $id(j)$.

For each type $(A, B, C, ...)$ of reproduction defined, the *<type>-parse* of x, denoted $type-P(x)$ is the (unique) decomposition of x in terms of those productions that are pinpointed through the greedy left to right scanning of the symbols of x. The production of x that is selected by actuating the hth step in this process represents the hth *phrase* in the parse. Since each phrase is also a production, we can associate with the parse of x its *translation* $\sigma(x)$, defined as the concatenation of the identifiers for the productions that are also phrases in the parse.

The paradigm of the procedure *parse* below encompasses most instantiations of the generative processes in [ZI, LZ, SZ1, SZ2, AGU]. We assume that the operation of *insert* is accompanied with the identification of the current (re)production via the auxiliary function *Lprefix*, and by the insertion of an auxiliary endmarker node whenever needed for possible later reference.

```
    procedure parse (x,q)
    ## produces σ(x) from inputs x and characteristic function q ##
1.  begin i:=1; j:=1; h:=1; T_1:={suf_1};
          phrase_1 := prod_1 := x[1]; σ := id(1) := <x[1]>;
2.  while i<n do ## produce next phrase ##
3.     begin i:=i+1; j:=j+1; h:=h+1;
4.     T_j := insert(T_{j-1},suf_i)
5.     phrase_h := prod_j := Lprefix(suf_i);
6.     σ =σ ·id(j);
7.     if i+|rep_j|<n then
              ## generate intermediate (re)productions ##
              begin
8.              m:=j
9.              with k ∈ q(i,|rep_j|) do
10.             begin m:=m+1; T_m:=insert(T_{m-1},suf_k) end
              end
11.         i:=i+|rep_j|
       end
12.    σ(x):=σ
    end.
```

The loop of lines (9,10) enriches the vocabulary between 'active' parsing steps by inserting extra suffixes according to some given characteristic function q. The two extreme cases are when q exhausts all intermediate positions (i.e., $k=i+1, i+2$, etc.), and when it neglects them all. In this latter case it results in $j=h$ at all times during *parse*. One expects the number of phrases in the parse to decrease as the number of intermediate insertions increases. However, there is a subtle interplay between the number of intermediate insertions and the sizes resulting for identifiers, which might offset this benefit. For example, let:

$$x = 11111111111111111101110111011101010101010001000100$$

The A-parse is characterized as follows:

Lprefix(suf_i) - coincides with the longest prefix of suf_i that matches some past production (=phrase), extended by concatenation of the next symbol of suf_i.

Example: $A-P(x)$ = 1-11-111-1111-11111-11110-1110-11101-110- 111010-10-101-0-00-100-01-00$

phrases = 17

$|id(j)| \tilde{=} [\log j] = [\log h]$ bits [SZ 1] (i.e., roughly the bits needed to identify one among $h-1$ previous phrases plus the empty phrase λ.)

The B-parse is as follows:

Lprefix(suf_i) - is given by the longest prefix of suf_i that matches the concatenation of two past phrases followed by the terminal symbol as above, or else it is as per scheme A if no such pair of phrases exists.

Example: $B-P(x)$ = 1-11-1111-1111111-11111-0-1110-11101-110- 111010-10-101-010-00-010-001-00$

phrases = 16

$|id(j)| \tilde{=} [\log(3j)] = [\log(3h)]$ [SZ 2] (roughly, the current phrase is identified by selecting one of the h possible simple phrases, plus the $h-2$ pairs followed by an incoming 1, plus as many pairs followed by a 0).

The C-parse and the D-parse are closely related. For the first one we have:

Lprefix(suf_i) - is chosen as the longest concatenation of past phrases, ending perhaps in a prefix of a past phrase, followed by the new symbol as above.

Example: $C-P(x)$ = 1-11-1111-11111111-1110-11101-110-1110111010- 10-101-0-00-10001-00$

phrases = 14

$|id(j)| \tilde{=} [\log h] + [\log i] + 1$ ([$\log h$] bits are needed to identify the first past phrase, [$\log i$] bits contain the length of the current phrase and the last bit is needed for the terminal symbol).

In the D-parse, we waive the requirement that the copying process be terminated during some *past* phrase, i.e., we have now:

$rep_j = head_j$

Example: $D-P(x)$ = 1-1111111111111111110-1110-111011101110-10- 1010100-0-100-0100$

phrases = 9

$|id(j)| \tilde{=}$ [log$h$] + [log($n-i$+1)] + 1 (this has an interpretation similar to that of scheme C, except that the length of the current phrase then exceed the i bits).

The suffix in the E-parse is exactly the same as for the D-parse except that it is now $r>n$. It follows that it now $rep_j = rep_i = head_i$.

Example: $E-P(x)$ = 1-1111111111111111110-111011101110111010- 1010100-01000-100$

phrases = 6

$|id(j)| \tilde{=}$ [log$i$] + [log($n-i$+1)] + 1 (the copying process may now start at any past position).

It is readily seen that the instantiations $A-D$ of *parse* can be set up to run in linear time.

Other variations and applications are discussed elsewhere in this book [MW,LZ1], along with a broader survey of data compression [ST], and novel compression methods [FK] for sparse bit strings. Intermediate characterizations for the set Q were introduced in [AGU]. Efficient ways of dealing with buffers of limited sizes [ZL] are presented in [RPE].

3. SQUARES IN A WORD

A *square* of x is a word on the form ww, where w is a *primitive* word, i.e., a word that cannot be expressed in any way as v^k with $k>1$. Square free words, i.e., words that do not contain any square subwords have attracted attention since the early works by A. Thue in 1912 [TH]. A copious literature, impossible to report here, has been devoted to the subject ever since.

By keeping special marks to all nodes leading to suf_1 it is possible to spot all square prefixes of x as a byproduct of the construction of T_x. The same straightforward strategy can be used for square suffixes. On the other hand, devising efficient algorithms for the detection of (all) squares has required more efforts [ML,CR,AP]. The number of distinct occurrences of squares in a word can be $\Theta(n \log n)$, which sets a lower bound for all algorithms that find all squares [CR]. For instance, infinitely many *Fibonacci words*, defined by:

$$w_0 = b ; w_1 = a$$

$$w_{m+1} = w_m w_{m-1} \quad for \; m > 1$$

have $O(n \log n)$ distinct occurrences of square subwords. Interestingly enough, by following the proof in [CR] as a guideline and making use of the fact that cyclic permutations of a primitive word are also primitive, it is not difficult to show that, for $m \geq 4$, the number S_m of different square subwords in w_m is such that $S_m \geq 1/12$

($|w_m| \log |w_m|$). This fact is of some consequence in trying to assess the space needed for the allocation of the statistics without overlap of all subwords of a text-string [AP1]. We show now that the E-parse $\text{prod}_1\text{prod}_2\cdots\text{prod}_k$ of a string x, can be used nicely as a filter to spot the leftmost occurring nontrivial square of x. Our approach is similar to the one in [CR1]. In this context, a square is *trivial* if it is a suffix of prod_j for some $j \in \{1,2,...k\}$ (which takes, trivially, overall linear time to spot), or if it is detected following the situation described below.

For $j \in \{1,2,...,k\}$, let $\text{prod}_j = \text{rep}_j \cdot a$ with $a \in I \cup \{\$\}$. Now prod_1 is obviously squarefree. Assume $\text{prod}_1\text{prod}_2\cdots\text{prod}_{j-1}$ square free and let l be its length. Then if $|\text{rep}_j| \geq l$, there is a square in $\text{prod}_1\cdots\text{prod}_{j-1}\text{rep}_j$, due to two occurrences of rep_j that either overlap or are contiguous. This circumstance can be easily detected on line with carrying out the E-parse of x, hence in linear time, and we shall say that such square is trivial too.

A few more definitions are needed in order to illustrate the full criterion. We say that two subwords w and $w\cdot$ of x satisfy the *left* (*right*) *property*, denoted $l(w,w\cdot)$ ($r(w,w\cdot)$), if $ww\cdot$ are squarefree but $ww\cdot$ embeds a square vv centered to the left (right) of $w\cdot$. Let x be a string with no trivial square. Then:

> *x is not squarefree iff there is* $i \in (1,2,...,k-1)$ *such that:* $l(\text{prod}_i\,\text{prod}_{i+1})$ *or* $r(\text{prod}_i\,\text{prod}_{i+1})$ *or* $r(\text{prod}_1\text{prod}_2...\text{prod}_{i-1}, \text{prod}_i\,\text{prod}_{i+1})$.

To prove this claim, let yvv be the shortest non squarefree prefix of x and let j be the smallest index for which yvv is a prefix of $\text{prod}_1\cdots\text{prod}_{j+1}$. Under our assumptions, it suffices to show that the second occurrence of v must fall entirely within $\text{prod}_j\,\text{prod}_{j+1}$. But this follows at once from the definition of rep_j. Indeed, if the second occurrence of v does not fall within $\text{prod}_i\,\text{prod}_{j+1}$ then rep_j would be contained in the second occurrence of v without being a suffix of v, a contradiction.

The left and right properties can be checked in overall linear time with the aid of auxiliary 'local' subword trees, or simply by resorting to the 'failure function' [AH]. We leave this as an exercise for the reader. An alternative procedure for testing squarefreeness [ML1] and a simple and elegant probabilistic algorithm for this problem [RA] are both discussed elsewhere in this book.

We turn now to the problem of finding all squares in a word. The use of subword trees in this task is brought up by the following fact [AP].

> *x contains a square occurrence at position i iff there is a primitive word* $w \in V_x$ *and a vertex* α *in* T_x *such that* i *and* $j = i + |w|$ *are consecutive leaves in the subtree of* T_x *rooted at* α *and furthermore* $|w(\alpha)| \geq (i-j)$.

The algorithmic criterion provided by the above condition is implemented straightforwardly in a bottom up computation. Starting from the leaves of T_x, for each interior vertex visited we construct the sorted list of the labels of its leaves. The sorted list of any such vertex is obtained by merging the sorted lists of its offspring vertices. The strategy runs in $O(n \log n)$ time if T_x is nearly balanced or completely unbalanced. Optimal handling of intermediate cases involves pebbling of T_x with an *ad hoc* data structure suited to the efficient repeated merging of integers in a known range [AP].

We devote the remainder of this section to highlight that the structure of T_x may help disclosing general properties about power subwords in a string [AA]. For instance, unlike the number of squares, the number of distinct cube subwords of any string x is bounded by n. To show this, we introduce the notion of cube constrained word (CCW) as follows: we say that $ww \in V_x$ is *cube constrained* if $w^3 \in V_x$. It is seen [AA] that the number of distinct CCW's in any string x is bounded by n. In order to prove this fact, one first uses the definition of T_x to show that if $w^{k+1}(k \geq 1)$ is a subword of x, then w^k and w^{k+1} have distinct loci in T_x. Next one uses this in conjunction with the *periodicity lemma* [LS] to show that if w^2 and v^2 are distinct CCW's of x, then they must have distinct loci in T_x. The assertion follows then from the fact that the number of interior vertices of T_x is bounded by n.

4. STATISTICS WITHOUT OVERLAPS

The (primitive rooted) squares in V_x have consequences on the amount of storage needed to allocate the statistics without overlap of all substrings of x [AA,AP1], which leads us to another application of T_x. Consider the *weighted vocabulary* (V_x, C') where C' associates, with each $w \in V_x$, the maximum number k of distinct occurrences of w such that it is possible to write $k = w_1 w w_2 w w_3 \cdots w w_{k+1}$ with w_d possibly empty $(d = 1,2,...,k+1)$.

The construction of (V_x, C') requires in general augmenting T_x [AP1] by inserting auxiliary nodes of degree 1. The role of such nodes in the augmented tree is to function as proper loci for subwords whose loci in the original tree T_x would not report the actual number of their nonoverlapping occurrences. To be more precise, assume that all nodes in the tree of Fig. 1 are weighted with their associated C' values. Now ab occurs 8 times in w_7, the word of Fig. 1; but the locus α of ab has $w(\alpha) = aba$ with a $C' = 5$. In order for the tree to report the appropriate C' value for ab we have to split an edge and create the proper locus for this subword. Let $\tilde{T}_x$ be the minimal (i.e., with the least auxiliary nodes) augmented subword tree. The following fact gives a handle in establishing where the auxiliary nodes should be inserted in T_x in order to produce $\tilde{T}_x$ [AP1].

If α is an auxiliary node of $\tilde{T}_x$, then there are subwords u, v in x and an integer $k \geq 1$ such that $w(\alpha) = u = v^k$ and there is an $w \in V_x$ such that $w = v^m v \cdot$ with $v \cdot$ a prefix of v and $m \geq 2k$.

An $O(n \log n)$ upper bound on the number of auxiliary nodes needed in $\tilde{T}_x$ can be readily set, based on the above fact and on the upper bound on the number of positioned squares in a word. However, it seems to be an interesting open question whether there are words whose minimal augmented suffix trees do in fact attain that bound. The insertion of candidate auxiliary nodes can be carried out during the brute force construction of T_x, after which redundant nodes can be removed through one visit of the structure. Hence $\tilde{T}_x$ can be obtained in $O(n^2)$ time, almost straightforwardly. A more efficient construction is also more elaborate [AP2], and we shall not attempt at reporting it here.

5. CONCLUDING REMARKS

Since subword trees embody remarkably structured information about the word(s) they are built out of, it is not surprising that they can be used in a variety of tasks that either aim at retrieving some such information or make crucial use of it in answering disparate queries. Sometimes there are better methods than those based on such trees, however, no digital index seems to outperform subword trees in versatility and elegance.

For instance, the subword tree associated with $y = x\#x^r\$$ can be used to detect all palindrome subwords of x, in $O(nlogn)$, by repeated bottom up merging of leaves (as with the detection of squares) and by making use of the fact that any palindrome in V_x must have a proper locus in T_y, as the reader may check for himself. As is well known, there are linear time solutions for this problem (see for instance [MA]).

Similarly, the subword tree associated with a set of m words of total length l can be adapted to test the *unique decipherability* of the code consisting of those words in $O(m \cdot l)$ time [RO]. However, the same performance can be achieved by a simpler construction, based on pattern matching machines [AC], as shown in [AG]. The subject of unique decipherability testing is also addressed elsewhere in this book [CH]. The relation between subword trees and pattern matching machines is investigated in [CR2].

Acknowledgements

Joel Seiferas kindly supplied some of the reference material. Zvi Galil and Sam Wagstaff made many very helpful comments on a preliminary version of this paper.

References

AC Aho, A., and Corasick, M.J., Efficient String Matching: An Aid to Bibliographic Search, *CACM* 18, 335-340 (1975).

AH Aho, A., Hopcroft, J.E., Ullman, J.D., *The Design and Analysis of Computer Algorithms*, Addison Wesley, Reading (1974).

AA Apostolico, A., On Context Constrained Squares and Repetitions in a String, *R.A.I.R.O. Journal Theoretical Informatics* 18, 2, 147-159 (1984).

AG Apostolico, A., Giancarlo, R., Pattern Matching Machine Implementation of a Fast Test for Unique Decipherability, Inf. Proc. Letters 18, 155-158 (1984).

AGU Apostolico, A., Guerrieri, E., Linear Time Universal Compression Techniques for Dithered Images Based on Pattern Matching (extended abstract), *Proceedings of the 21st Allerton Conference on Communication, Control and Computing*, 70-79 (1983).

AP Apostolico, A., Preparata, F.P., Optimal Off-line Detection of Repetitions in a String, *Theoretical Computer Science*, 22, 297-315 (1983).

AP1 Apostolico, A., Preparata, F.P., The String Statistics Problem, Tech. Report, Purdue Univ. CS Dept. (1984). A preliminary version: A Structure for the Statistics of all Substrings of a Textstring With and Without Overlap,

Proceedings of the 2nd World Conference on Math. at the Service of Man, 104-109 (1982).

BL Blumer, A., Blumer, J., Ehrenfeucht, A., Haussler, D., McConnell, R., Building a Complete Inverted File for a Set of Text Files in Linear Time, *Proceedings of the 16th ACM STOC*, 349-358 (1984).

BE Blumer, A., Blumer, J., Ehrenfeucht, A., Haussler, D., McConnel, R., Building the Minimal DFA for the Set of All Subwords of a Word On-line in Linear Time, *Springer-Verlag Lecture Notes in Computer Science* 172, 109-118 (1984).

CH Capocelli, R.M., Hoffmann, C.H., Algorithms For Factorizing and Testing Subsemigroups, *Combinatorial Algorithms on Words* (A. Apostolico and Z. Galil, eds.) Springer-Verlag (1985).

CR Crochemore, M., An Optimal Algorithm for Computing the Repetitions in a Word, *Inf. Proc. Letters* 12, 5, 244-250 (1981).

CR1 Crochemore, M., Recherche Lineaire d'un Carre dans un Mot, *C.R. Acad. Sc. Paris*, t.296, Serie I, 781-784 (1983).

CR2 Crochemore, M., Optimal Factor Transducers, *Combinatorial Algorithms on Words* (A. Apostolico and Z. Galil, eds.) Springer-Verlag (1985).

CS1 Chen, M.T., Seiferas, J., Additional Notes on Subword Trees, unpublished lecture notes (1982).

CS2 Chen, M.T., Seiferas, J., Efficient and Elegant Subword Tree Construction, *Combinatorial Algorithms on Words* (A. Apostolico and Zvi Galil, eds.), Springer-Verlag (1985).

FK Fraenkel, A. S., Klein, S. T., Novel Compression of Sparse Bit Strings, *Combinatorial Algorithms on Words* (A. Apostolico and Z. Galil, eds.) Springer-Verlag (1985)

KMP Knuth, D.E., Morris, J.H., Pratt, V.R., Fast Pattern Matching in Strings, *SIAM Journal on Computing* 6, 2, 323-350 (1977).

KMR Karp, R.M., Miller, R.E., Rosenberg, A.L., Rapid Identification of Repeated Patterns in Strings, Trees, and Arrays, *Proceedings of the 4th ACM STOC*, 125-136 (1972).

KN Knuth, D.E., *The Art of Computer Programming*, Vol. 3: *Sorting and Searching*, Addison-Wesley, MA (1973).

LS Lyndon, R.C., Schützenberger, M.P., The Equation $a^M = b^N c^P$ in a Free Group, *Michigan Math. Journal* 9, 289-298 (1962).

LZ Lempel, A., Ziv, J., On the Complexity of Finite Sequences, *IEEE TIT* 22, 1, 75-81 (1976).

LZ1 Lempel, A., Ziv, J., Compression of Two-dimensional Images, *Combinatorial Algorithms on Words* (A. Apostolico and Z. Galil, eds.) Springer-Verlag (1985).

MA Manacher, G., A New Linear-time On-line Algorithm for Finding the Smallest Initial Palindrome of a String, *JACM* **22, 346-351 (1975).**

MR Majster, M.E., Reisner, A., Efficient On-line Construction and Correction of Position Trees, *SIAM Journal on Computing* 9, 4, 785-807 (1980).

MC McCreight, E.M., A Space Economical Suffix Tree Construction Algorithm, *JACM* 23, 2, 262-272 (1976).

ML Main, M.G., Lorentz, R.J., An O(nlogn) Algorithm for Finding all Repetitions in a String, *Journal of Algorithms*, 422-432 (1984).

ML1 Main, M.G., Lorentz, R.J., Linear Time Recognition of Square-Free Strings, *Combinatorial Algorithms on Words*, (A. Apostolico and Z. Galil, eds.) Springer-Verlag (1984).

MO Morrison, D.R., PATRICIA – Practical Algorithm to Retrieve Information Coded in Alphanumeric, *JACM* 15, 4, 514-534 (1968).

MW Miller, V.S., Wegman, M.N., Variations on a Theme by Ziv and Lempel, *Combinatorial Algorithms on Words* (A. Apostolico and Z. Galil, eds.) Springer-Verlag (1985).

PR Pratt, V.R., Improvements and Applications for the Weiner Repetition Finder, unpublished manuscript (1975).

RA Rabin, M.O., Discovering Repetitions in Strings, *Combinatorial Algorithms on Words* (A. Apostolico and Z. Galil, eds.), Springer-Verlag (1985).

RO Rodeh, M., A Fast Test for Unique Decipherability Based on Suffix Trees, *IEEE TIT* 28, 648-651 (1982).

RPE Rodeh, M., Pratt, V.R., and Even, S., Linear Algorithms for Data Compression via String Matching, *JACM* 28, 1, 16-24 (1981).

SE1 Seiferas, J., Subword Trees, unpublished lecture notes (1977).

SL Slisenko, A.O., Detection of Periodicities and String Matching in Real Time, *Journal of Soviet Mathematics* 22, 3, 1316-1387 (1983).

ST Storer, J.A., Textual Substitution Techniques for Data Compression, *Combinatorial Algorithms on Words* (A. Apostolico and Z. Galil, eds.) Springer-Verlag (1985).

SZ1 Seery, J.B., Ziv, J., A Universal Data Compression Algorithm: Description and Preliminary Results, Bell Labs TM77-1212-6/77-1217-6 (1977).

SZ2 Seery, J.B., Ziv, J., Further Results on Universal Data Compression, Bell Labs, TM78-1212-8/78-1217-11 (1978).

TH Thue, A., Uber Die Gegenseitige Lage Gleicher Teile Gewisser Zeichenreichen, Skr. Vid. Kristiana I. Math. Naturv. Klasse 1, 7-67 (1912).

WE Weiner, P., Linear Pattern Matching Algorithms, *Proceedings of the 14th Annual Symposium on Switching and Automata Theory*, 1-11 (1973).

ZI Ziv, J., Coding Theorems for Individual Sequences, *IEEE TIT* 24, 4, 405-413 (1978).

ZL Ziv, J., Lempel, A., A Universal Algorithm for Sequential Data Compression, *IEEE TIT* 23, 3, 337-343 (1977).

ZL1 Ziv, J., Lempel, A., Compression of Individual Sequences On Variable Length Encoding, *IEEE TIT* 24, 5, 530-536 (1978).

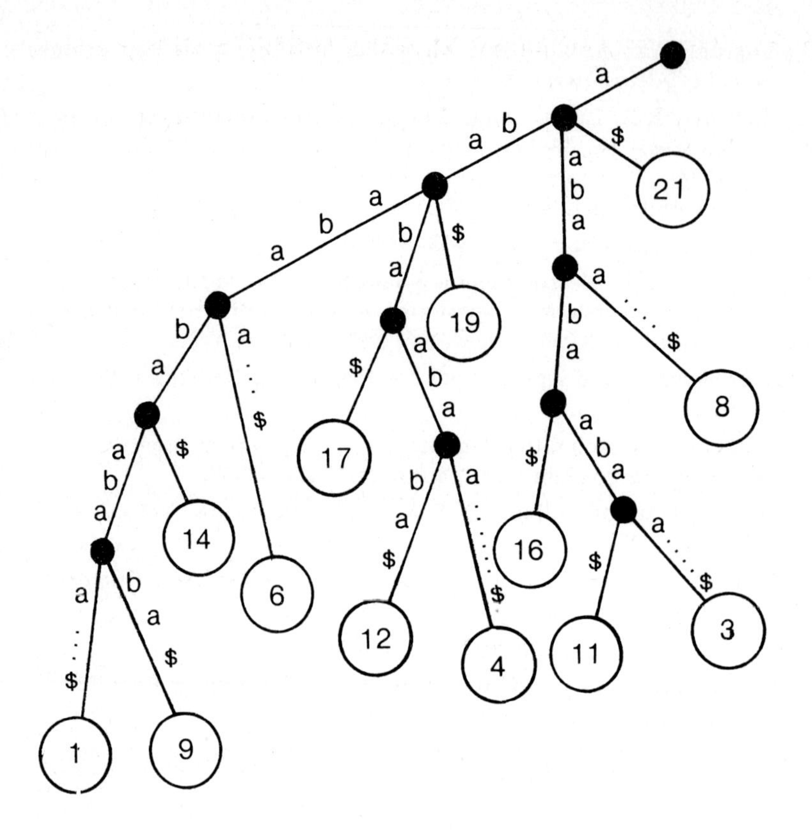

Figure 1

A partial view (all suffixes starting with *a*) of the subword tree of the string *abaababaabaababaababa*.

EFFICIENT AND ELEGANT SUBWORD-TREE CONSTRUCTION†

M. T. Chen
Department of Computer Science
University of Nanjing
People's Republic of China

Joel Seiferas
Department of Computer Science
University of Rochester
Rochester, New York, U. S. A. 14627

Abstract. A clean version of Weiner's linear-time compact-subword-tree construction simultaneously also constructs the smallest deterministic finite automaton recognizing the *reverse* subwords.

Introduction

Any finite set S of words that is prefix-closed (i.e., $xy \in S \Rightarrow x \in S$) has a *prefix tree* with node set S, ancestor relation "is a prefix of", and father relation "is obtained by dropping the last letter of". The set S of all subwords of a text string is a prefix-closed set whose prefix tree, the text's *subword tree*, is particularly useful. For example, it lets us test arbitrary words for membership in S in time proportional to their own lengths, regardless of how long the entire text is. Even more useful is the subword tree for $¢w\$$, where $¢$ and $\$$ are delimiting symbols not occurring in w. This, for example, lets us test easily whether a word is a prefix or suffix of w. In one appropriate walk through the tree, we can easily augment each node with such information as the count of its leaf descendents. Then it becomes convenient to tell *how many* times a word appears, *where* a word first or last appears, what is the longest repeated subword, and more. As an example, the subword tree for $¢aabab\$$ is shown in Figure 1.

The number of distinct subwords of a text string of large length n can be very large (proportional to n^2 for $a^{n/2}b^{n/2}$, for example), so subword trees can have prohibitively many nodes. Fortunately, however, there are compact but functionally equivalent data structures that can even be *built* in time proportional to just n. Weiner [13], McCreight [5], Pratt [6, 7, 8], and Slisenko [12, Section 2] have described such data structures and algorithms. (See also [1, Section 9.5], [4].) Each of their algorithms is complicated by the maintenance of additional auxiliary structure along with the developing compact subword tree. (In Slisenko's case, more ambitious applications account for an extra measure of additional structure [9–12].) In this report, we describe a version with auxiliary structure that is unusually clean and clearly desirable in its own right.

---

† Part of this work was done while the first author visited the University of Rochester. The work of the second author was supported in part by the National Science Foundation under grant MCS-8110430.

NATO ASI Series, Vol. F12
Combinatorial Algorithms on Words
Edited by A. Apostolico and Z. Galil

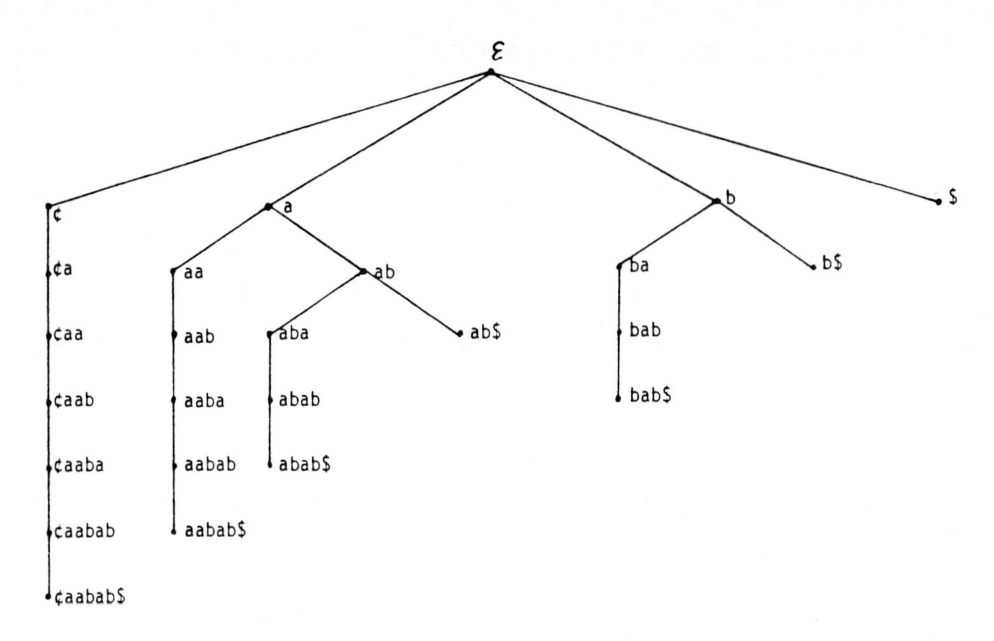

Figure 1. Node-labeled subword tree for *¢aabab$*

Compact Representations

The most obvious way to represent a subword tree compactly is to omit interior nodes of degree 1, replacing them with through edges. The string corresponding to each remaining node can be represented by a (not necessarily unique) pair of pointers into the text string; or, alternatively, the incremental substring corresponding to each edge can be represented by such a pair. Resulting representations for *¢aabab$* are shown in Figure 2. Either way, no information is lost, and the degree of each remaining node continues to be bounded by the alphabet size. The size of this representation is thus proportional to the number of nodes. And the number of nodes is proportional to the length of the text, because the number of leaves is so bounded (one for each suffix of the text) and because, in a tree without interior nodes of degree 1, the number of interior nodes is bounded by the number of leaves.

We can obtain a quite different compact representation by identifying (edge-)isomorphic subtrees. (The edge labels to be preserved by each such isomorphism are the incremental *letters*, not their indices in the text.) The edge-labeled version of the subword tree for *¢aabab$* (shown in Figure 3), for example, has isomorphic subtrees below the subwords *b* and *ab*. The result of making all such identifications is shown in Figure 4. (For expositional clarity in our figures, we revert to explicit edge labels. For later reference, in addition, parenthesized capital letters have been arbitrarily assigned as names for the nodes in Figure 4.) Except for omission of the one nonaccepting state, from which all transitions are self-loops, this directed acyclic graph is just the smallest deterministic finite automaton recognizing the set of subwords of the text. We will see that the size of this representation is again only proportional to the length of the text,

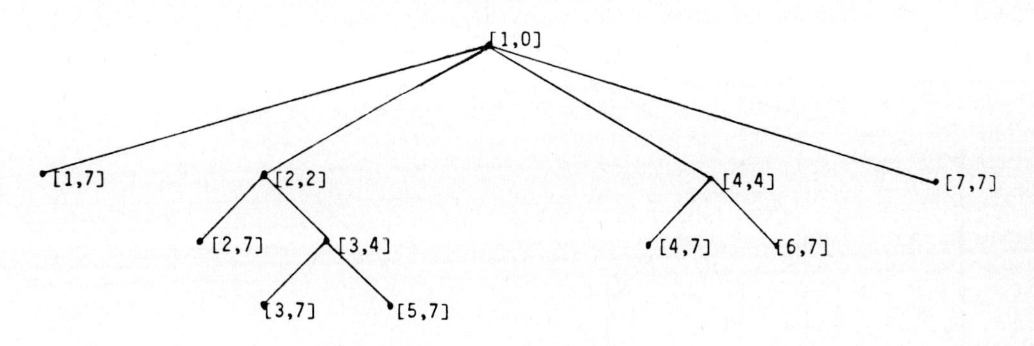

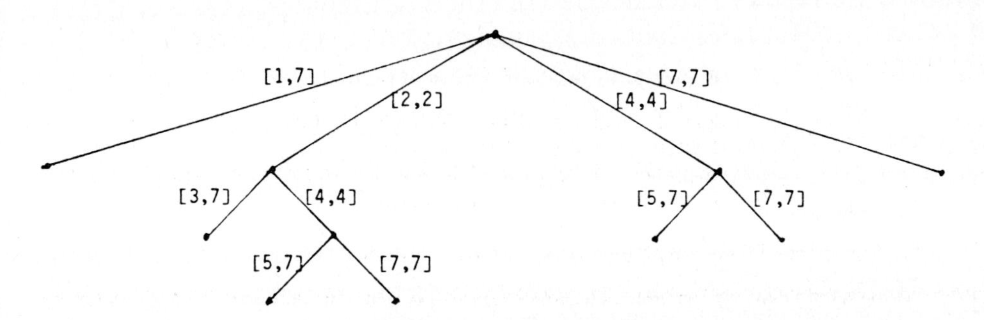

Figure 2. Compacted subword tree for *¢aabab$*, node-labeled and edge-labeled

and for essentially the same reason.

Approaches to Construction

The main objective of Weiner's algorithm is to build the first compact representation of the subword tree. To do this, it maintains auxiliary information that happens to include explicitly some (but not all) of the son links from the ***uncompacted***, ***reverse*** subword tree. (Such terminology is unambiguous, because the tree of reverse subwords of the text is identical to the tree of ordinary subwords of the reverse text.) Similarly, McCreight's algorithm happens to maintain ***father*** links from the uncompacted reverse subword tree.

Pratt's and Slisenko's algorithms are essentially the same as Weiner's algorithm, but with a different viewpoint. The main objective is to build that portion of the uncompacted subword tree whose

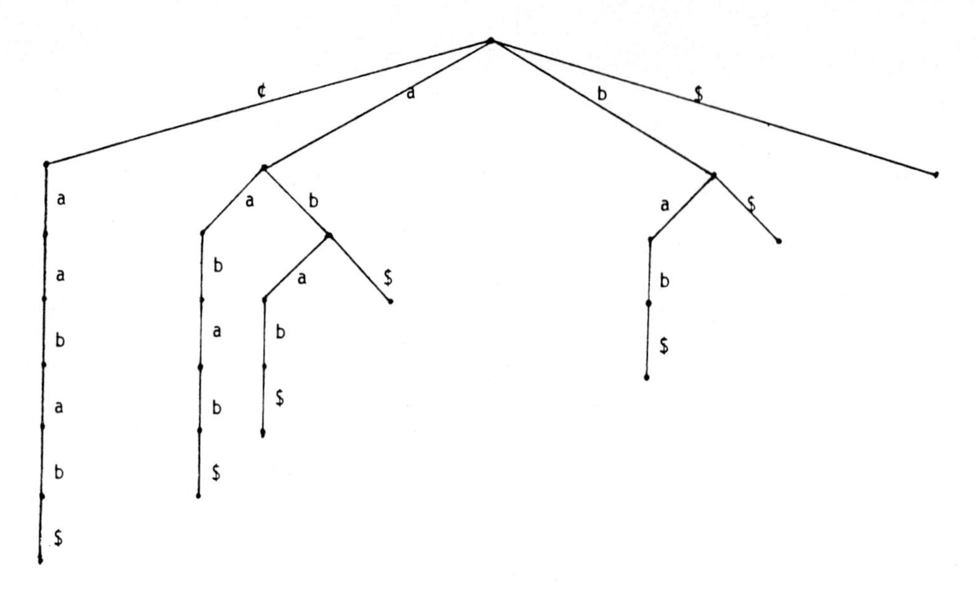

Figure 3. Edge-labeled version of Figure 1

nodes are the text's "longest repetitions". With a few additional patches, this turns out to be as useful as the entire subword tree.

The concept of a "longest repetition" is a dynamic or historical one, based on successive consideration of longer and longer text prefixes. In each prefix, the longest suffix that also occurs elsewhere in the prefix is a "longest repetition". Fortunately, a simple *static* definition is easily seen to be equivalent: A string x is a longest repetition if and only if there are distinct letters a and b for which both ax and bx occur as subwords. From this redefinition, it is clear that the longest repetitions correspond precisely to the interior nodes of degree greater than 1 in the reverse subword tree. The auxiliary structure from Weiner's algorithm turns out to include enough of the uncompacted reverse subword tree so that Pratt and Slisenko can simply run that algorithm on the reverse of their texts, regarding Weiner's bathwater as their baby (and his baby as their bathwater).

Our new observation is that there is a much more natural choice of the "few additional patches" maintained above. Instead of information of some new kind, we can add additional edges (but *no* additional nodes) to the fragment of the uncompacted reverse subword tree to get a directed acyclic graph that, for the reverse subword tree, is the *second* compact representation described above. The key coincidence is that the nodes in the first compact representation of the subword tree correspond to the reverses of the longest representatives of the isomorphism classes of the words in the uncompacted reverse subword tree. To see this, first observe that, whenever one subword u is a prefix of another subword uv *and occurs only in that context*, the subtrees below the reverses of the two strings in the reverse subword tree must be isomorphic; i.e., for every z, zu is a subword if and only if zuv is. (From a reverse perspective, this is the reason for the isomorphism below b and ab in Figure 3, for example.)

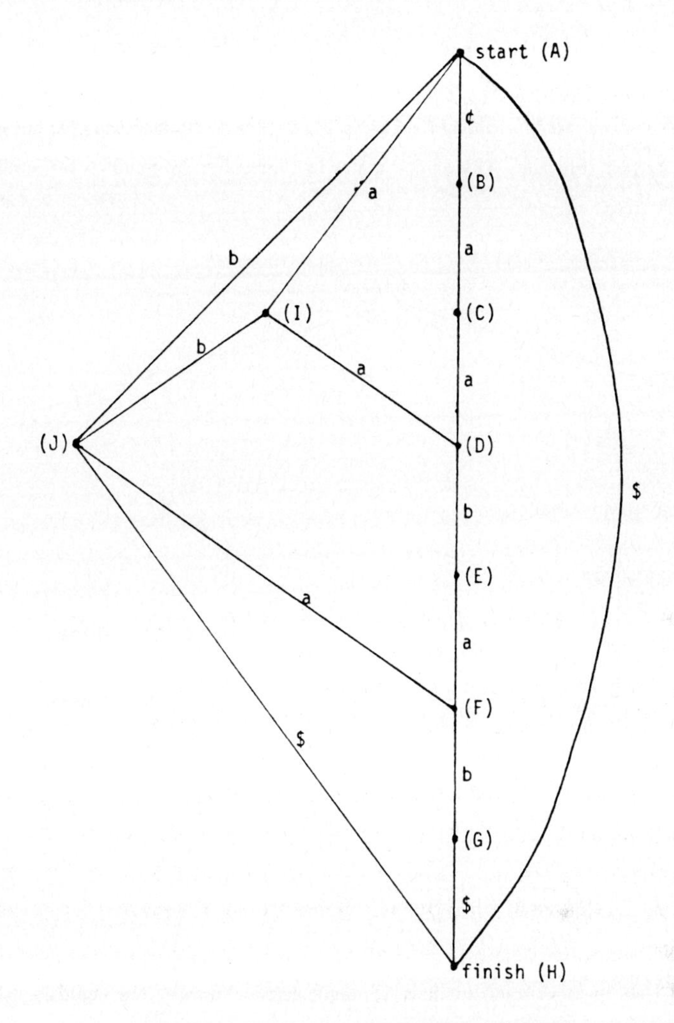

Figure 4. Smallest subword DFA for *¢aabab$*

And observe conversely that, for a text with the assumed unique left endmarker, this is the ***only*** possible reason for subtree isomorphism in a reverse subword tree.

Note that the last point can be false for a text with no left endmarker. In the text *aab*, for example, $\{z|za \text{ is a subword}\} = \{z|zab \text{ is a subword}\} = \{\epsilon, a\}$, but the first occurrence of *a* is ***not*** as a prefix of *ab*. When there is a left endmarker ¢, however, the starting positions of each subword *u* in the text are completely ***determined*** by $\{z|zu \text{ is a subword}\}$.

Continuing with the left-endmarked case, we have concluded that the subtrees below the reverses of two text subwords are isomorphic if and only if one of the subwords is a prefix of the other and occurs only in that context. Equivalently, they are isomorphic if and only if, in the ordinary subword tree, the longer subword can be reached from the shorter one via a path with no branching. This is the claimed coincidence. Hence, we will have the second compact representation for the reverse subword tree if we direct the a-edge from each node x to the shortest extension axy of ax that is also a node. The compact representation of the subword tree for *¢aabab$* shown in Figure 4 was obtained from the subword tree for *$babaa¢* (Figure 5) by just this rule. The named nodes in Figure 5 correspond to the similarly named nodes in Figure 4.

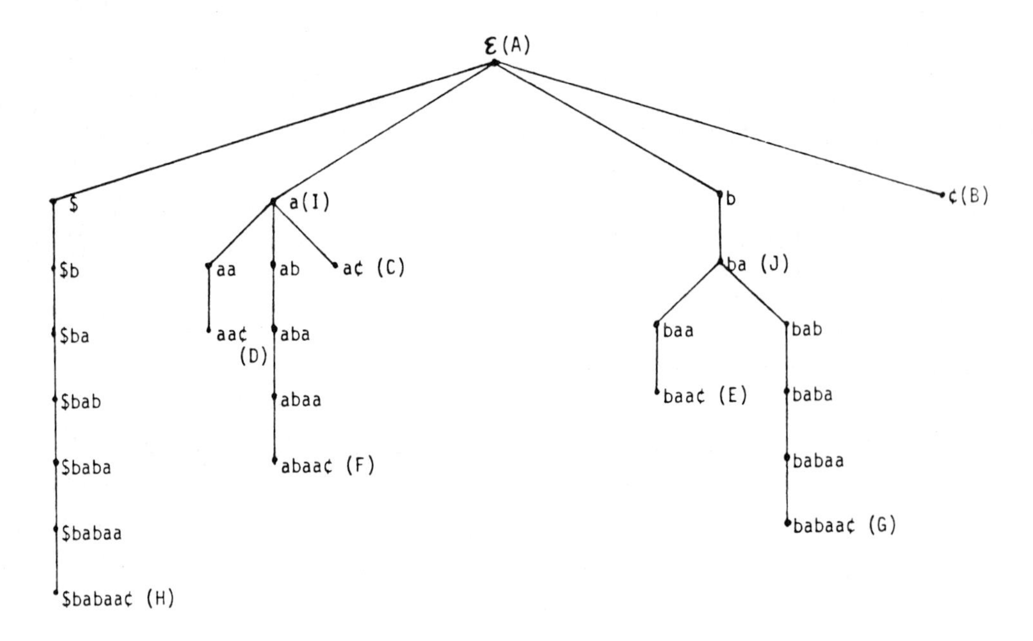

Figure 5. Node-labeled subword tree for *$babaa¢*

In the case that the text does *not* have a unique left endmarker, our condition above is still *sufficient* for isomorphism. For such a text, therefore, we will just get a linear-size deterministic finite automaton that is not necessarily the smallest possible. If the text is *aab$*, for example, we get an automaton with one extra state. In a later section we will sketch a neat way to patch the construction to eliminate extra states.

Our Algorithm

Before we spell out our variant of the algorithm, let us summarize the specifications for the cleaned up data structure the algorithm is supposed to construct. The text w is a character string, and $\$$ is a right endmarker not occurring in w. There is a node for each suffix of $w\$$ (including the empty subword ϵ) and a node for each subword x occurring in two distinct immediate right contexts (i.e., with distinct letters a and b for which both xa and xb are subwords of $w\$$). From each node x, for each letter a, there is an a-*extension* link to the shortest node (if any) with prefix xa and an a-*shortcut* link to the shortest node (if any) with prefix ax. From each node, there is a *prefix* link to its longest proper prefix node (if any). (Noting that prefix links are just the reverses of extension links, we will leave setting them implicit whenever we create or change extension links.) Finally, for each node, there is a pair of indices into $w\$$ to identify (one instance of) the corresponding subword. (We will also leave implicit the setting of these indices. Each new node added will be a prefix of the entire text so far considered; it will be either that entire text or a prefix 1 longer than the known length of some older node.)

As our terminology suggests, we will describe the algorithm from Weiner's viewpoint: The main objective is to build the extension and prefix links, and the other links provide time-saving "shortcuts". Even so, the algorithm works from *right to left* in the text, with no need to look "ahead" (to the left), provided we index the letters of $w\$$ from right to left.

The structure for $w\$ = \$$ is trivial (two nodes) and can obviously be built in constant initial time. To build the structure for $aw\$$, we assume inductively that we have the structure for $w\$$ and that we have pointers to the root node and to the node $w\$$, where we finished the previous step. The new subwords will be the "sufficiently long" prefixes of $aw\$$; we will have to install $aw\$$ as a new extension from the longest prefix y of $aw\$$ that is already a subword of $w\$$. If the root (corresponding to the empty subword) does not yet have an a-extension, then it serves as y. If the root does already have an a-extension, we could still find y by "following $aw\$$ along extension links" down from the root until continuation would leave the tree; but for a string like $a^n\$$, this would accumulate to time proportional to n^2. Instead, noting that y's least-proper suffix x ($y = ax$) must already be a node in the structure for $w\$$ and that it must be a prefix of $w\$$, we can trace *up* along *prefix* links from $w\$$, watching for the node x; it will be the first node with an a-shortcut. Following that shortcut will lead to $ax = y$ if it is already a node, or to its shortest extension axz that is a node otherwise. In the latter case, we will have to install y as a new node between axz and its prefix parent, initially with the same shortcut links axz has. The shortcuts *to* the new node will be directed from the nodes lying on the prefix path from x up through the last node x' not already having an a-shortcut link to a *proper prefix* of ax.

With y found and properly installed, we can install $aw\$$ as an extension below it as required, initially without any extension or shortcut links. Shortcut links should be directed *to* this new node from the nodes lying on the prefix path from $w\$$ up through the last node not already having an a-shortcut link. Both these and the shortcut links redirected to y in the case that y had to be installed can be set in a traversal of the prefix path from $w\$$ up through x'; so the time to obtain the structure for $aw\$$ from the one for $w\$$ is proportional to some constant plus the number of nodes on the prefix path from $w\$$ to x'.

To see that this time bound accumulates only to linear time, we look at the node depth of each successive text suffix in the developing extension tree. The key observation is that, except for some small additive constant, the depth of $aw\$$ within its structure is *reduced* from the depth of $w\$$ within its structure at least enough to compensate for the time-indicative number of nodes on the prefix path from $w\$$ to x' above. (See Figure 6.) To see this, first note that the depth of x' is certainly so reduced.

Then note that, if ax'' is any node on the prefix path above y, x'' must be a node on the prefix path above x'. The consequence of the observation is that to spend more than linear total time would require more than linear total depth reduction, which is impossible since the greatest possible *increase* in depth is constant for each iteration.

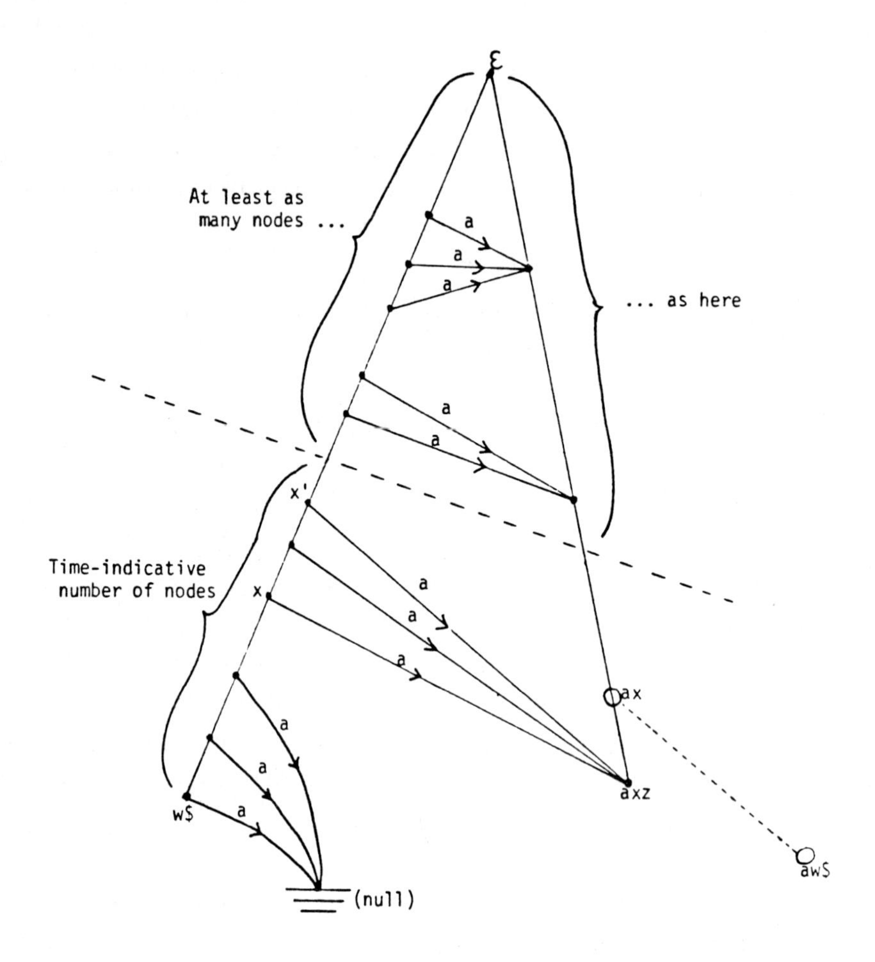

Figure 6. How depth reduction compensates for time

Significance of Endmarkers

Of course subword trees and their two compact representations exist even for texts without endmarkers. A satisfactory way to obtain the first compact representation is to affix auxiliary endmarkers, to perform the entire construction described above, and then finally to remove the endmarkers' effects from the result. Suppose we have the first compact representation of the subword tree for $\not{c}w\$$. To obtain the representation for w is quite easy: Delete the $\not{c}$-subtree below the root, and replace each leaf $u\$$ with u, omitting the leaf entirely if u was already the parent of $u\$$. This yields a compact subword tree for w, but it might introduce interior nodes of degree 1; to get the most compact tree, straightforwardly replace each such node with a through edge.

Now we turn to the second compact representation. From the reverse perspective, we have described an *on-line*, linear-time algorithm to construct linear-size deterministic finite automata recognizing the subwords of the successive prefixes of a left-endmarked text, in such a way that the automaton is minimal for each text prefix ending in a previously unseen symbol ("endmarker"). Following [2], we can modify the construction so that the automaton is minimal in *every* case, without even giving up on-line operation.

The smallest finite automaton equivalent to a given one can always be obtained by identifying the states in each equivalence class [3, Section 3.4]. In the case of our construction, it turns out that each nontrivial equivalence class consists of just two states in the constructed automaton, joined by a prefix/extension link, and that we can easily modify the construction to keep track of which pairs are equivalent. Still in linear time and on-line, we can also maintain a version of the automaton with these pairs actually merged.

To see all this inductively, consider what happens when we go from the structure for $w\$$ to the structure for $aw\$$ (Figure 6). From our reverse perspective, two states (nodes) are equivalent if and only if they correspond to text subwords that occur in precisely the same left contexts. When we go from the structure for $w\$$ to the structure for $aw\$$, pairs of nodes that were already left-distinguishable will continue to be left-distinguishable. The new node $aw\$$ initially occurs only in the empty left context, so it cannot be equivalent to any other node. If there is a new node ax, however, it will be initially left-indistinguishable from axz, the only left extension of its one new occurrence being the empty word.

It turns out to be remarkably easy to determine when left-indistinguishable pairs of nodes in the structure for $w\$$ become distinguishable in $aw\$$. For any parent-child pair of nodes u and uv that are left-indistinguishable in the structure for $w\$$, there must be an occurrence in $w\$$ of u not having right context v. Let t be the full left context of the ***rightmost*** such occurrence. Because there is no node properly between u and uv in the structure for $w\$$, there can be no node properly between tu and tuv either. Because we are looking at the rightmost occurrence of u not having right context v, it follows that tu is a node with exactly two children, $w\$$ and some extension of tuv. If the node tu has no a-shortcut, then the left context at will distinguish u and uv in $aw\$$. Conversely, if tu does have an a-shortcut, then at will not left-distinguish u and uv in $aw\$$, and u and uv will continue to be left-indistinguishable. (If some further-left occurrence of u not having right context v became left-distinguished from uv, then a similar occurrence would have occurred and been so distinguished as part of the shortcut occurrence of atu within just $w\$$.) In summary, therefore, each of the left-indistinguishable node pairs for $w\$$ will remain left-indistinguishable in $aw\$$ if and only if the prefix parent of $w\$$ already has an a-shortcut in the structure for $w\$$.

Finally, to complete restoration of the induction hypothesis, we must be sure that the one possible new node ax does not lie on the extension edge between any pair (u, uv) that continues to be left-indistinguishable, and that the one possible new indistinguishable pair (ax, axz) is disjoint from each pair (u, uv) that continues to be left-indistinguishable. It suffices to note that axz will be strictly longer than uv if any old indistinguishability continues. For the latter to occur, we have noted that the parent of $w\$$ must already have an a-shortcut (and thus serve as x in Figure 6), be of the form tu, and have exactly one other child, some extension of tuv. Therefore, atu occurs only with right context v in $w\$$, and axz must be some extension of $atuv$, which is certainly longer than uv.

To maintain a version of the automaton with equivalent pairs actually merged is relatively straightforward. Initially create each new node ax as mere "auxiliary structure". Install temporary copies of any links to and from it also to and from the equivalent node axz if needed, as part of the desired reduced automaton. For each equivalent pair, maintain a list of all such temporary designations and structure, so that they can be found and modified or erased if the nodes later become distinguishable. Whenever such a pair of nodes does become distinguishable, redesignate the corresponding temporarily auxiliary structure as now part of the desired reduced automaton, and erase all the corresponding temporary links to and from the formerly equivalent extension node. Each node ax can go through this change only once, so the additional time will be only proportional to other time already counted in our earlier analysis, and the total running time will remain linear.

The remaining endmarker $\$$ is easier to do without. If we have the smallest deterministic finite automaton recognizing the subwords of $\$w^R$, then we can get a deterministic finite automaton recognizing the subwords of just w^R by simply eliminating the $\$$-transitions. Any two distinct states that are still accessible must be accessible via subwords u and v of just w^R for which

$$\{z|uz \text{ is a subword of } \$w^R\} \neq \{z|vz \text{ is a subword of } \$w^R\}.$$

Unless one of u and v is the empty word, this yields

$$\{z|uz \text{ is a subword of } w^R\} \neq \{z|vz \text{ is a subword of } w^R\},$$

since the first two sets respectively equal the second two sets in this case. If one of u and v is the empty word, on the other hand, then the two sets are unequal because w^R is in their symmetric difference. If we remove inaccessible states, therefore, we finally get the smallest deterministic finite automaton recognizing the subwords of w^R.

This last adaptation gives up on-line operation, but again such a concession is not necessary. The extraneous states are always those nodes accessible only via shortcut links the first of which is a $\$$-shortcut. These states always correspond to an initial segment of the chain of nonempty prefixes of the left-endmarked text. The construction can be modified simply to view these states and the shortcut links to them as additional auxiliary structure. To maintain identification of the ***significant*** (***non***auxiliary) nodes on-line is easy: Initially, the root (ϵ) is significant and the one leaf ($\$$) is auxiliary. Each subsequent node we create is initially auxiliary; but each time we create a shortcut link from a significant node to an auxiliary one, the latter becomes significant, and so do any auxiliary nodes further along the shortcut chain from there. The total additional cumulative time to count is thus at most proportional to the length of that entire text-length chain.

Conclusion

A clean version of Weiner's algorithm provides natural but very different indices of the subwords and the reverse subwords of a text. It is remarkable that such functional symmetry is efficiently achieved by such an integrated asymmetric contruction.

References

1. A. V. Aho, J. E. Hopcroft, and J. D. Ullman, *The Design and Analysis of Computer Algorithms*, Addison-Wesley, Reading, Massachusetts, 1974.

2. A. Blumer, J. Blumer, A. Ehrenfeucht, D. Haussler, and R. McConnell, *Building the minimal DFA for the set of all subwords of a word on-line in linear time*, Eleventh International Colloquium on Automata, Languages and Programming (Antwerp, Belgium, 1984), to be presented, July 18, 1984.

3. J. E. Hopcroft and J. D. Ullman, *Introduction to Automata Theory, Languages, and Computation*, Addison-Wesley, Reading, Massachusetts, 1979.

4. M. E. Majster and A. Reiser, *Efficient on-line construction and correction of position trees*, SIAM Journal on Computing **9**, 4 (November 1980), 785–807.

5. E. M. McCreight, *A space-economical suffix tree construction algorithm*, Journal of the Association for Computing Machinery **23**, 2 (April 1976), 262–272.

6. V. R. Pratt, *Improvements and applications for the Weiner repetition finder*, unpublished manuscript (May 1973, October 1973, and March 1975).

7. M. Rodeh, V. R. Pratt, and S. Even, *A linear algorithm for finding repetitions and its applications in data compression*, Technical report no. 72, Department of Computer Science, Technion—Israel Institute of Technology, Haifa, Israel (April 1976).

8. M. Rodeh, V. R. Pratt, and S. Even, *Linear algorithm for data compression via string matching*, Journal of the Association for Computing Machinery **28**, 1 (January 1981), 16–24.

9. A. O. Slisenko, *String-matching in real time*, Preprint P-7-77, The Steklov Institute of Mathematics, Leningrad Branch (September 1977) (Russian).

10. A. O. Slisenko, *String-matching in real time: Some properties of the data structure*, Mathematical Foundations of Computer Science 1978 (Proceedings, 7th Symposium, Zakopane, Poland, 1978) (Lecture Notes in Computer Science **64**), Springer-Verlag, Berlin, 1978, pp. 493–496.

11. A. O. Slisenko, *Determination in real time of all the periodicities in a word*, Soviet Mathematics —Doklady **21**, 2 (March–April 1980), 392–395.

12. A. O. Slisenko, *Detection of periodicities and string-matching in real time*, Journal of Soviet Mathematics **22**, 3 (June 11, 1983), 1316–1387; translated from Zapiski Nauchnykh Seminarov Leningradskogo Otdeleniya Matematicheskogo Instituta im. V. A. Steklova AN SSSR **105** (1980), 62–173.

13. P. Weiner, *Linear pattern matching algorithms*, 14th Annual Symposium on Switching & Automata Theory (Iowa City, Iowa, 1973), IEEE Computer Society, Long Beach, California, 1973, pp. 1–11.

3 - DATA COMPRESSION

Textual Substitution Techniques for Data Compression

James A. Storer
Computer Science Dept.
Brandeis University
Waltham, MA 02254

June, 1984

Abstract: With many types of networks (e.g., distributed computing, electronic mail, etc.) communication channels are relatively slow. The ability to put large amounts of processing power on a single chip promises to make sophisticated data compression algorithms truly practical. A data encoding/decoding chip can be placed at the ends of every communication channel, with no computational overhead incurred by the communicating processes. Similarly, secondary storage space can be increased by hardware that (invisible to the user) performs data compression. For the purposes of this paper, *data compression* refers to transforming a string of characters to another (presumable shorter) string, from which it is possible to recover (exactly) the original string at some point later in time. This paper surveys research on data compression methods that employ textual substitution.

1. Introduction

A powerful (and practical) approach to compression of a string of characters is that of *textual substitution,* a technique studied by this author (Storer [1984, 1984b, 1982, 1982b, 1979, 1977, 1977b], Gonzalez and Storer [1982], Storer and Szymanski [1982], Maier and Storer [1977], and Gallant, Maier, and Storer [1980]) and by many others (e.g., Miller and Wegman [1984], Choueka, Fraenkel, and Perl [1982], Pechura [1982], Rodeh, Pratt, and Even [1981], Langdon [1981], Apostolico [1979], Sheifler [1977], Vasyukova [1977], Seery and Ziv [1977,1978], Lempel and Ziv [1976,1977,1978], Rubin [1976], Visvalingham [1976], Mayne and James [1975], Hahn [1974], Morris and Thompson [1974], Wagner [1973], McCarthy [1973], Hagamen, Linden, Long, and Weber [1972], Ruth and Kreutzer [1972], Lesk [1970], and Marron and DeMaine [1967]).

We use the term *textual substitution* data compression method to refer to any data compression method that compresses text by identifying repeated substrings and replacing some substrings by *pointers* to another copy. Thus, the input to a data compression algorithm employing textual substitution is a sequence of characters over some *source alphabet* Σ and the output is a sequence of characters from Σ interspersed with pointers. We make the distinction between pointers and characters only for convenience. Since any well defined compression algorithm must be able to take as input a string consisting of a single character, it must be possible to represent any single character with whatever coding scheme is used to represent the output from the algorithm. Hence, characters can be viewed as a pointer to a string of exactly one character and the output of the compression algorithm a simply a sequence of pointers, where each pointer specifies some string of ≥ 1 characters of Σ.

Example 1.1: To compress English text, we construct a *dictionary* consisting of the 128 ASCII characters together with the 37640 most common substrings of (of two or more characters) of printed English text†. The compression algorithm somehow partitions its input into strings contained in the dictionary and then outputs 15 bit pointers for each one. Here, "characters" in the output are

† See Storer [1984], Kucera and Francis [1967], and Shannon [1951] for such statistics.

NATO ASI Series, Vol. F12
Combinatorial Algorithms on Words
Edited by A. Apostolico and Z. Galil

simply those pointers which indicate one of the first 128 entries. ○

Example 1.2: To compress English text, we construct a *dictionary* consisting of the 37768 most common substrings of (of two or more characters) of printed English text. The compression algorithm somehow partitions its input into strings that are either contained in the dictionary or are single ASCII characters and then outputs for each string consisting of a single ASCII character a 0 followed by a 7 bit code and for strings of two or more characters a 1 followed by a 15 bit code. Here, pointers have two sizes (8 and 16 bits) and "characters" in the output are the 8 bit pointers. ○

The above two examples show different ways of encoding pointers for what are essentially identical methods. We assume that it is always possible to determine where one pointer ends and the next one begins. Besides this assumption, the precise manner in which pointers are encoded into bits is not important for most of the theoretical results to be discussed in this paper. The exact assumptions regarding pointer representations and lengths will be made primarily for convenience (e.g., a problem that is NP-complete will remain so for virtually any reasonable assumptions) and will vary slightly among the sections to follow. However, when discussing some practical algorithms in Section 3, we will address this issue more carefully; for example, when the distribution of pointer values is not uniform, Huffman codes can be used to represent pointers.

Section 2 considers *off-line* models of data compression where the entire input string can be read before any of the output is produced; this section is based on the work of Storer and Szymanski [1982], Storer [1979], and Storer [1977]. Off-line models are not only powerful, but also convenient for examining theoretical issues. Section 3 considers *on-line* models of data compression where output must be continuously produced as input is read in; this section is based on the work of Storer [1984]. On-line models are useful for practical applications such as communications. Section 4 then considers parallel implementations of Algorithms presented in Section 3 which may be appropriate for high performance hardware in the future; this section is based on the work of Gonzalez and Storer [1982], Storer [1982], and Storer [1982b]. Section 5 outlines current areas of research by this author concerning data compression by textual substitution.

2. Off-Line Compression: The Macro Model

2.1. Definitions

We shall treat the source data as a finite string over some alphabet. With *external macro schemes,* a source string is encoded as a pair of strings, a *dictionary* and a *skeleton*. The skeleton contains characters of the input alphabet interspersed with pointers to substrings of the dictionary. The dictionary is also allowed to contain pointers to substrings of the dictionary. The source string is recovered by substituting dictionary strings for pointers. With *internal macro schemes,* a string is compressed by replacing duplicate instances of substrings with pointers to other occurrences of the same substrings. The result is a single string of characters and pointers.

For the purposes of this section, we let $p \geq 1$ be an integer denoting the implementation dependent size of a pointer. If x is a string containing pointers, the *length* of x, denoted $|x|$, is defined to be the number of characters in x plus p times the number of pointers in x. We shall treat a pointer as an indivisible object which, in some unspecified fashion, uniquely and unambiguously identifies some string which is referred to as the *target* of that pointer. The way a pointer is written is not important; for simplicity, we shall write a pointer as a pair (m,n) where m indicates the position of the first character in the target, n indicates the length of the target, and $|(m,n)|$ is the pointer size p. Note that the first argument m can be either an absolute location or a displacement.

For example, with internal schemes, it is likely to be the distance from the pointer to its target in practice. However, for conceptual simplicity, in this section we will always assume the first argument to specify an absolute location.

Example 2.1.1: Let $p = 1$ and consider the string

$$w = \text{aaBccDaacEaccFacac}$$

which might be encoded under the external macro model as

$$x = \text{aacc\#}(1,2)\text{B}(3,2)\text{D}(1,3)\text{E}(2,3)\text{F}(2,2)(2,2)$$

where # separates the dictionary from the skeleton; for convenience, we assume $|\#| = 0$. The compression achieved by the string x (i.e., the ratio $|x| / |w|$) is 14/18. Using the internal macro model, w could be encoded as

$$y = \text{aaBccD}(1,2)\text{cEa}(4,2)\text{Fac}(13,2)$$

achieving a compression of 15/18. ○

At this point, we digress briefly with some comments about our assumptions concerning pointers. As mentioned in the introduction, in practice, the output of a textual substitution compression algorithm can be viewed simply as a sequence of pointers, rather than as a sequence of characters interspersed with pointers. In this context, we have really allowed two pointer sizes: unit size (which can only indicate single characters) or size p. The motivation for doing this is to simplify theoretical results by having a uniform integer pointer size p (which can be arbitrarily large); however, it is also convenient to be able to represent characters without a large increase in size, so we in addition allow a unit cost for single characters. This "2-step" scheme of characters and pointers of size p is already more general than is required to model many practical methods, since $p = 1$ models a method that outputs a sequence of uniform size pointers. For example, 12 bit pointers (which may be subsequently Huffman coded) work well for some of the practical methods discussed in Section 3; here, the true length in bits of the output of the compression algorithm is simply 12 times the length as defined in this section. It is possible to generalize the research discussed in this section to allow for a continuous range of pointer sizes between 1 and p (including non-integer sizes) and to allow p to be a function of the size of the source string. However, it does not seem to lend significant additional insights; the model of characters interspersed with uniform integer sized pointers seems to be a nice compromise for examining theoretical issues pertaining to off-line data compression by substitution.

Implementation considerations motivate us to describe a number of variations on our basic models. A scheme is *recursive* if a pointer target is allowed to contain pointers. Two pointers *overlap* if their targets overlap. Whether overlapping pointers are permitted in the external model depends highly on the implementation chosen for the dictionary. In fact, implementation considerations can lead to the placement of various restrictions on the kinds of overlapping permitted. Some of these restrictions are considered in Storer [1977, 1979, 1982] and Gonzalez and Storer [1982] and by other authors such as Choueka, Fraenkel, and Perl [1982] (who consider prefix overlappings). An *original pointer* is one which denotes a substring of the original source string whereas a *compressed pointer* denotes a substring of the compressed representation itself. The string y of the previous example contains compressed pointers. Using original pointers we could encode w as

$$z = \text{aaBccD}(1,2)\text{cEa}(4,2)\text{F}(8,2)(8,2)$$

achieving a compression of 14/18. Original pointers are more natural for one-pass decoding. Compressed pointers allow the recovery of portions of the source string without requiring the implicit decompression of the entire string. A *left* (or *right) pointer* is one which denotes a substring occurring earlier (respectively, later) in the string. Considering the strings x, y, and z presented above, only x uses overlapping pointers, only z uses recursion, and none of these strings use right pointers. By using both left and right pointers it is possible to save additional space over the use of just one direction. For example, using both right and left pointers, the compressed forms y and z presented above could be replaced by

$$y = (5,2)B(10,2)DaacEaccF(6,2)(6,2)$$
$$z = (7,2)B(12,2)DaacE(8,2)cF(8,2)(8,2)$$

achieving a compression of 14/18 and 13/18 respectively. We discuss recursion in relation to original pointers primarily to study the "power" of various methods. With original pointers, a pointer is recursive if all or part of the string it represents is represented by a pointer.

Cycles cannot occur with compressed pointers but using original pointers, cycles can often make sense. For example, the compressed form

$$ab\,(5,2)\,a\,(1,3)$$

uniquely determines the string *abaaaaba* even though the two pointers (5,2) and (1,3) form a cycle in the sense that each points to a portion of the string represented by the other. An example of a degenerate cycle is given by the compressed form $a\,(1,n)$ which uniquely determines the string a^{n+1}. Schemes which allow recursion but which do not allow cycles are said to have *topological recursion*. From the above discussion it should be clear that topological recursion is not necessary for a compressed form to be uniquely decodable. However, it can be useful to consider topological recursion for three reasons: First, authors in the past have assumed this. Second, study of such schemes leads to a deeper understanding of the power of original pointers. Third, topological recursion may model some practical considerations in the design of efficient original pointer compression methods.

The above discussion leads us to formally define four basic *macro schemes* and three types of *restrictions* which may be placed on any of these schemes. Throughout this section, Σ will denote the underlying alphabet from which the data in question is constructed, and for convenience, we assume that $|\#|=0$.

Definition 2.1.1: A *compressed form* of a string s using the *EPM* (external pointer macro) *scheme* is any string $t=s_0\#s_1$ satisfying

(1) s_0 and s_1 consist of characters from Σ and pointers to substrings of s_0.

(2) s can be obtained from s_1 by performing the following two steps:

A: Replace each pointer in s_1 with its target.

B: Repeat step A until s_1 contains no pointers.

Definition 2.1.2: A *compressed form* of a string s using the *CPM* (compressed pointer macro) *scheme* is any string t satisfying:

(1) t consists of characters from Σ and pointers to substrings of t.

(2) s can be obtained from t by forming the string $t\#t$ and then decoding as with the EPM scheme. ○

Definition 2.1.3: A *compressed form* of a string s using the *OPM* (original pointer macro) *scheme* is any string t satisfying:

(1) t consists of characters from Σ and pointers representing substrings of s.

(2) s can be obtained from t by replacing each pointer (n,m) by the sequence of pointers

$(n,1)$, $(n+1,1)$, ..., $(n+m-1,1)$ and then decoding as with the CPM scheme with the stipulation that pointers are considered to have length 1. ○

Definition 2.1.4: A *compressed form* of a string s using the *OEPM* (original external pointer macro) *scheme* is any string $t = s_0 \# s_1$ satisfying:

(1) t consists of characters from Σ and pointers.

(2) s_0 may be decoded using the OPM scheme to produce a string r. Furthermore, pointers in s_1 point to substrings of r.

(3) s may be obtained by replacing each pointer in s_1 with its target in r. ○

A string may have more than one minimal length compressed form. For formal discussions, we assume that there is some lexicographic ordering two distinguish among compressed forms of the same length, and we henceforth refer to *the* minimal length compressed form. The *contraction* of a string s for pointer size p according to a given scheme is the minimal length compressed form of s using that scheme with pointer size p. The contraction of a string s using a scheme X is denoted by $\Delta_X(s)$. We shall refer to the process of replacing a string r by a pointer as *factoring out* r and often refer to a string that is a target or potential target as a *factor*.

Definition 2.1.5: A CPM (OPM) pointer q_1 *depends on* pointer q_2 if the target of q_1 contains q_2 (all or part of the string represented by q_2) or if there is a pointer q_3 such that q_1 depends on q_3 and q_3 depends on q_2. A macro scheme is restricted to *no recursion* if dependent pointers are forbidden and *topological recursion* if no pointer may depend on itself; that is, there exists no circular sequence of pointer dependencies.

Definition 2.1.6: Two pointers *overlap* if their targets overlap. A macro scheme is restricted to *no overlapping* if overlapping pointers are forbidden. ○

Definition 2.1.7: A CPM (OPM) pointer q *points to the left* if the leftmost character of its target is to the left of q (the leftmost character of the string represented by q). A *right pointer* is similarly defined. A macro scheme is restricted to *unidirectional pointers* if all pointers must point in the same direction (of course, with the EPM and OEPM schemes, this only applies to the external dictionary). As a special case of this, we can restrict a macro scheme to have only *left* or *right* pointers. ○

The different combinations of the four basic macro schemes we have defined and the recursion, overlapping, and pointer direction restrictions provide us with a large number of data compression methods. The combinations are sufficiently general to cover a large number of the textual substitution schemes proposed in the literature.

2.2. Inherent Power of the Model

Simply examining how a data compression method compresses a string consisting of a single repeated character (e.g. a string of zeros) can say much about the basic "power" of the method, even though such "easy" strings to compress are unlikely to occur in practice. The theory of *program size complexity* (e.g., Storer [1983], Katseff and Sipser [1977], Chaitin [1976, 1975, 1969, 1969b, 1966], Daley [1976, 1974, 1973], Kamae [1973], Kolmogorov [1969, 1965], Loveland [1969, 1969b], Blum [1967], Martin-Lof [1966]) guarantees arbitrarily powerful compression methods, including methods for which finding the contraction is undecidable.

Example 2.2.1: Suppose that $F(i)$ is some very fast growing function; for the purposes of this example, let $F(i) = ACKERMAN(i,i)$[†]. If for an integer i, we let $BIN(i)$ denote the string obtained by writing i in binary, then the following is clearly a well-defined and computable compression algorithm:

if $s = BIN(F(i))$ for some i

then $\Delta(s) := 0BIN(i)$

else $\Delta(s) := 1s$

Although the above method is unlikely to have any practical value, it does compress infinitely many strings by an amount that cannot be specified by any primitive recursive function (and does not increase the length of any string by more than one bit). O

Theorem 2.2.1: For all strings s, if only topological recursion is allowed, then (assuming s is compressible) both $|\Delta_{EPM}(s)|$ and $|\Delta_{OEPM}(s)|$ are:

(A) $\geq p * \log_2(|s|/p) + 1.9p$.

(B) $\geq 3p * \log_3(|s|/p) - .02p$ when overlapping is forbidden.

(C) $\geq 2(p\,|s|)^{1/2}$ when recursion is forbidden.

(D) $\geq 2(p\,|s|)^{1/2}$ when both recursion and overlapping are forbidden.

(E) A through D hold even if pointers are required to be unidirectional.

If non-topological recursion is allowed, then

(F) The bounds of (A) through (E) hold for the EPM scheme.

But:

(G) $|\Delta_{OEPM}(s)| \geq 2p+1$ independent of what overlapping and pointer direction restrictions are made.

Furthermore, all of the bounds in (A) through (G) are tight; that is, each is attained for infinitely many strings s[††].

Proof: We include only a proof of (B). Since $|\Delta_{OEPM}(s)| \leq |\Delta_{EPM}(s)|$, it is sufficient to show that the OEPM scheme satisfies this bound and the EPM scheme can attain it infinitely often. We can assume that $s = a^{|s|}$. Since overlapping is forbidden, the pointers of $\Delta_{OEPM}(s)$ can be divided into a sequence of sets $S_1, \ldots, S_m$ such that the pointers in S_i, $1<i\leq m$, have targets whose compressed representation consists of pointers in some s_j, $j<i$; in fact, since we are concerned only with worst-case performance[†††], we can assume that $j = i-1$. We can also assume that S_i, $1\leq i\leq m$, contains at most 3 pointers. This is because for any $k\geq 4$, there is an i and j such that $2i+3j\leq k\leq 2^i 3^j$ and so we can replace a set S_i of 4 or more pointers by a sequence of sets having at most 3 pointers where the last set in the sequence will represent a string at least as long as that represented by the original sequence of 4 or more. Hence, for some $x\leq 2$, we can assume that for all $i>x$, S_i contains exactly 3 pointers. This is because we can assume that the sets of 2 come first and a

† For a nice introduction to Ackerman's function, see Henie [1977].

†† Actually, the bound in (A) is just an approximation for the expression $\mathrm{MIN}\{p\lceil\log_2(|s|/i)\rceil + i + p : p < i \leq 2p\}$ which is attained exactly infinitely often. Similarly, the bound in (B) is just an approximation for the expression $\mathrm{MIN}\{3p\lceil\log_3(|s|/i)\rceil + i : 2p < i \leq 4p\}$ which is attained exactly for infinitely many strings.

††† It is not necessary to consider a compressed form of length x for a string a^y if there is a compressed form of length $<$x for a string a^z where $z\geq y$.

sequence of three 2 pointer sets can be replaced by two 3 pointer sets. Given this, it can be assumed that for some $2p<k\leq 4p$, $0\leq M\leq L\leq 1$, and n, $\Delta_{OEPM}(s)$ is of the form:

$$a^k(q_0^2)^L(q_1^2)^M(\prod_{i=2}^{n} q_i^3)\#q_{n+1}^3$$

where q_0 points to a^k; q_1 points to q_0^2; q_2 points to a^K, q_0^2, or q_1^2 depending on the the values of L and M; and for $3\leq i\leq n+1$, q_i points to q_{i-1}^3. Thus we have:

$$|\Delta_{OEPM}(s)| \geq k+3np+2Lp+2Mp$$

$$\geq 3p\lceil\log_3(|s|/k\,2^L\,2^M)\rceil+k+2Lp+2Mp$$

$$\geq 3p\lceil\log_3(|s|/k)\rceil+k+3p(\frac{2}{3}-\log_3 2)(L+M)$$

$$\geq 3p\lceil\log_3(|s|/k)\rceil+k$$

$$\geq \mathrm{MIN}\{3p\lceil\log_3(|s|/i)\rceil+i:\ 2p<i\leq 4p\}$$

$$\geq 3p*\log_3(|s|/p) + \mathrm{MIN}\{p(h-3\log_3 h):\ 2<h\leq 4\}$$

$$> 3p*\log_3(|s|/p) - .02p$$

For any $2p<i\leq 4p$ and $n>1$, the bound of $\mathrm{MIN}\{3p\lceil\log_3(|s|/i)\rceil+i:\ 2p<i\leq 4p\}$ is achieved using the EPM scheme with no overlapping on the string:

$$s = a^{i3^n} \quad \bigcirc$$

The above theorem (and a similar theorem for internal schemes) says that macro schemes are not as "wild" as methods like Example 2.2.1 but, if recursion is allowed, do have the ability to "count"; that is, a string of n zeros can be represented with $O(log(n))$ bits. Bounds (C) and (D) show that recursion is in some sense a more serious restriction than overlapping. Bound (G) shows the limitations of a model that allows p to be independent of $|S|$; clearly, if p is constant, a pointer cannot represent an infinite number of strings.

Another way to examine the power of the macro model is to consider worst-case performance ratios between schemes. For example, the following can be shown about the relationship between internal and external schemes.

Theorem 2.2.2: For all strings s,

$$\text{(A)}\ \frac{2}{3}|\Delta_{CPM}(s)| < |\Delta_{EPM}(s)| \leq |\Delta_{CPM}(s)|+p$$

$$\text{(B)}\ \frac{1}{2}|\Delta_{OPM}(s)| < |\Delta_{OEPM}(s)| \leq |\Delta_{OPM}(s)|+p$$

regardless of whether topological recursion is assumed. Furthermore, these bounds are tight. ○

For the proof of the above theorem and related results concerning pointer direction, overlapping, etc., see Storer and Szymanski [1982] and Storer [1979].

2.3. Two NP-Complete Encoding Problems

Storer [1977] presents a large number of NP-completeness results concerning encoding for the macro model. In addition, some additional results are contained in Gallant [1982], Maier [1977], Maier and Storer [1977], and Gallant, Maier, and Storer [1980]. Proofs for these results include reductions from the various restricted forms of the node cover problem (Maier and Storer [1977], Storer [1977]), and the superstring problem† (Maier and Storer [1977], and Gallant, Maier, and Storer [1980]). In this section, we present two "sample" theorems. These two theorems parallel two linear time algorithms (for restricted versions of these problems) to be presented in the next section. Section 2.5 then gives an informal characterization of what makes an encoding problem for the macro model NP-complete.

Theorem 2.3.1: Given a string s and an integer K, it is NP-complete to determine whether $|\Delta_{EPM}(s)| \leq K$ or $|\Delta_{OEPM}(s)| \leq K$ when both recursion and overlapping forbidden. Furthermore, this is true regardless if p is part of the problem input or is constrained to be a fixed integer ≥ 1.

Proof: We consider only the case $p > 1$; the case $p = 1$ is shown in Storer [1977]. Note that when both recursion and overlapping are forbidden, the EPM and OEPM schemes are equivalent. Let

$$G = (V = \{v_1, \ldots, v_n\}, E = \{e_1, \ldots, e_m\}), K$$

be an instance of the node cover problem and let $p = p_0$. Let \$ be a special symbol and let @ denote a new, distinct symbol each time it occurs. For v_i in V, let $V_i = \$v_i^{p-1}\$$ and for $e_i = (v_j, v_k)$ in E, let $E_i = \$v_j^{p-1}\$v_k^{p-1}\$$. Now let:

$$s = (\prod_{i=1}^{p}\prod_{j=1}^{n} V_j @)(\prod_{i=1}^{m} E_i @)$$

We claim that G has a node cover of size K if and only if $|\Delta(s)| \leq |s| + K - m$.

First suppose that G has a node cover X of size K. We shall construct a compressed form t for s (having length $|s| + K - m$) where t is of the form $s_0\#(\prod_{i=1}^{p}\prod_{j=1}^{n}\overline{V}_j @)(\prod_{i=1}^{m}\overline{E}_i @)$ where s_0 contains those V_i for which v_i is in X, and $\overline{V}_j$ is V_j if v_j is not in X and a pointer to v_j in s_0 if v_j is in X. If E_i is $\$v_j^{p-1}\$v_k^{p-1}\$$, then $\overline{E}_i$ is either $rv_k^{p-1}\$$ or $\$v_j^{p-1}q$ where r is a pointer to v_j in s_0 and q is a pointer to v_k in s_0. Since X is a node cover, this can always be done. If we now compute the length of t, $|s_0| = K(p+1)$, $|\prod_{i=1}^{p}\prod_{j=1}^{n}\overline{V}_j @| = |\prod_{i=1}^{p}\prod_{j=1}^{n} V_j @| - pK$, and $|\prod_{i=1}^{m}\overline{E}_i @| = |\prod_{i=1}^{m} E_i @| - m$. Hence, $|\Delta(s)| \leq |t| = |s| + K(p+1) - pK - m = |s| + K - m$, as was to be shown.

Conversely, suppose that $|\Delta(s)| \leq |s| + K - m$. We shall show that G has a node cover of size at most K. First observe that since overlapping of pointer targets is forbidden, no pointer in $\Delta(s)$ can refer, for any strings x and y, to a string of the form $xv_i\$v_j y$, $x@y$, or $x@y$ since such a string can occur at most once in s and no gain can be achieved by factoring it out. Thus, $\Delta(s)$ is of the form $s_0\#(\prod_{i=1}^{p}\prod_{j=1}^{n}\overline{V}_j @)(\prod_{i=1}^{m}\overline{E}_i @)$ where s_0 is a dictionary of macro bodies and the $\overline{V}_j$'s and $\overline{E}_i$'s are the shortest compressed forms of the V_j's and E_i's respectively using s_0. As mentioned earlier, without loss of generality we are assuming throughout this section that every pointer references a string of length at least $p+1$. Thus, since $|V_i| = p+1$, we can infer that each $\overline{V}_i$ is either V_i itself or a pointer to an occurrence of v_i in s_0. Similarly, since $|E_i| = 2p+1$, each $\overline{E}_i$ must either consist of E_i itself or else be a string of the form $rv_j^{p-1}\$$ or $\$v_j^{p-1}r$ where r is a pointer to some V_i

† The superstring problem is: Given a set S of strings and an integer K, is there a string of length K that contains all strings of S as a substring.

in s_0. Now let L be the number of $\overline{E}_i$'s such that $\overline{E}_i = E_i$; that is, the number of $\overline{E}_i$'s that have not had a factor removed. Then $|\prod_{i=1}^{m} \overline{E}_i @| = |\prod_{i=1}^{m} E_i @| - (m-L)$ since removing a factor from an E_i saves one character. Let J be the number of V_i's in s_0. Then $|s_0| = J(p+1)$ and $|\prod_{i=1}^{p}\prod_{j=1}^{n} \overline{V}_j @| = |\prod_{i=1}^{p}\prod_{j=1}^{n} V_j @| - Jp$ because each v_i that is replaced by a pointer saves one character. Thus $|\Delta(s)| = J(p+1) + |s| - Jp - (m-L) = |s| + J + L - m$, and so $J+L \leq K$. We now claim that G has a node cover of size $J+L$ formed by taking the J nodes represented in s_0 and one node from each of the L edges not factored in $\Delta(s)$. Therefore G has a node cover of size K, as was to be shown. O

Theorem 2.3.2: Given a string s and an integer K, it is NP-complete to determine whether $|\Delta_{CPM}(s)| \leq K$ or $|\Delta_{OPM}(s)| \leq K$. Furthermore, this is true regardless if p is part of the problem input or is constrained to be a fixed integer ≥ 1. O

See Storer [1977] and Gallant [1982] for a proof of the above theorem. Note that in the reduction used in the proof of Theorem 2.3.1, the alphabet size is allowed to depend on the size of the NP-complete problem in question. However, all of the above results hold even when the alphabet is fixed.

2.4. Two Linear-Time Encoding Algorithms

In this section, we present linear time encoding algorithms for restricted versions of the problems shown NP-complete in the last section. We start with a restricted version of the EPM scheme where *the dictionary is specified in advance.*

Given a dictionary, a particular string may have many different compressed forms. Wagner [1973] gives a simple dynamic programming algorithm to find a minimal length compressed form, when the dictionary is given. In fact, although his algorithm does not assume variable length pointers, the ability to point to a substring of a dictionary element, or pointers within dictionary elements, it is easily generalized to accommodate these features. For D a given dictionary of strings, and $s = s_1, \ldots, s_m$ a string to be compressed, let:

$$q_{i,j} = \begin{cases} \textit{IF } (i=j) \textit{ THEN } s_i \\ \textit{ELSE IF (exists a pointer with expanded target } s_i \cdots s_j \textit{) THEN the shortest one} \\ \textit{ELSE } \infty \end{cases}$$

The representation and lengths of pointers is unimportant, but this must be specified before $q_{i,j}$ can be defined. For a set of strings S, MIN(S) denotes a shortest one.

For each $1 \leq i \leq m$, Algorithm 2.4.1 computes $SHORT(i)$, a shortest compressed form of $s_1, \ldots, s_i$ (using dictionary D). Algorithm 2.4.1 can clearly be implemented in polynomial time. Moreover, with some reasonable assumptions about relative pointer sizes and by using appropriate data structures, this algorithm can be implemented in linear time.

In Storer and Szymanski [1982], it is shown that a similar approach (dynamic programming from right to left) works for the OPM/L scheme (OPM scheme restricted to left pointers). The key to implementing this algorithm in linear time is to be able to quickly compute for a position in the string what is the longest substring that both starts at this position and occurs earlier in the string. Rodeh, Pratt, and Even [1976] give an algorithm for computing this information (which they use to implement a greedy algorithm for a model similar to the OPM/L scheme) that is based on the work of McCreight [1976] (see also Chen and Seiferas [1984], Majster [1979], and Weiner [1973]).

SHORT $(m+1)$:= **<the empty string>**

for $i = m$ **to** 1 **by** -1 **do**

SHORT (i) := MIN$\{q_{i,j}$, *SHORT* $(j+1)$: $i \leq j \leq m\}$

Algorithm 2.4.1
(Wagner's Algorithm, slightly modified)

2.5. A Characterization of NP-Completeness

The difference between the compression methods considered in the last two sections is that with the schemes considered in Section 2.4, the set of legal pointers does not depend on the string being compressed, whereas this was not the case in Section 2.3. For example, suppose positions 50 through 59 are identical to positions 100 through 109 of a string. Then with the OPM scheme, a pointer (100,10) may or may not be legal, depending on whether it causes a cycle of pointer dependences. However, this pointer is always legal for the OPM/L scheme. The following is an informal characterization of when encoding for the macro model is NP-complete:

If for a macro scheme the set of legal pointers does not depend on the string being compressed, then the scheme has a polynomial time encoding algorithm. Otherwise, encoding is NP-complete for this scheme. ○

3. On-Line Compression: Practical Techniques

3.1. Definitions

A practical on-line model for data compression by textual substitution is to have a *encoder* and *decoder,* each with a fixed, finite amount of local memory, which we refer to a a *local dictionary,* and we let M denote the maximum number of entries it may contain. We assume that the two local dictionaries may be initialized at the beginning of time to contain identical information. The encoder forever repeats the following steps:

Encoder's Algorithm:

(1) Let t be the longest prefix of the input stream that matches an entry of the local dictionary, advance the input stream forward by $|t|$ characters, and transmit a code of $\log_2(M)$ bits corresponding to t.

(2) Update the dictionary:
   a. If the dictionary is full then delete one of its entries.
   b. Add a new entry.

Similarly, the decoder forever repeats the following steps:

Decoder's Algorithm:

(1) Receive a $\log_2(M)$ bit code and expand it to a string t by a dictionary lookup.

(2) Same as Step 2 of the encoder's algorithm.

In practice, there are a host of details which we do not address here, such as allowing for several pointer sizes (instead of just $\log_2(M)$ as described above), error recovery, etc. In particular, the output of the encoder can be passed through the on-line Huffman encoding algorithm of Gallager [1978] (see also Huffman [1952] and Karp [1960]) to correct for a non-uniform usage of the pointers. With fixed length pointers the Huffman coder would view the source alphabet as characters of $\log_2(M)$ bits; although Gallager's algorithm can also be modified to accommodate variable length pointers.

It should also be noted that Step 1 of the encoder amounts to a greedy algorithm for parsing the input stream. For some implementations we will consider, it it possible to achieve better worst-case performance by incorporating some amount of look-ahead. However, we will not address this issue in this here.

This basic model can be viewed as a practical realization of the work of Lempel and Ziv [1976] and Ziv and Lempel [1977,1978]. The sub-sections to follow can be viewed as successively more general ways of interpreting the above informal descriptions of the encoder's and decoder's algorithms.

3.2. Static Dictionary Method

With the *static dictionary* method, Step 2 of the encoder and decoder algorithms is null. Compression is based only on the initial value of the local dictionaries. The static dictionary model is useful when the source is known in advance (e.g., a dictionary of common English substrings such as described in Examples 1.1 and 1.2). It also has nice properties with respect to error recovery, to be discussed in Section 5.3. In Storer [1982], worst-case bounds between the greedy approach of the encoder and Wagner's dynamic programming approach are considered:

Theorem 3.2.1: Let D be a dictionary and s a string. Let $GRE(D,s)$ and $DYN(D,s)$ denote the compressed forms obtained using the greedy approach and the dynamic programming approach, respectively. Then, assuming that pointers may indicate any suffix of a dictionary element:

$$\frac{|DYN(D,s)|}{|GRE(D,s)|} \geq \frac{p}{2p-1}$$

Furthermore, even if D is restricted to have only 2 elements and s is written over a 2 symbol alphabet, this bound may be obtained arbitrarily closely. That is, for any real $h>0$, there are infinitely many strings such that:

$$\frac{|DYN(D,s)|}{|GRE(D,s)|} < \frac{p}{2p-1} + h \quad \bigcirc$$

It should be noted that although the above worst-case bound suggests a a difference of as much as a factor of 2, in practice, the difference appears to be at most a few percent (Storer [1984]).

3.3. Sliding Dictionary Method

The OPM scheme restricted to left pointers can be used for the on-line model as follows. First, we write pointers as (m,n) where m is the *displacement* back from the current position and n is the length of the target. Second, we pick an integer k and use k bits for n and $\log_2(M)-k$ bits for m. Third, we restrict m to be less than or equal to

$$2^{\log_2(M)-k} - \frac{|\Sigma|}{2^k}$$

to allow $|\Sigma|$ "extra" pointer values so that a single character can always be represented. Step 2 of the encoder and decoder algorithms amounts to sliding a window of length given by the above formula.

A typical practical choice of parameters for compressing sources like English or programming language text might be to use 16 bit pointers with a 3 bit length field and a 13 bit displacement field; here, assuming the 128 character ASCII alphabet, $1 \leq n \leq 8{,}176$ and $2 \leq m \leq 9$.

If all pointers output by the encoder are the same length as we are assuming in this paper, then the greedy approach of the encoder can easily be shown to yield an optimal length parse of the input stream. However, if pointers of different lengths are used (e.g., short ones for characters and long ones for strings), then a similar theorem to the one in the preceding section can be shown for worst-case performance. Similarly, in practice, the difference appears to be at most a few percent (Storer [1984]).

The sliding dictionary method is somewhat "wasteful" because a given substring may occur more than once within a given window, thus unnecessarily consuming more than one pointer value. Using the appropriate data structures, one can store only the unique substrings contained in the window, yielding an effectively longer window. In practice, the improved performance does not appear to be significant for many types of data (Storer [1984]), but this issue is theoretically interesting and leads us naturally to Section 3.4.

3.4. Adaptive Dictionary Method

In the last section, the dictionary was updated in a very restricted fashion. *Adaptive dictionary* methods employ some heuristic for Step 2 of the encoder and decoder algorithms. There are at least two natural heuristics for Step 2a:

(1) Don't delete; once the dictionary becomes full, convert to the static dictionary model. For sources with reasonable stationary statistics, this method works well in practice. One variation of this approach is to periodically delete the dictionary and restart the process. Another is to use double the memory and maintain a duel dictionary to which all additions are made; once the duel dictionary becomes full, it is substituted for the main dictionary and then reset to be empty. Experimentation with this approach has been performed by Lempel and Ziv [1984b], Welch [1984], and Seery and Ziv [1977,1978].

(2) Delete the least recently used entry. Experimentation with this approach has been performed by Miller and Wegman [1984] and Storer [1984].

For Step 2b, it seems natural to base the choice of a new entry on the most commonly used entries in the dictionary. There has been very little research to date on good heuristics. There are, however, two simple heuristics for forming new dictionary entries for which a significant experimental research has been performed: (e.g., Miller and Wegman [1984] and Storer [1984]).

(1) Concatenate the last match with the first character of the current match. This can

be viewed as a direct practical implementation of the Ziv-Lempel algorithm. Experimentation with this approach has been performed by Miller and Wegman [1984] and Storer [1984].

(2) Concatenate the last match and the current match. The motivation behind this heuristic is that it allows for faster "adaptation" to a changing source. Miller and Wegman [1984] were the first to formally propose this heuristic and perform experiments with it; these experiments show this heuristic be a significant improvement over (1) for highly compressible sources such as terminal screen refreshes. In addition, Miller and Wegman [1984] present a space efficient data structure for this heuristic; although its worst-case time performance is poor, it can be implemented to run in expected linear time. Further experiments with this heuristic can be found in Storer [1984].

4. Parallel Implementations

4.1. The Computational Model

We implement our algorithms in VLSI with *systolic arrays,* a model of parallel computation studied by many authors (see Section 8.3 of Mead and Conway [1980]). The idea is to lay out a regular pattern of processing elements that have a simple interconnection structure; idealy, each element is connected only to adjacent elements. In contrast to this model of computation, one could simply implement a serial data compression algorithm using, for example, a micro-processor chip with some memory. However, a practical advantage of the systolic array implementation is speed, making the resulting data compression chip appropriate for a wider range of applications. In addition, the through-put of systolic structures is independent of their size; this is important from both a practical and theoretical standpoint.

4.2. Static Dictionary Method

One way of implementing the static dictionary method with a systolic array is a pipe which can be viewed as 3 processing elements for each string of the dictionary. The dictionary strings themselves are stored in the middle row, one dictionary string per element. If pointers within dictionary elements are present, it will always be that they point to a dictionary element to the left. Data to be encoded enters from the left, is piped along the bottom row, and encoded as it progresses. Data to be decoded enters from the right, is piped along the top row, and decoded as it progresses.

One advantage of the this structure is that any number of these can be combined to produce a larger one (by simply sticking them together end to end). Another advantage, both practical and theoretical, concerns maximum edge (wire) lengths. There has been much debate as to the right model of time for a VLSI signal to propagate along a wire (see Bilardi, Pracchi, and Preparata [1981]), and some authors have argued that as much as $O(n^2)$ should be charged to an edge of length n. Here, no edges depend on the length of the pipe. Thus, from the standpoint of VLSI layout, this array structure is quite desirable:

⇒ There are 3 basic elements from which the entire array is constructed.

⇒ Each array element has a fixed size; strings that are too long to fit into a single array element can be represented via recursive pointers.

⇒ Each array element is connected only to its nearest neighbors in the pipe.

⇒ The structure has a linear area layout with all edges having length $O(1)$ (it can be folded in a "snake" fashion to fit in a rectangular area).

⇒ The layout strategy is independent of the number of chips used.

Ironically, decoding can be a problem if compression is too great, because the pipe can "back up". This issue and many other are addressed in Storer [1982].

4.3. Sliding Dictionary Method

In this section, we briefly overview systolic implementations of the sliding dictionary method which are presented in Gonzalez and Storer [1982]. The idea is to store the last n characters processed (i.e., the sliding dictionary) in a systolic pipeline (one character per processing element). The dictionary is updated by sliding old characters left in order to bring in new ones. We can think of this dictionary as a window of width n which is sliding through the input stream. As new characters are read in, we compare them with each of the elements of the dictionary (i.e., compare them with the n characters preceding them). To each character we will attach a pair of values [*location*, *length*]; which indicate the location and length of the longest substring of the dictionary that matches a prefix of the portion of the input that starts with this character (when the character gets shifted from element to element, its [*location*, *length*] values get shifted too). The [*location*, *length*] information is updated any time a longer match is found. After a character has been compared to the n elements, its [*location*, *length*] information is examined to decide whether data compression should take place.

Since the dictionary needs to be constantly updated so that each character can be compared to exactly the n characters preceding it, our implementation uses an array of $2n$ processors; that is, instead of just storing the last n characters, we store the last $2n$ characters. The systolic pipeline can be thought of as having two processors for each of the $2n$ array elements, which we refer to as the top and bottom rows. The top row will serve as the dictionary consisting of processors numbered 1 through $2n$, from right to left. The bottom row will consist of the data elements being processed. Each element will be processed by the processor directly above it. The comparisons are done in blocks of n characters at a time. While these characters are processed, new characters are read in, and characters that have already been processed are transmitted (or pointers to them if compression took place). Note that although we view the systolic array as operating on blocks of n characters at a time, this is invisible to the outside world; that is, the systolic array accepts input characters in real time.

Although there are many details that must be addressed, the hardest problem is obtain the information [*location*, *length*]. In Gonzalez and Storer [1982], two schemes for doing this are presented, which are appropriate for different assumptions concerning the underling technology. Layout issues involving these schemes are addressed in Storer [1982b].

5. Current Areas of Research

5.1. Beyond the Macro Model

There are several ways that the macro model can be generalized. One of the most interesting is to allow the addition of arguments to pointers which can specify modifications to be made on a target before it is substituted. It is not clear whether such generality would have much practical value for standard sources such as English text, but this generality is clearly useful and necessary for special applications. For example, the subsequence and supersequence compression methods discussed in Maier [1977]. Other ways of extending the macro model are discussed in Gallant [1982].

5.2. 2-Dimensional Substitution Techniques

Most current systems for (lossless) compression of 2-dimensional data (e.g., digital images) somehow scan the 2-dimensional data to produce a string, which is then compressed with 1-dimensional techniques such as those discussed in this paper. In fact, by scanning the image via a space filling curve, such an approach can lead to optimal compression algorithms from an information-theoretic point of view (Lempel and Ziv [1984]). However, an interesting 2-dimensional approach to this problem is to generalize the concept of textual substitution to that of pointers to 2-dimensional objects.

5.3. Error Recovery for On-Line Compression

A problem that we did not address in Section 3 is what to do about errors that occur along the communication line between the encoder and decoder. For many applications, the errors themselves are not a problem. A much more serious problem, considered by Reif and Storer [1984], is that of the encoder and decoder's dictionaries becoming different, resulting in a single error to cause all subsequent text to be garbled.

5.4. Adaptation Heuristics

There is much more research needed concerning the adaptive dictionary method of Section 3.4. One area is the development of more powerful heuristics for dictionary updating; the experimental work of Storer [1984] suggests that there may be room for improvement over the two mentioned here. Another interesting direction for both theoretical and experimental research is heuristics for letting the size of the dictionary adapt with the data. This also applies to the sliding dictionary model (i.e., let the maximum displacement and maximum length parameters adapt) and the static dictionary model (i.e., have a very large dictionary sorted in order of what entries are considered most likely to occur and let the initial portion that is used adapt) Experimental evidence contained in Storer [1984] suggests that optimal values for these parameters differ significantly for different sources.

5.5. Parallel Algorithms

One interesting area of research is parallel algorithms for the adaptive dictionary method, perhaps along the lines of the algorithms overviewed in Sections 4.2 and 4.3 for the static and sliding models. Another is sub-linear time parallel algorithms for data compression; this research would be applicable to a model where more than one character can be read and written per unit of time.

5.6. Parsing (Approximation) Algorithms for Subsequent Huffman Coding

For practical reasons, it is often convenient use fixed length pointers and the correct for non-uniform usage later by passing the pointers through a Huffman algorithm, as mentioned in Section 3.1. At least from a theoretical point of view, there are problems in choosing among equivalent pointer representations in order to improve the performance of the Huffman coding. Storer [1984b], several versions of this problem are shown NP-complete. Approximation algorithms for this sort of problem is an interesting area for research.

6. References:

A. Apostolico [1979]. "Linear Pattern Matching and Problems of Data Compression ", Proc. IEEE International Symposium on Information Theory.

G. Bilardi, M. Pracchi, and F. P. Preparata [1981]. "A Critique and Appraisal of VLSI Models of Computation", Conference on VLSI Systems and Computations, Carnegie-Mellon U., 81-88.

M. Blum [1967b]. "On the Size of Machines", *Information and Control* 11, 257-265.

G. J. Chaitin [1966]. "On the Length of Programs for Computing Finite Binary Sequences", *JACM* 13:4, 547-569.

G. J. Chaitin [1969]. "On the length of Programs for Computing Finite Binary Sequences; Statistical Considerations", *JACM* 16:1, 145-159.

G. J. Chaitin [1969b]. "On the simplicity and Speed for Computing Infinite Sets of Natural Numbers", *JACM* 16:3, 407-422.

G. J. Chaitin [1975]. "A Theory of Program Size Formally Identical to Information Theory", *JACM* 22:3, 329-340.

G. J. Chaitin [1976]. "Information-Theoretic Characterizations of Recursive Infinite Strings", *Theoretical Computer Science* 2, 45-48.

M. T. Chen and J. Seiferas [1984]. "Efficient and Elegant Subword-Tree Construction", Technical Report, Dept. of Computer Science, U. Rochester.

Y. Choueka, A. S. Fraenkel, and Y. Perl [1982]. "Polynomial Construction of Optimal Prefix Tables for Text Compression", draft.

R. P. Daley [1973]. "An Example of Information and Computation Trade-Off", *JACM* 20:4, 687-695.

R. P. Daley [1974]. "The Extent and Density of Sequences Within the Minimal-Program Complexity Hierarchies", *JCSS* 9, 151-163.

R. P. Daley [1976]. "Noncomplex Sequences: Characterizations and Examples", *Journal of Symbolic Logic* 41:3, 626-638.

R. G. Gallager [1978]. "Variations on a Theme by Huffman", *IEEE Transactions on Information Theory* 24:6, 668-674.

J. Gallant [1982]. "String Compression Algorithms", Ph.D. Thesis, Dept. EECS, Princeton University.

J. Gallant, D. Maier, and J. A. Storer [1980]. "On finding Minimal Length Superstrings", *JCSS* 20, 50-58.

Gonzalez and Storer [1982]. "Parallel Algorithms for Data Compression", Technical Report CS-82-109, Computer Science Department, Brandeis University.

W. D. Hagamen, D. J. Linden, H. S. Long, and J. C. Weber [1972]. "Encoding Verbal Information as Unique Numbers", *IBM Systems Journal* 11.

B. Hahn [1974]. "A New Technique for Compression and Storage of Data", *CACM* 17:8, 434-436.

F. Henie [1977]. *Introduction to Computability,* Addison Wesley, Reading, MA, 226-236.

D. A. Huffman [1952]. "A Method for the Construction of Minimum-Redundancy Codes", *Proceedings of the IRE* 40, 1098-1101.

T. Kamae [1973]. "On Kolmogorov's Complexity and Information", *Osaka Journal of Mathematics* 10, 305-307.

R. M. Karp [1960]. "Minimum Redundancy Coding for the Discrete Noiseless Channel", IRE Transactions on Information Theory, 27-38.

H. P. Katseff and M. Sipser [1977]. "Several Results in Program Size Complexity", Proceedings IEEE 18th Annual Symposium on Foundations of Computer Science, Providence, R.I.

A. N. Kolmogorov [1965]. "Three approaches to the Quantitative Definition of Information", *Problems of Information Transmission* 1, 1-7.

A. N. Kolmogorov [1969]. "On the Logical Foundation of Information Theory", *Problems of Information Transmission* 5, 3-7.

H. T. Kung and C. E. Leiserson [1978]. "Systolic Arrays (for VLSI)", Technical Report CMU-CS-79-103, Dept. of Computer Science, Carnegie-Mellon University.

G. Langdon [1981]. "A Note on the Ziv-Lempel Model for Compressing Individual Sequences", Technical Report RJ3318, IBM Watson Research Laboratory.

H. Kucera and W. N. Francis [1967]. *Computational Analysis of Present-Day American English,* Brown University Press., Providence, RI.

A. Lempel and J. Ziv [1976]. "On the Complexity of Finite Sequences", *IEEE Transactions on Information Theory,* 22:1, 75-81.

A. Lempel and J. Ziv [1984]. "Compression of Two-Dimensional Data", draft.

A. Lempel and J. Ziv [1984b]. Private communication.

M. E. Lesk [1970]. "Compressed Text Storage", Bell Laboratories Technical Report, Bell Laboratories, Murray Hill, NJ.

D. W. Loveland [1969]. "A Variant of the Kolmogorov Concept of Complexity", *Information and Control* 15, 510-526.

D. W. Loveland [1969b]. "On Minimal-Program Complexity Measures", Proceedings First Annual

ACM Symposium on Theory of Computing, Marina Del Rey, California, 61-65.

D. Maier [1977]. "The Complexity of Some Problems on Subsequences and Supersequences", Proc. Conference on Theoretical Computer Science, University of Waterloo, Waterloo, Ontario, Canada.

D. Maier and J. A. Storer [1977]. "A Note Concerning the Superstring Problem", Proc. 1978 Conference on Information Sciences and Systems, Baltimore, MD.

M. E. Majster [1979]. "Efficient On-Line Construction and Correction of Position Trees", Technical Report 79-393, Dept. of Computer Science, Cornell University.

B. A. Marron and P.A.D. DeMaine [1967]. "Automatic Data Compression", *CACM* 10:11, 711-715.

P. Martin-Löf [1966]. "The Definition of Random Sequences", *Information and Control* 9, 602-619.

A. Mayne and E. B. James [1975]. "Information Compression by Factorizing Common Strings", *The Computer Journal* 18:2, 157-160.

J. P. McCarthy [1973]. "Automatic File Compression", International Computing Symposium (North Holland).

E. M. McCreight [1976]. "A Space-Economical Suffix Tree Construction Algorithm", *JACM* 23:2, 262-272.

C. Mead and L. Conway [1982]. *Introduction to VLSI Systems,* Addison-Wesley, Reading, MA.

V. S. Miller and M. N. Wegman [1984]. "Variations on a Theme by Lempel and Ziv", Technical Report, IBM Watson Research Laboratory.

R. Morris and K. Thompson [1974]. "Webster's Second on the Head of a Pin", Bell Laboratories Technical Report, Bell Laboratories, Murray Hill, NJ.

M. Pechura [1982]. "File Archival Techniques Using Data Compression", *CACM* 25:9, 605-609.

J. Reif and J. A. Storer [1984]. Draft.

M. Rodeh, V. R. Pratt, and S. Even [1981]. "Linear Algorithms for Data Compression Via String Matching", *JACM* 28:1, 16-24.

F. Rubin [1976]. "Experiments in Text File Compression", *CACM* 19:11, 617-623.

S. S. Ruth and P. J. Kreutzer [1972]. "Data Compression for Large Business Files", *Datamation* 18:9, 62-66.

J. B. Seery and J. Ziv [1977]. "A Universal Data Compression Algorithm: Description and Preliminary Results", Technical Memorandum 77-1212-6, Bell Laboratories, Murray Hill, N.J.

J. B. Seery and J. Ziv [1978]. "Further Results on Universal Data Compression", Technical Memorandum 78-1212-8, Bell Laboratories, Murray Hill, N.J.

C. E. Shannon [1951]. "Prediction and Entropy of Printed English", *Bell System Technical Journal* 30, 50-64; Reprinted in *D. Slepian (ed.) [1973]. Key Papers in the Development of Information Theory,* IEEE Press, New York, NY, 42-46.

R. W. Sheifler [1977]. "An Analysis of Inline Substitution for a Structured Programming Language",

CACM 20:9, 647-654.

J. A. Storer [1977]. "NP-Completeness Results Concerning Data Compression", Technical Report 234, Dept. of Electrical Engineering and Computer Science, Princeton University.

J. A. Storer [1977b]. "PLCC- A Compiler-Compiler for PL1 and PLC Users", Technical Report 236, Dept. of Electrical Engineering and Computer Science, Princeton University.

J. A. Storer and T. G. Szymanski [1978]. "The Macro Model for Data Compression", Proceedings Tenth Annual ACM Symposium on Theory of Computing, San Diego, C. A.

J. A. Storer [1979]. "Data Compression: Methods and Complexity Issues", Ph. D. Thesis, Dept. of Computer Science, Princeton University.

J. A. Storer [1983]. "Toward an Abstract Theory of Data Compression", *TCS* 24, 221-237.

J. A. Storer [1982]. "Data Compression Arrays to Reduce VLSI Communication Traffic", Technical Report CS-82-101, Dept. of Computer Science, Brandeis University.

J. A. Storer [1982b]. "Combining Pipes and Trees in VLSI", Technical Report CS-82-107, Dept. of Computer Science, Brandeis University.

J. A. Storer and T. G. Szymanski [1982]. "Data Compression Via Textual Substitution", *JACM* 29:4, 928-951.

J. A. Storer [1984]. "Experiments with On-Line Data Compression of Digital Text Using Dictionaries", draft.

Storer [1984b]. Draft.

N. D. Vasyukova [1977]. "On the Compact Representation of Information", *Mathematika i Kibernetika* 4, 90-93.

M. Visvalingam [1976]. "Indexing with Coded Deltas - A Data Compaction Technique", *Software - Practice and Experience* 6, 397-403.

R. A. Wagner [1973]. "Common Phrases and Minimum-Space Text Storage", *CACM* 16:3, 148-152.

P. Weiner [1973]. "Linear Pattern Matching Algorithms", Proceedings 14th Annual Symposium on Switching and Automata Theory, 1-11.

T. A. Welch [1984]. "A Technique for High-Performance Data Compression", *IEEE Computer* 17:6, 8-19.

J. Ziv [1978]. "Coding Theorems for Individual Sequences", *IEEE Transactions on Information Theory* 24:4, 405-412.

J. Ziv and A. Lempel [1977]. "A Universal Algorithm for Sequential Data Compression", *IEEE Transactions on Information Theory* 23:3, 337-343.

J. Ziv and A. Lempel [1978]. "Compression of Individual Sequences Via Variable-Rate Coding", *IEEE Transactions on Information Theory* 24:5, 530-536.

Variations on a theme by Ziv and Lempel

VICTOR S. MILLER
MARK N. WEGMAN

IBM, THOMAS J. WATSON RESEARCH CENTER
YORKTOWN HEIGHTS, NY 10598

Abstract: **The data compression methods of Ziv and Lempel are modified and augmented, in three ways in order to improve the compression ratio, and hold the size of the encoding tables to a fixed size. The improvements are in the area of dispensing with any uncompressed output, ability to use fixed size encoding tables by using a replacement strategy, and more rapid adaptation by widening the class of strings which may be added to the dictionary. Following Langdon, we show how these improvements also provide an adaptive probabilistic model for the input data. The issue of data structures for efficient implementation is also addressed.**

The Compression methods of Ziv and Lempel

Ziv and Lempel [6] , [7] , [8] propose two related methods for data compression which are motivated by the theory of algorithmic complexity on information-lossless finite-state transducers, rather than more traditional probablity based approaches. The second, simpler, method can be implemented straightforwardly and efficiently, and produces satisfactory compression on most data. However, the method has some shortcomings, which we remedy in the next section. Below we use the term encoder for the compression program, and decoder for the inverse operation. Ziv and Lempel's original algorithm is given in Figure 2 on page 2. It will henceforth be referred to as the ZL algorithm.

I is the input string.
I_P denotes the P-th character in the input string.
P is the number of characters which have been compressed.
A is the size of the input alphabet.
D is the dictionary: an array of strings.
n is the size of the dictionary.
We use the notation $\lg x$ to denote $\log_2 x$
match(I,P,D) finds the longest string from D which
matches I starting at P.
output(x,y) outputs x in y bits.
input(k) returns the next k bits as an integer.
replacelast(l,c) replaces the last character of D_l by c
write writes its argument at the end of the output.
chr(i) is the i-th character in the alphabet.

Figure 1. Notation

NATO ASI Series, Vol. F12
Combinatorial Algorithms on Words
Edited by A. Apostolico and Z. Galil

```
D_1 ← Λ
P ← 1
n ← 1
While P ≤ |I| begin
  k ← match(I,P,D)
  S ← D_k
  output(k,max(1,lg n))
  P ← P + |S| -- Advance the cursor
  c ← I_P
  output(c,lg (A))
  n ← n + 1
  D_n ← Sc
  P ← P + 1
  end
```

Figure 2. Ziv and Lempel's original algorithm

Before discussing the problems with the ZL algorithm we will point out some of its strong points. These points are carried over in our modifications.

- It is adaptive. Thus, it can work on a wide variety of different file types and still get acceptable preformance.
- It can be thought of as a general model for compression. For example, Rissanen and Langdon [5] , and Gilbert and Momna [1] propose schemes which are optimal (or nearly optimal) once a specific model is fixed (e.g. First order Markov model, or a "good" set of n-grams). However, there does not seem to be a general strategy for constructing the model in question.
- Both the encoder and the decoder maintain a dictionary of strings, and at every point those dictionaries change in lock-step, so that no extra data is needed to transmit the dictionary. This is possible because the encoder only uses information that has been transmitted to adapt its encoding tables.
- It records probable events and does not waste space with information about improbable events. To construct a Huffman code compressing all pairs of 8-bit characters, it would be necessary to have a table of 65,536 frequencies. The resulting code would only have information about digram frequencies, even though some trigram frequencies might be more useful (This defect is also remedied in the suggested algorithms of Langdon and Rissanen [5]) With the ZL algorithm if there is a 100-gram which is more common than a single letter, the 100-gram would be "learned" first. It might be more important to store information that runs of blanks occur more frequently than the letter "z".
- A fourth advantage is that there is a good representation for the dictionary. It is possible to keep a fixed number of bits for each dictionary entry, no matter how long the string of characters represented by that dictionary entry is.

Improving Ziv and Lempel's encoding

The first problem with the above method is that part of the output consists of uncompressed symbols from the input. While this is of little consequence asymptotically, it is of significant consequence for practical data. For example, if a file is partitioned into 20,000 sequences by the above algorithm, over one-third of the output will consist of uncompressed characters. The solution proposed here avoids transmitting any uncompressed data: The entire output consists of a sequence of identifying numbers.

The reason the original algorithm needs to output uncompressed characters, is to ensure that the cursor actually moves, as it does in the line marked "Advance the cursor". This may be done by ensuring that **match** always finds a non-empty string. Instead of initializing the dictionary to contain the empty string, we initialize the dictionary to contain all strings of length 1. Thus, a match is always found. The dictionary is augmented by adding the string S concatenated with the first character of the next string which is matched.

We can achieve an average of 4 characters (on English text) for each dictionary entry, when there are 4000 dictionary entries. The ratio does not improve much as the dictionary grows beyond this point. Thus, the uncompressed character occupies 8 bits out of twenty in the ZL algorithm, with the remaining four characters occupying about three bits apiece. Even if we compress a file of 2^{40} characters by 3 to 1, the extra characters take up over 10% of the compressed file. Asymptotically, the one character may not matter (which is what Ziv and Lempel prove), but it will continue to matter even when compressing more text than has been written in the history of mankind.

for $i \leftarrow 1$ **to** A **begin** $D_i \leftarrow$ chr(i) **end**
$P \leftarrow 1$
$n \leftarrow A + 1$
While $P \leq |I|$ **begin**
  $k \leftarrow$ **match**(I,P,D)
  $S \leftarrow D_k$
  output(k,lg n)
  $P \leftarrow P + |S|$
  $c \leftarrow I_P$
  $n \leftarrow n + 1$
  $D_n \leftarrow Sc$
  end

Figure 3. The character extension improvement

```
for i ← 1 to A begin D_i ← chr(i) end
n ← A + 1
l ← 0
Do forever begin
  r ← input(lg n)
  If end of input then stop
  cc ← first(D_r)
  -- The first character of D_r is always known
  If l ≠ 0 then replacelast(l,cc)
  write(D_r)
  l ← r
  n ← n + 1
  D_n ← D_r?
  -- The ? will be replaced
  --  during the next iteration
  end
```

Figure 4. The decoding algorithm for the character extension improvement

The ZL algorithm does not put a bound on the size of the dictionary. Because this is not practical, Ziv and Lempel recommend blocking the input into those segments whose dictionaries just fill up available space, compressing each block independently. However it is more advantageous to replace individual strings in the dictionary. We discard strings which have been in some sense least recently used. We have done this in two ways, and believe that all reasonable ways of interpreting least recently used will work. One way defines a string as "used" if it is matched, or is a prefix of a match. The other definition associates a reference count with each string in the dictionary. The reference count of a string S is the number of strings in the dictionary which are of the form $S\alpha$ for some symbol α Among those strings with reference count 0, the one whose reference count has been 0 the longest is the next string which will be replaced (It is that string which has been unused for the longest time, therefore it is the "least recently used".) The strategy for discarding entries has been modified in order to make the implementation easier. In the implementation, we have pointers from strings to their prefixes, and if the prefixes were to be discarded, the data structure would not be consistent. Note, that we cannot discard the single characters with which we have initialized the dictionary. This strategy may be used with the original ZL algorithm or with either of the other two improvements given here.

In the figures below we assume that slots are filled or marked **available**. There is a procedure **avail** which returns the number of an available slot, if one is available, or deletes the least recently used entry of D and returns its entry number. The dictionary has a fixed number, 2^r, of entries.

```
for i ← 1 to A begin D_i ← chr(i) end
for i ← A + 1 to 2^r
  begin
    D_i ← available
  end
P ← 1
While P ≤ |I| begin
  k ← match(I,P,D)
  output(k,r)
  j ← avail(D)
  P ← P + |S|
  D_j ← D_k I_P
  end
```

Figure 5. The replacement strategy

Intuitively, the dictionary should be filled with the natural units of the file being compressed. In English this would mean that most entries in the dictionary would be words, or even phrases. However, the method used to build up larger units must go through transitions which are not units. So, one might have the a unit composed of one word, plus the first half of another word. It is less likely to have the last part of that word in the dictionary, than the whole word. Moreover, if there are pieces of words in the dictionary, the dictionary may be bigger than it should be. For example if three fourths of the dictionary has strings which are never, or rarely used then two extra bits will be used in the transmission of each bit so that these useless entries can be referred to. Furthermore, it takes a long time to adapt to long strings. All these problems can be eliminated or ameliorated by adding into the dictionary entries which are the concatenation of two matches rather than the concatenation of the first match and the first character of the second match.

This strategy may be used with the replacement strategy previously given (as in the figure), or with a growing dictionary, as with the original ZL algorithm. When using the replacement strategy with the above algorithm, it is necessary to redefine the meaning of reference count: The reference count of a string S, is the number of strings in the dictionary which were constructed as ST, or TS, for some T (SS is counted twice).

```
for i ← 1 to A begin D_i ← chr(i) end
for i ← A + 1 to 2^r
  begin
    D_i ← available
  end
P ← 1
S' ← Λ
while P ≤ |I| begin
  k ← match(I,P,D)
  S ← D_k
  P ← P + |S|
  output(k,r)
  j ← avail(D)
  D_j ← S'S
  S' ← S
  end
```

Figure 6. The String extension improvement

Combining these three ideas we get a powerful encoder, whose implementation we will next discuss. A million characters of English text can be compressed in less than 20 seconds on an IBM/370 model 3081. Characters after compression will average two and two thirds bits each. On other material we often do better. Program source, modules, and listings all get three to one compression. Facsimile images of text get around ten to one. Terminal sessions get (by some measures) thirty to one.

Implementation

The major difficulties in obtaining a practical implementation are in finding a good data structure for the dictionary of strings. This structure should be small, yet allow rapid searches. We will first describe a data structure sufficient for all the encodings save the last that we presented. Then we will extend it to that case. The size of all these structures is proportional to the number of strings in the dictionary, and does not depend on the size of the strings.

In the first encodings all strings, S, in the dictionary are either one character longer than a prefix, P, of S, with P being in the dictionary, or S is one character long. Thus the dictionary resembles a tree. The root of the tree is the empty string. Each node (except for the root) is the child of the node representing the string labeling the node with the last symbol omitted. In the encoding algorithm the recognition is accomplished by first recognizing a prefix and then seeing if it has a child which matches the next character. Let n be a node with parent, P, C being the last character of the string corresponding to n (the character which is not in the string corresponding to P). A hash table indexed by the pair (P,C) returns n if there is such a child. Thus, given a node, we can quickly tell if there is a longer match. The decoding algorithm works similarly, but instead of a hash table a simple pointer from n to P suffices.

The least recently used encoding presents little problem. The only thing worth noting is that holes cannot be left in the dictionary. Thus the string 'abc' cannot stay in the dictionary when the string 'ab' is thrown out. The simplest strategy is to only place strings on the LRU list when they are leaves of the tree. Thus, a new string or one whose children have been deleted become candidates for deletion.

The algorithms given in Figure 3 on page 3 and Figure 5 on page 5 obviously have running time linear in the input using the above data structures.

The string concatenation algorithm is more difficult. We must maintain two data structures. One structure we call the *discriminator tree*, and is used to find candidates for matches. This structure resembles the above dictionary, and is similar to a *PATRICIA* tree of Morrison [2] . The other data structure is called the *pair forest* and it allows us to choose between the candidates.

The pair forest succinctly represents all strings. Each string is represented by a node. A node is either a character, or two pointers to other nodes. Thus the concatenation of two strings in the dictionary can be represented by a node which points to both. All nodes in the forest are placed in an array, the ith element of which points to the ith string in the dictionary.

The discriminator tree maintains the property that the parent, P , of a node, n, corresponds to a prefix of the string represented by n. However the prefix may be more than one character shorter. Moreover, not all nodes in the tree neccessarily correspond to strings in the dictionary; they may be prefixes of strings in the dictionary.

All strings in the dictionary which are not prefixes of other strings in the dictionary correspond to leaves in the discriminator tree. All other strings in the dictionary are also placed in the tree as internal nodes. If S is the prefix of two or more strings, T and U, which are in the dictionary and S is the longest such string, then a node, N, corresponding to S is in the tree, even if S is not in the dictionary. Since we cannot store S efficiently, we store in N a pointer to either T or U in the pair forest. Given T or U, and the length of S it is possible to re-create S.

With each node is stored the length of the string that it matches. A hash table, as before, allows us to find the appropriate child from a parent. So, if we are trying to match a string which is a leaf in the tree, we start at the root, and hash the first character to find the next node. From this node we find the length of the string it matches, and hence the character beyond it. We hash the node and that character to obtain the next node. This process is repeated to find a candidate match. A problem arises if the candidate match does not match the text on a character which was not used in the hashing process.

We know at this point that the real match in the dictionary must either be the string found by the discriminator tree or a prefix of it. The pair forest is used to find the longest prefix of the candidate match that matches the text being compressed. As this process goes on, the discriminator tree is scanned to find the longest string in the dictionary which corresponds to this prefix. That string is the correct match.

Maintaining these structures can be difficult when elements are to be deleted, using the LRU strategy. However, if the only elements that are deleted are those which are not pointed at by any other elements in the pair tree it becomes relatively easy. Using the definition that an element is used if it is pointed at, makes the implementation much easier and experiments show has no harmful effect on the compression.

It should be pointed out, that in the string extension algorithm the possibility arises that when adding a string to the discriminator tree, that it is already there. In that case no changes should be made to the discriminator tree.

The remaining multitudinous details of implementation are left as an exercise to the reader.

The Symbolwise Equivalent Model

There is a weakness with the above algorithms. When the identifying number is transmitted, it is encoded with a very simple encoding: each possible identifying number is considered to be equally probable, and lg n bits are used to encode it. Langdon has already pointed out that there may be a gain in compression in the original Ziv-Lempel algorithm by estimating the probabilities of the identifying numbers by means of frequency counting, and then using Arithmetic Coding to encode the output. For those readers familiar with the terminology in [3] , we may further interpret the two new algorithms above in terms of a symbolwise Markov Model (the objects which are encoded are the symbols in the original alphabet, as opposed to "extended" symbols, in our case the indices into to dictionary).

It should be pointed out that the assumption of equiprobability is very close to being true: Any unit occuring often enough will very likely only occur subsequently inside of a longer unit.

We may estimate the probabilities of occurence of the identifying numbers from their frequency of occurence, and then use Arithmetic Coding to encode them. This may be further improved by using frequencies of digrams of pairs: Prob($S \mid S'$) is estimated by Prob(**first**(S) $\mid$**last**(S')) $\times$ Prob($S \mid$**first**(S)) where the latter two probabilities are estimated by relative frequencies (Prob($A \mid B$) means the conditional probability of A given B, **first**(S) means the first character of S, and **last**(S) means the first character of S.).

Following Langdon [3] we show that like the original algorithm of Ziv and Lempel, we may interpret our improvements in terms of a symbolwise Markov model, without using alphabet extension.

The states of the model correspond to the nodes of the discriminator tree, plus the "phantom" nodes on paths between nodes. The latter nodes are the nodes that would be in the full tree of all prefixes of all elements in the dictionary, but are not in the discriminator tree. The probabilities are obtained by a counting technique as described in [3] . Namely, the estimate of the probability for seeing a symbol in a given state is the ratio of the frequency of occurences of that symbol in that state, to the total frequency of visits to that state. Note that the "comma" of Ziv and Lempel must be counted as a symbol in the alphabet (the comma is considered to be a new symbol which indicates the end of a parsed string. The comma is necessary in a symbolwise model so that the decoder can estimate the probabilities in the same context as the encoder). However, in the case of the string extension, unlike the case of the single character extension, no coding space needs to be allocated to the comma, unless the node of the discriminator tree corresponds to a string represented in the pair forest (i.e. in the former contexts, the comma has probability 1). In addition, no coding space is needed at any phantom node, because only one symbol is possible at those nodes.

It remains to give the probability estimates for the single characters in the discriminator tree. One might, as in [3] take the 0-order probabilities for their occurences in that context. This ignores, however, any serial correlation between parsed strings. Better results may be obtained by conditioning the first character on the last character of the previous string (i.e. obtain the probability estimates by tabulating the digram statistics for the pairs (last character,first character)). Using this, the string encoder calculates a theoretical entropy for a number of files containing English text and formatting commands to be 2.0 to 2.1 bits per character.

Extension and Possible Future work

One might obtain some of the benefits of the arithmetic coding mentioned in the last section without the attendant work (many multiplications) in the following way: Keep all the strings in the dictionary organized into two (or more) "buckets", with successively bigger sizes. A replacement strategy as outlined above would be used for each bucket, but when a string is deleted from a bucket, it would be inserted into the bucket down one level in the hierarchy. The top level bucket would be relatively small, corresponding to the most frequently used strings. One might be able to estimate the probabilities of referring to a string in a given bucket statically, and use a simple Huffmann code as a prefix to the node reference. For example, suppose that the first bucket in the hierarchy had a capacity of 1024, and the second bucket had a capacity of 4096. Then one would only need 11 bits to refer to a string in the first bucket, but 13 bits to refer to a string in the second. If the probablity of refering to a string in the first bucket were greater than 1/2, then one would need less than 12 bits per string for a reference. Notice that $\lg(1024+4096)=12+\lg(5/4)$, which is greater than 12.

Acknowledgements

We would like to thank John Cocke, who first got us interested in the Ziv-Lempel algorithm, and Glen Langdon, who made many constructive comments, and supplied us with the DAFC program [4] which we used for comparison.

Results on some test files

	A1	A1'	A2	A2'	DAFC	0ORD	1ORD
1.	1.99	1.65	1.52	1.05	2.40	4.29	2.10
2.	2.91	2.44	2.23	1.91	3.69	5.35	3.01
3.	3.06	2.47	2.73	2.38	3.55	4.41	2.75
4.	1.63	1.33	1.33	1.19	1.73	2.45	1.67
5.	0.34	0.29	0.16	0.14	0.65	1.52	0.56
6.	0.96	0.81	0.61	0.55	1.22	2.20	1.28
7.	3.45	2.92	2.73	2.41	4.13	5.02	3.59
8.	0.84	0.72	0.26	0.23	1.36	2.52	1.24
9.	2.49	1.97	2.11	1.73	2.75	3.96	2.20
10.	2.67	2.20	2.35	1.96	2.91	3.78	2.53
11.	5.06	4.20	4.43	3.86	5.03	6.55	4.98
12.	3.04	2.50	2.72	2.36	3.20	4.64	2.93

All of the above figures are expressed in output bits/input character (which are 8 bits in this case). This gives a measure which is independent of the length of the input character set, which for example is 7 bits for ASCII but 8 bits for EBCDIC. To get compression ratios, each of the above figures should be divided by 8.

1. Formatted manual..333K
2. History file of changes to compiler..................182K
3. High level source 36K
4. Assembler source..347K
5. Assembler source of encoding tables...............480K
6. Cross reference of a compiler..........................3482K
7. SCRIPT file of technical paper........................224K
8. Screens sent to a terminal in one day's work...1200K

9. Database from Customer.................................229K
10. Online telephone directory 1202K
11. Object code module for formatting language ..619K
12. APL workspace 440K

A1 TERSE algorithm incorporating "character extension" and replacement strategy. Dictionary size 4096 entries.

A1' TERSE algorithm with theoretical entropy for model in section "The Symbolwise Equivalent Model" on page 8.

A2 TERSE algorithm incorporating "string extension" and replacement strategy. Dictionary size 4096 entries.

A2' Algorithm A2 with theoretical entropy for model in section "The Symbolwise Equivalent Model" on page 8.

DAFC Algorithm modeling input as a sixteen state Markov source with runs, using adaptive arithmetic coding.

0ORD The theoretical zeroth-order entropy.

1ORD The theoretical first-order entropy.

REFERENCES

[1] E. Gilbert and C. Momna, *Multigram Codes*, IEEE Trans. Inform. Theory, *IT-28* (1982), 346-348

[2] D. Morrison, *PATRICIA -- A Practical Algorithm to Retrieve Information Coded in Alphanumeric*, J. ACM, *15* (1968), 514-534

[3] G. Langdon, *A Note on the Ziv-Lempel Model for Compressing Individual sequences*, IEEE Trans. Inform. Theory, *IT-29* (1983), 285-287

[4] G. Langdon, *A Double-Adaptive File Compression Algorithm*, IEEE Trans. Commun., *COM-31* (1983), 1253-1255

[5] J. Rissanen and G. Langdon, *Universal Modeling and Coding*, IEEE Trans. Inform. Theory, *IT-24* (1981), 12-23

[6] J. Ziv, *Coding Theorems for Individual Sequences*, IEEE Trans. Inform. Theory, *IT-24* (1978), 405-412

[7] J. Ziv and A. Lempel, *A Universal algorithm for Sequential Data Compression*, IEEE Trans. Inform. Theory, *IT-23* (1976), 75-81

[8] J. Ziv and A. Lempel, *Compression of Individual Sequences via Variable-rate Coding*, IEEE Trans. Inform. Theory, *IT-24* (1978), 530-536

COMPRESSION OF TWO-DIMENSIONAL IMAGES

Abraham Lempel
Computer Science Department
Technion
Haifa, Israel

and

Jacob Ziv
Electrical Engineering Department
Technion
Haifa, Israel

Abstract

Distortion-free compressibility of individual pictures, i.e., two-dimensional arrays of data, by finite-state encoders is investigated. For every individual infinite picture I, a quantity $\rho(I)$ is defined, called the compressibility of I, which is shown to be the asymptotically attainable lower bound on the compression-ratio that can be achieved for I by any finite-state, information-lossless encoder. This is demonstrated by means of a constructive coding theorem and its converse that, apart from their asymptotic significance, might also provide useful criteria for finite and practical data-compression tasks. The proposed picture-compressibility is also shown to possess the properties that one would expect and require of a suitably defined concept of two-dimensional entropy for arbitrary probabilistic ensembles of infinite pictures. While the definition of $\rho(I)$ allows the use of different machines for different pictures, the constructive coding theorem leads to a universal compression-scheme that is asymptotically optimal for every picture. The results of this paper are readily extendable to data arrays of any finite dimension. The proofs of the theorems will appear in a forthcoming paper.

NATO ASI Series, Vol. F12
Combinatorial Algorithms on Words
Edited by A. Apostolico and Z. Galil

I. INTRODUCTION

In a recent paper [1], the compressibility of individual sequences was defined and shown to be the asymptotically attainable lower bound on the compression ratio that can be achieved by any finite-state encoder.

In this paper, we present similar results for individual pictures, i.e., two-dimensional arrays of data. We define and investigate the compressibility of individual pictures with respect to the class of information-lossless encoders that allow for any finite-state encoding. As in [1], no distortion is allowed and the original data must be fully recoverable from its compressed image. This accounts for the requirement of the information-losslessness [2] property.

In our model, an encoder E is defined by a septuple (S,A,B,D,g,t,f), where S is a finite set of states; A is a finite (input) alphabet of α picture symbols and an extra 'end-of-picture' symbol; B is a finite set of output words, over a finite output alphabet, including the null-word, i.e., the 'word' of length zero; D is a finite set of incremental displacement vectors that determine the move of the reading head to the next input letter; g is the next-state function that maps $S \times A$ into S; t is the incremental-displacement function that maps $S \times A$ into D; and f is the output function that maps $S \times A$ into B.

When a picture is scanned and processed by such an encoder $E=(S,A,B,D,g,t,f)$, the encoder reads an (input) sequence $x=x_1x_2x_3\ldots$, $x_i \in A$; writes an (output) sequence of words $y=y_1y_2y_3\ldots$, $y_i \in B$; advances according to a sequence of incremental displacements $\vec{d}=\vec{d}_1\vec{d}_2\vec{d}_3\ldots$, $\vec{d}_i \in D$; and goes through a sequence of states $z=z_1z_2z_3\ldots$, $z_i \in S$, in accordance with

$$y_i = f(z_i, x_i)$$

$$\vec{d}_i = t(z_i, x_i)$$

$$z_{i+1} = g(z_i, x_i).$$

The position $\vec{\Delta}_{i+1}$ of the scanner when it reads the $(i+1)$–*st* input symbol is given by

$$\vec{\Delta}_{i+1} = \vec{\Delta}_0 + \sum_{j=1}^{i} \vec{d}_j,$$

where $\vec{\Delta}_0$ is the initial position of the scanner.

The finite segment $x_i x_{i+1} \cdots x_j$, $1 \le i \le j$, of x will be denoted by x_i^j; similar notation will, naturally, apply to finite segments of other sequences. Occasionally, we shall write $f(z_1, x_1^n)$ for y_1^n and $\hat{g}(z_1, x_1^n)$ for z_{n+1}.

An encoder is said to be *information-lossless* (IL) if for all $z_1 \in S$ and all $x_1^n \in A^n$, $n \ge 1$, the triplet $[z_1, f(z_1, x_1^n), \hat{g}(z_1, x_1^n)]$ uniquely determines x_1^n (and, therefore, also z_1^n and d_1^n).

To simplify the discussion, without any real loss of generality, we assume throughout that the output alphabet of the encoder is binary; the initial state z_1 is some fixed member of S; and that $\vec{\Delta}_0 = (0,0)$, the coordinates of the top-left picture-cell of the given picture, with the first coordinate increasing downward and the second rightward.

Furthermore, since every finite state encoder can be simulated by one whose incremental-displacement set D consists only of the elementary steps required to reach adjacent picture-cells, we shall assume, again without loss of generality, that the set D does indeed consist of exactly those elements which can move the scanner from any cell to any of the eight adjacent cells, or can make it stay in place.

Given an encoder E and an $N \times N$ picture I_N (i.e., a square array of N^2 picture-cells) the *compression-ratio* for I_N with respect to E is defined by

$$\rho_E(I_N) \triangleq \frac{L(y_1^n)}{N^2 \log\alpha} \tag{1}$$

where $\alpha=|A|-1$ is the number of picture symbols, n is the number of steps it takes E to complete the scan of I_N, and $L(y_1^n)$ is the length, in bits, of the compressed image y_1^n of I_N. Note that n is not necessarily equal to N^2 and that

$$L(y_1^n) = \sum_{i=1}^{n} l(y_i).$$

where $l(y_i)$ is the length of the output word y_i.

Let $E(s)$ denote the class of all finite-state IL encoders with $|S|\le s$, and let

$$\rho_{E(s)}(I_N) = \min_{E\in E(s)} \{\rho_E(I_N)\}. \tag{2}$$

For the infinite picture I, obtained by letting N tend to infinity, we define

$$\rho_{E(s)}(I) \triangleq \limsup_{N\to\infty} \rho_{E(s)}(I_N) \tag{3}$$

and

$$\rho(I) \triangleq \lim_{s\to\infty} \rho_{E(s)}(I). \tag{4}$$

It should be clear that for every infinite picture I

$$0\le\rho(I)\le 1.$$

This normalized quantity $\rho(I)$, which depends solely on I, will be referred to as the *compressibility* or *complexity* of I.

Our main results are presented in the form of two theorems. In Theorem 1, the converse-to-coding theorem, we derive a lower bound on $\rho(I)$. In Theorem 2, the coding theorem, we demonstrate, constructively, that the bound of Theorem 1 is asymptotically attainable by a universal compression scheme which combines a specific scanning path with the authors' compression algorithm for individual sequences [1].

The remainder of this paper consists of two parts: descriptive part (Section II) where all the results are stated and discussed, including one constructive proof (Theorem 2) which is essential for the mainstream of Section II, and a conclusion (Section III) where further interpretation and possible generalizations are discussed. The formal proofs of the theorems will appear elsewhere in a forthcoming paper.

II. STATEMENT OF RESULTS

The first result establishes a lower bound on the compression-ratio attained by any encoder $E \in E(s)$ for a given finite, $N \times N$, picture I_N over an alphabet of α picture symbols and an extra end-of-picture symbol.

Let $q = 2^j$, $j = 0,1,2,\ldots$, let $N^* = \lfloor \frac{N}{q} \rfloor q$ and let I_{N^*} be the $N^* \times N^*$ picture occupying the upper-left corner of I_N. Consider the partition of I_{N^*} into $\left(\frac{N^*}{q}\right)^2$ subpictures, of size $q \times q$, containing q^2 picture-cells each, and denote these subpictures by $I_q(1), I_q(2), \ldots, I_q\left(\left(\frac{N^*}{q}\right)^2\right)$. For $1 \le m \le \left(\frac{N^*}{q}\right)^2$, let

$$\delta(I_q(m), W) = \begin{cases} 1 & \text{if } I_q(m) = W \\ 0 & \text{otherwise.} \end{cases}$$

Then, the relative frequency of occurrence of the $q \times q$ array W in the partitioned I_{N^*} is given by

$$P(I_{N^*}, W) = \left(\frac{q}{N^*}\right)^2 \sum_{m=1}^{(N^*/q)^2} \delta(I_q(m), W). \tag{5}$$

This relative frequency can be interpreted as a 'probability measure' of $W \in A^{q \times q}$, relative to I_{N^*}, where $A^{q \times q}$ is the set of all $q \times q$ arrays over A. The corresponding normalized 'entropy' of I_N is then defined by

$$H_q^*(I_N) \triangleq \frac{-1}{q^2 \log \alpha} \sum_{W \in A^{q \times q}} P(I_{N^*}, W) \log P(I_{N^*}, W), \tag{6}$$

where here, and elsewhere in the sequel, log h means $\log_2 h$.

$$LS: (1,1)\to(1,0)\to(0,0)\to(0,1);$$

$$US: (1,1)\to(0,1)\to(0,0)\to(1,0);$$

$$DS: (0,0)\to(1,0)\to(1,1)\to(0,1).$$

For $k\geq1$, the basic and standard scans are defined by (see Fig. 3):

If k is odd, SS is RS and if k is even, SS is DS;

RS for I_q begins with DS for $I_q(0,0)$, followed by RS for $I_q(0,1)$, followed by RS for $I_q(1,1)$, and ends with US for $I_q(1,0)$;

DS for I_q begins with RS for $I_q(0,0)$, followed by DS for $I_q(1,0)$, followed by DS for $I_q(1,1)$, and ends with LS for $I_q(0,1)$;

US for I_q begins with LS for $I_q(1,1)$, followed by US for $I_q(0,1)$, followed by US for $I_q(0,0)$, and ends with RS for $I_q(1,0)$;

LS for I_q begins with US for $I_q(1,1)$, followed by LS for $I_q(1,0)$, followed by LS for $I_q(0,0)$, and ends with DS for $I_q(0,1)$.

The standard scan for I_{2^k}, $k=1,2,3,4$, is shown in full detail in Fig. 4.

The SS is readily recognized as the Peano-Hilbert plane-filling curve [3],[4] whose main property, in the context of this paper, is that for each k it covers I_{2^k} quadrant-by-quadrant, i.e., never leaves a quadrant before visiting each of its cells.

Given a picture I_{2^k}, let $SS(I_{2^k})$ denote the sequence of 2^{2k} picture symbols read from I_{2^k} under SS. For $SS(I_{2^k})=x_1^n$, $n=2^{2k}$, let $m=2^{2j}$, $j=0,1,\dots,k$, and let

$$\delta(x_{im+1}^{(i+1)m},w) = \begin{cases} 1 & \text{if } x_{im+1}^{(i+1)m}=w \\ 0 & otherwise \end{cases}, \quad 0\leq i\leq \frac{n}{m}-1.$$

Then, the relative frequency of occurrence of w in x_1^n is

$$P(x_1^n,w) = \frac{m}{n}\sum_{i=0}^{n/m-1} \delta(x_{im+1}^{(i+1)m},w), \tag{10}$$

and the corresponding normalized 'entropy' of x_1^n is defined by

$$\hat{H}_m(x_1^n) \triangleq -\frac{1}{m\log\alpha} \sum_{w\in A^m} P(x_1^n,w)\log P(x_1^n,w). \tag{11}$$

In comparing the quantities defined by (6) and (11) for a given picture I_{2^k}, it is important to make the following observations.

(i) $H_q^*(I_{2^k})$ is the entropy of the $2^k\times 2^k$ *picture* I_{2^k} as induced by the distribution of its $q\times q$, $q=2^j$, subpictures.

(ii) $\hat{H}_m(x_1^n)$ is the entropy of the *sequence* $x_1^n=SS(I_{2^k})$, obtained from I_{2^k} under SS, as induced by the distribution of its subsequences of length $m=2^{2j}$. Each of these subsequences, in turn, is obtained from a $2^j\times 2^j$ subpicture under one of the four basic scans that constitute SS.

(iii) Since only two bits are required to distinguish between the four basic scans, it follows that when $m=q^2$, the numerical difference between $H_q^*(I_{2^k})$ and $\hat{H}_m(SS(I_{2^k}))$ is bounded by two bits per m input symbols.

Hence, we can write

$$\hat{H}_m(SS(I_{2^k})) \le H_q^*(I_{2^k}) + \frac{2}{q^2\log\alpha}, \quad m=q^2. \tag{12}$$

Letting k, and then m, tend to infinity we define

$$\hat{H}_m(SS(I)) = \limsup_{k\to\infty} \hat{H}_m(SS(I_{2^k})) \tag{13}$$

and

$$\hat{H}(SS(I)) = \lim_{m\to\infty} \hat{H}_m(SS(I)). \tag{14}$$

From (7), (12), and (13) we obtain

$$\hat{H}_m(SS(I)) \le H_q^*(I) + \frac{2}{q^2\log\alpha}, \quad m=q^2, \tag{15}$$

and from (8), (14), and (15) we have

$$\hat{H}(SS(I)) \le H^*(I). \tag{16}$$

Now, let $\rho(SS(I))$ denote the compressibility of the sequence $x=SS(I)$ as defined by equation (4) of [1]. Using Huffman's coding scheme for input blocks of length m, it is easy to show [5] that

$$\rho(SS(I)) \leq \hat{H}_m(SS(I)) + \frac{1}{m}$$

which, when m tends to infinity, becomes

$$\rho(SS(I)) \leq \hat{H}(SS(I)) \tag{17}$$

for all I.

Combining (16) and (17) with Theorem 1 yields the following result.

Lemma 1: For every infinite picture I

$$\rho(I) \geq \rho(SS(I)). \tag{18}$$

An interesting application of Lemma 1 is the use of (18) to identify certain infinite pictures that are incompressible by any finite-state IL encoder and can, therefore, be viewed as 'random' pictures. To illustrate this point, let $u(k)$ denote the binary sequence of length $k2^k$ that lists, in lexicographic order, all the 2^k binary words of length k, and let

$$u = u(1)u(2)u(3).....$$

It was shown in [1] that u is a 'random' sequence satisfying

$$\rho(u) = 1. \tag{19}$$

Since the picture-to-sequence transformation SS is invertible, one can readily construct the picture $U=SS^{-1}(u)$ which, by (18) and (19), satisfies

$$\rho(U) \geq \rho(SS(U)) = \rho(u) = 1.$$

Our next result combines the authors' algorithm [1] for the compression of sequences with a finite-state version of SS to demonstrate the existence of a universal compression scheme under which, for every picture I, the compression-ratio attained for the top-left corner I_{2^k} of I tends to $\rho(I)$ as k tends to infinity.

First, let SS_q denote the restriction of SS to a finite $q \times q$ picture, where $q=2^j$. Note that if j is odd, SS_q is a right-scan beginning at $\vec{\Delta}_0=(0,0)$ and ending at $\vec{\Delta}=(q-1,0)$, while if j is even, SS_q is a down-scan beginning at $\vec{\Delta}_0=(0,0)$ and ending at $\vec{\Delta}=(0,q-1)$.

Given a picture I_{2^k}, $k\geq j$, we define the *finite scan* FS_q of I_{2^k} as follows. $FS_q(I_{2^k})$ consists of repeated applications of SS_q to successive $q\times q$ subpictures, or blocks, of I_{2^k}. Starting at the top-left block, FS_q proceeds to the one below if j is odd (the one to the right if j is even), and so on until an 'end of picture' symbol is encountered. Having thus completed a column (row) of blocks, FS_q begins the next column (row) of blocks from bottom to top (right to left), and so on until all of I_{2^k} has been scanned (Fig. 5).

Theorem 2 (Coding Theorem):

For every $\varepsilon>0$ and every positive integer j there exists a finite-state IL encoder ξ with $s(j)$ states such that the compression-ratio $\rho_\xi(I_{2^k},j)$ attained for the top-left corner I_{2^k}, $k\geq j$, of a given picture I satisfies

$$\limsup_{k\to\infty} \rho_\xi(I_{2^k},j)\leq\rho(I)+\delta_\varepsilon(I,j), \tag{20}$$

where

$$\lim_{j\to\infty} \delta_\varepsilon(I,j)=\varepsilon.$$

Proof: The proof is constructive and employs FS_q, $q=2^j$, in conjunction with a universal, concatenated coding scheme for successive strings of length q^2. In a previous paper ([1], Theorem 2), the authors have presented a construction of a finite-state IL encoder ξ such that for every $\varepsilon>0$, the compression-ratio $\rho_\xi(x,n)$ attained by ξ for a given infinite sequence x, when fed by successive subsequences of length n, satisfies

$$\rho_\xi(x,n) \leq \rho(x)+\hat{\delta}_\varepsilon(x,n), \tag{21}$$

where

$$\lim_{n\to\infty} \hat{\delta}_\varepsilon(x,n)=\varepsilon.$$

Consider the top-left corner I_{2^k} of a given picture I. Apply FS_q, followed by the algorithm ξ to successive $q\times q$ blocks of I_{2^k} where each such $q\times q$ block is rotated at 0°, 90°, 180° or 270° so as to yield the smallest compression ratio attained by ξ for that block. Denote the resulting scan by $F'S_q$ and let $X(1),X(2),\cdots,X(r)$ be the $r=2^{2(k-j)}$ successive strings each of length $q^2=2^{2j}$, obtained under $F'S_q$ from the successive $q\times q$ blocks of I_{2^k}. Then, $F'S_q(I_{2^k}=X(1),X(2),\cdots,X(r)=X_1^{rq^2}$.

Clearly,

$$\rho_\xi(F'S_q(I_{2^k}),q^2)\le\rho_\xi(SS(I_{2^k}),q^2)+\frac{2}{q^2\log\alpha}$$

where the last term results from the extra two bits which convey information about the rotation phase of each of the r $q\times q$ blocks.

Thus it follows from (21) that

$$\limsup \rho_\xi(F'S_q(I_{2^k}),q^2)\le\rho(SS(I))+\hat{\delta}_\varepsilon(SS(I),q^2)$$

where

$$\lim_{q\to\infty}\hat{\delta}_\varepsilon(SS(I),q^2)=\varepsilon$$

By Lemma 1, this becomes the result claimed in (20), where $\delta_\xi(I,j)$ is substituted for $\hat{\delta}_\varepsilon(SS(I),q^2)$ and where $s(j)$ is the number of states needed to realize $F'S_q$ and ξ (for $n=q^2$).

Q.E.D.

III. CONCLUSION

The concept of picture-compressibility defined in this paper leads to a suitable definition of entropy for an arbitrary ensemble of infinite pictures by taking the entropy to be the expected value of the compressibility. In addition to such results as the coding theorem, the generalized Kraft inequality, and the converse-to-coding theorem, it can be readily verified that the proposed concept of picture compressibility is invariant under both translation and rotation.

It is also important to note the significance of the result of Lemma 1. It shows that the compressibility of any picture is lower-bounded by that of the sequence derived from it under the standard scan. This, together with Theorem 2, means that, in the long run, the finite version of the standard scan is as good, or better, than any other finitely-implementable scheme. As mentioned earlier, in the context of this paper, the important property of SS is that it covers arrays of size $2^k \times 2^k$, $k \geq 1$, quadrant-by-quadrant, never leaving a quadrant before visiting each of its cells. Any other scan with this property could perform as well as SS. (See Fig. 3).

Finally, we would like to point out that all the results of this paper are readily extendable to data arrays of any finite dimension.

Acknowledgement

The proof of Lemma 3 has been derived jointly with Mrs. Dafna Sheinwald who pointed out an error in an earlier version of the proof.

References

[1] J. Ziv and A. Lempel, "Compression of Individual Sequences via Variable-Rate Coding", IEEE Trans. Inform. Theory, Vol. IT-24, pp. 530-536, Sept. 1978.

[2] S. Even, "Generalized Automata and their Information Losslessness", Switching Circuit Theory and Logical Design, AIEE Special Publ., S-141, pp. 144-147, 1961.

[3] B.B. Mandelbrot, Fractals: Form, Chance, and Dimension, San Francisco: W.H. Freeman and Co., 1977.

[4] M. Gardner, "Mathematical Games", Sc. Am., pp. 124-133, Dec. 1976.

[5] R.G. Gallager, Information Theory and Reliable Communication, New York: Wiley, 1968.

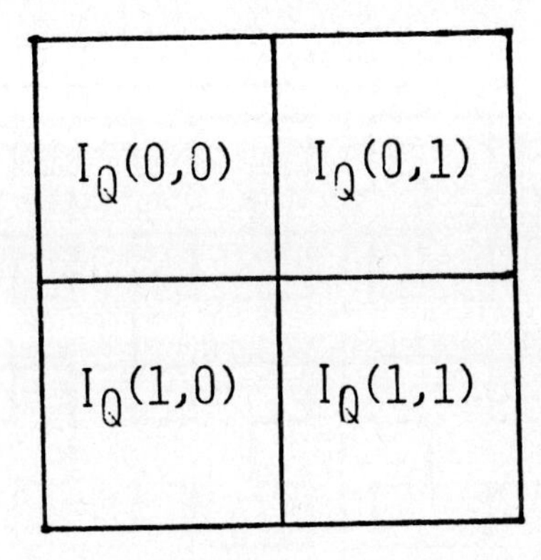

FIG. 1: PARTITION OF I_Q INTO QUANDRANTS.

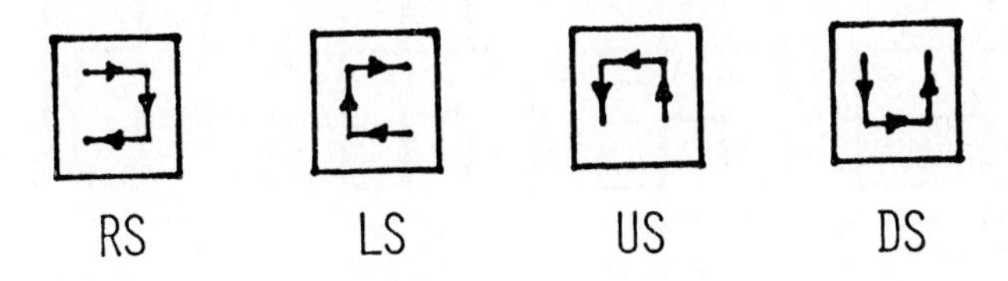

FIG. 2: BASIC SCANS FOR K=1.

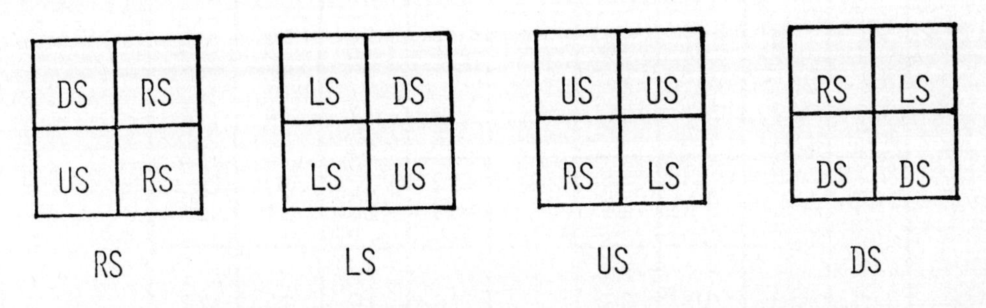

FIG. 3: RECURSIVE DEFINITION OF THE BASIC SCANS.

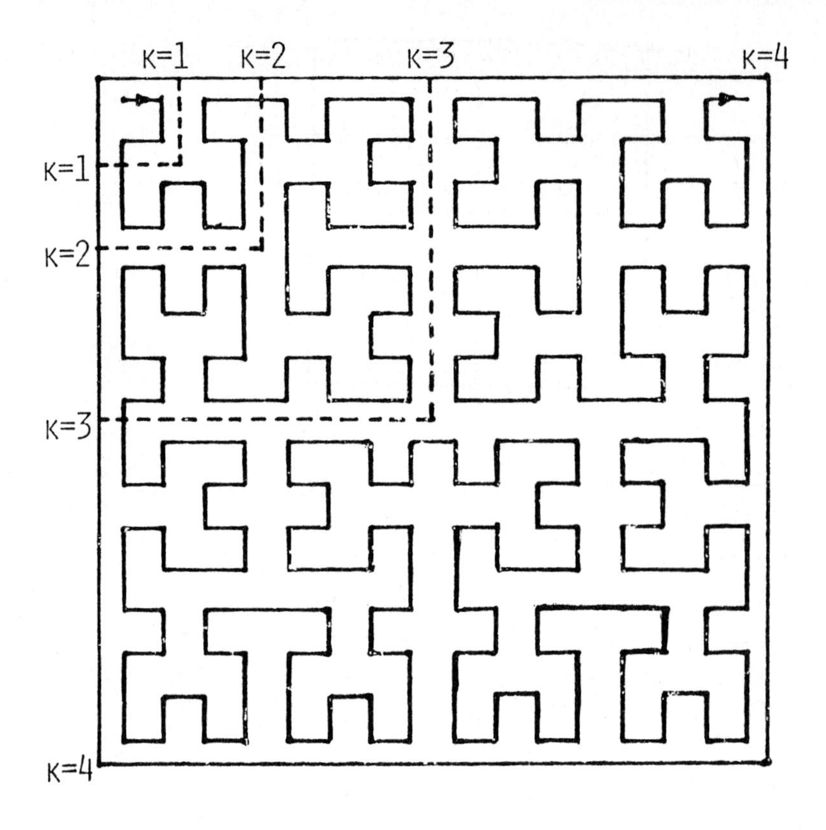

FIG. 4: THE STANDARD SCAN FOR I_{2K} , K=1,2,3,4.

1	8	9	16
2	7	10	15
3	6	11	14
4	5	12	13

(A)

1	2	3	4
8	7	6	5
9	10	11	12
16	15	14	13

(B)

FIG. 5: THE ORDERING OF $2^J \times 2^J$ BLOCKS IN THE FINITE SCAN FOR ODD J (A) AND FOR EVEN J (B).

OPTIMAL PARSING OF STRINGS

Alan Hartman and Michael Rodeh
IBM Israel Scientific Center
Technion City, Haifa 32000
Israel

1. INTRODUCTION

Parsing of strings has been suggested by Lempel and Ziv, as a tool to define the complexity of strings [LZ1] and to compress them [ZL1, ZL2]. Recently, they have shown that the scheme may be modified to handle two dimensional data [LZ2].

Strings may be parsed either with respect to themselves — leading to universal data compression algorithms [ZL1, ZL2, RPE] — or with respect to some reference dictionary. In this note we discuss both approaches: In Section 2 a real time algorithm is given which optimally parses an input string with respect to a reference tree. A non-on-line algorithm to generate the same parsing has been developed by Wagner [Wa]. In Section 3 Ziv and Lempel's algorithm [ZL2] is combined with McCreight's suffix tree construction algorithm to yield an online optimal parsing of strings relative to themselves.

The parsing technique described in [ZL1] has been implemented by Rodeh et al. [RPE] where an on-line linear algorithm has been proposed. That algorithm can be modified to yield an optimal parsing algorithm which is based on [LZ1], and which processes the input string in an on-line fashion. A non-on-line optimal parsing algorithm which gives the same result has been suggested by Storer and Szymanski [SS]. That algorithm processes the input string from right to left and parses it optimally in a way which is similar to Wagner's algorithm [Wa].

2. OPTIMAL PARSING OF STRINGS RELATIVE TO A REFERENCE TREE

Let $x = x_1, x_2 \ldots x_n$ be a string over the alphabet A and let T be a tree rooted at r with edges labelled by elements of A. Moreover, assume that for every element a of A and every vertex v of T at most one edge emanating from v is labelled a. For every vertex v of T let $\bar{v}$ denote the string spelled by the path from r to v. We shall sometimes identify v and $\bar{v}$.

NATO ASI Series, Vol. F12
Combinatorial Algorithms on Words
Edited by A. Apostolico and Z. Galil

An *optimal parsing* of x relative to T is the process of inserting a minimal number of commas into x such that every substring between two consecutive commas is a vertex of T.

For each v of T and each element a in A, let $child(v,a)$ denote the vertex of T connected to v by an edge labelled a; $child(v,a)$ is undefined if no such vertex exists in T.

To ensure that every string over A has some parsing relative to T we assume that $child(r,a)$ is defined for every a in A.

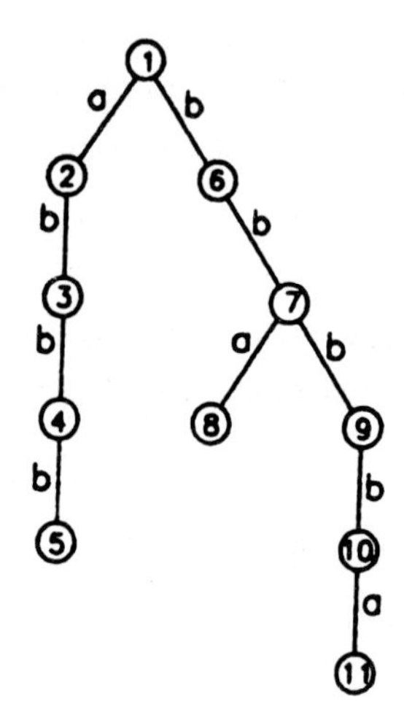

Figure 1. A tree T over $A = \{a,b\}$

<u>Example 1:</u>

A string x	$a\ b\ b\ b\ b\ b\ a$
A non optimal parsing	$a\ b\ b,\ b\ b\ b,\ a$
An optimal parsing	$a\ b,\ b\ b\ b\ b\ a$
Another optimal parsing	$a\ b\ b\ b,\ b\ b\ a$

□

Let $span_i$ denote the largest value of j such that $x_i \ldots x_j$ is a vertex of T.

<u>Example 2:</u>

	1	2	3	4	5	6	7
x_i	a	b	b	b	b	b	a
$span_i$	4	5	7	6	7	6	7

□

To find an optimal parsing notice that if a comma has been placed at position i (separating x_{i-1} from x_i) then the next comma may be placed in any location k, $(i < k \leq span_i + 1)$ for which $span_k$ is maximum. Let $next_i$ be the smallest such value of k. If $span_i = n$ then $next_i = \infty$.

Example 3:

	1	2	3	4	5	6	7
x_i	a	b	b	b	b	b	a
$next_i$	3	3	∞	5	∞	7	∞

□

Thus, one way to find an optimal parsing is to construct a sequence $c_1, c_2, \ldots, c_{m+1}$ so that $c_1 = 1$, $c_{i+1} = next_{c_i}$ and $c_{m+1} = \infty$. The complexity of x relative to T is m.

Example 4: For the string in the above examples, $c_1 = 1$, $c_2 = 3$, $c_3 = \infty$. Therefore, the complexity of x relative to T is 2.

□

2.1 Preprocessing the Reference Tree

A *suffix* of an non-empty string $a_1 a_2 \ldots a_k$ is a (possibly empty) string of the form $a_i a_{i+1} \ldots a_k$, where $i > 1$.

Let $v = a_1 \ldots a_k$ be a vertex of T, other than the root. The *suffix link* $sl(v,a)$ of v and a is the longest suffix $\bar{w}$ of $\bar{v}$, such that w is in T and $child(w,a)$ is defined.

Our definition of suffix links is slightly different from that of Weiner [We] and McCreight [M].

To construct the suffix links we process T in a breadth first search order. For each vertex v at level 1 $sl(v,a) = r$, for every element a of A. Assume that $sl(v,\cdot)$ has been defined for all the vertices v at level smaller than k, and let $u = child(v,a)$ be a vertex at level k. Let $w = child(sl(v,a),a)$. To define $sl(u,b)$ for each $b \in A$, use the following:

$$sl(u,b) = \begin{cases} w \text{ if } child(w,b) \text{ is defined.} \\ sl(w,b) \text{ otherwise.} \end{cases}$$

Since w is at level smaller than k, we have completed the construction of all suffix links. Fast access to the *child* information as well as to the old values of sl (e.g. by storing them in an array) leads to an $O(|A||V|)$ algorithm, where V is the set of vertices of T.

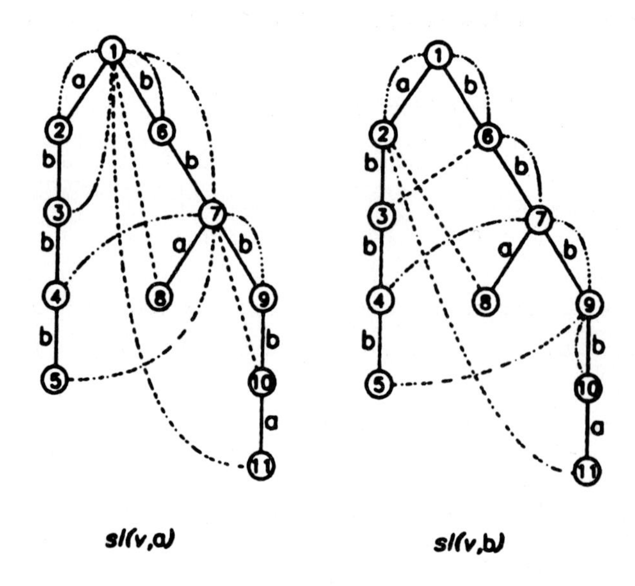

Figure 2. The tree *T* of Figure 1 with the various suffix links

2.2 A Real Time Algorithm

In this section we present a real-time algorithm to generate an optimal parsing of the input string relative to T. The algorithm uses the suffix links the construction of which is described in Section 2.1. The algorithm also uses $|\bar{v}|$ — the length of the path from the root of T to v.

The algorithm produces a sequence of integers each one of which is the number of characters between two consecutive commas. The length of the sequence is the smallest possible.

Algorithm 1: (to find an optimal parsing relative to a tree)

```
(compute span_1)
    v:=r;
    while the input is not empty do;
        a:= input character;
        if child(v,a) is undefined then go to L;
        v:=child(v,a);
        end;
    output (|v̄|);
    stop;
(compute next_i)
    L: c:=|v̄|; d:=0;
    while d ≤ c do;
        if d+|v̄| - |sl(v,a)‾| > c then do;
            output (d) ;
            go to L;
            end;
        d:=d+|v̄| - |sl(v,a)‾|;
        v:=sl(v,a);
        while child(v,a) is defined do;
            v:=child(v,a);
            if the input is empty then do;
                output (d);
                stop;
                end;
            a:=input character;
            end;
        end;
```

2.3 Application to Data Compression

Let T be a fixed reference tree. A parsing of a string x relative to T may be represented as a sequence $v_1 v_2 \ldots v_m$ where $x = \bar{v}_1 \bar{v}_2 \ldots \bar{v}_m$. To obtain this alternative representation we can follow the steps of Algorithm 1: when the position of the i-th comma is found we may initiate another scan of the tree which spells the characters between the $i-1$st and the i-th comma. The last vertex visited is v_i. Notice that buffering is needed here to be able to rescan the segment of the input bounded by the

$i-1$st and the i-th commas.

Example 5: The string *abbbbba* may be expressed as 5, 8 relative to the tree T of Figure 1.

□

Let $C_T(x)$ be the number of commas in an optimal parsing of x relative to T. Then x may be represented by $C_T(x) \cdot \lceil \log_2 |V| \rceil$ bits. Thus, if the paths of T "resemble" substrings of x, this will lead to a compact representation of x.

3. OPTIMAL SELF PARSING OF STRINGS

Ziv and Lempel [ZL2] suggested the following parsing scheme of a string $x = x_1 x_2 \ldots x_n$ into substrings $x = s_1 \ldots s_m$:

(i) $s_1 = x_1$

(ii) Assume that $s_1 s_2 \ldots s_{i-1} = x_1 x_2 \ldots x_p$ has been defined. s_i is defined to be the longest prefix of $x_{p+1} \ldots x_n$ which has the form $s_j a$ for some $j < i$ and $a \in A$.

Example 6:	A string x:	$a \; a \; b \; a \; b \; b \; a \; b \; b \; a \; b \; b \; a$
	Ziv and Lempel parsing:	$a, \; a \; b, \; a \; b \; b, \; a \; b \; b \; a, \; b, \; b \; a$

□

Ziv and Lempel's scheme is easy to implement by building a tree which contains s_i as vertices. Let T_i be the tree which contains s_j for all $j \leq i$.

Example 7: In Figure 3 we see how Ziv and Lempel's tree for the string in Example 6 is built up.

□

3.1 Optimality Considerations

During the build up of T_i, s_i is added to T_{i-1} by way of adding exactly one new vertex to T_{i-1}. Similarly, T_{i+1} differs from T_i by a single vertex. Therefore, the cost of processing $|s_i| + |s_{i+1}|$ characters in terms of the expansion of the tree, is the addition of two new vertices. Alternatively, we could try to process as many characters per new vertex as possible.

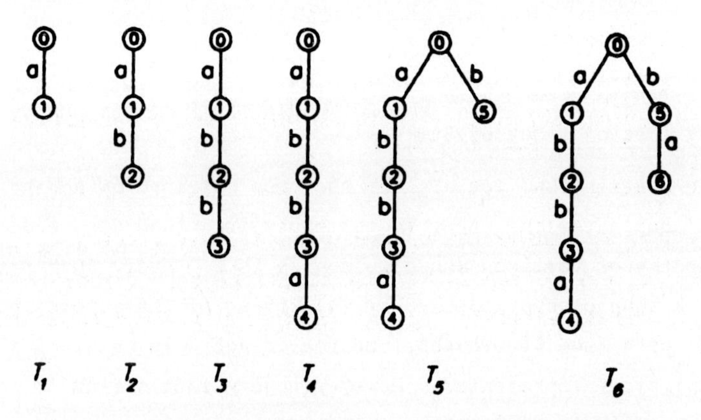

Figure 3. The evolution of Ziv and Lempel's tree for x = aababbabbabba

Example 8: Consider the string of Example 6 and assume that T_3 has already been built (see Figure 3). Instead of adding $s_4 = abba$ and $s_5 = b$ we could use $s_4' = abb$ and $s_5' = abba$ and end up with 5 commas instead of 6.

□

With respect to the tree build up, two variants should be investigated:

Variant 1: Maintain Ziv and Lempel's tree in a way which is identical to their original construction, but parse optimally with respect to its current value.

Variant 2: When the optimal parsing generates a substring s_i such that $s_i \in T_{i-1}$ then $T_i = T_{i-1}$. Otherwise T_i is obtained from T_{i-1} by adding s_i to it.

Example 9: For the string in Example 6, Variant 1 would yield T_5 as in Figure 3, while T_5 of Variant 2 is identical to T_4 of Figure 3.

□

It is clear that Variant 1 cannot yield a longer parsing than Ziv and Lempel's algorithm, while in many cases it may generate a smaller number of commas. The properties of Variant 2 are much less clear and require more study of both its information theoretic aspects and its practical properties.

As for the implementation of the two variants they may both be considered in the same framework: Let T_i represent $s_1, \ldots, s_i$ and let y be the suffix of the input string x which has not been processed yet. We have to decide where to locate the $i+1$st comma. Notice that this problem is similar to the problem discussed in Section 2. The additional difficulty here is that the tree T is not known before hand, and that it is dynamic. To resolve the issue, we shall use McCreight's suffix tree construction algorithm [M].

3.2 The Suffix Tree of a Set of Strings

Let s be a string and let T_s be the tree obtained by adding all the suffixes of s one after the other to the empty tree (see Figure 4a). T_s may have long paths of vertices whose outdegree is 1. Every such path may be replaced by a single edge labelled by the string of characters spelled by the original path (see Figure 4b). No real benefit is obtained if these strings are explicitly represented. However, every such string t may be represented by a pair $(p,|t|)$ so that $t = s_p, s_{p+1} \ldots s_{p+|t|-1}$ (see Figure 4c). The resultant tree is called the suffix tree of s.

The suffix tree of a string s may be built in linear time [M,We] McCreight's suffix tree construction algorithm uses *suffix links* which are slightly different from those of Section 2. A suffix link $sl(v)$ of a vertex v is the vertex w such that $\bar{w} = \bar{v}_2 \ldots \bar{v}_{|\bar{v}|}$. An interesting property of suffix trees is that the suffix link of all their vertices other than the root is always defined. (Notice that this property does not hold for arbitrary trees.)

Although the suffix links are usually considered as an implementation detail and are discarded when the algorithm terminates, they are important in our application. In the sequel we shall assume that the suffix links of all the vertices of a suffix tree are an integral part of the suffix tree itself (see Figure 4d).

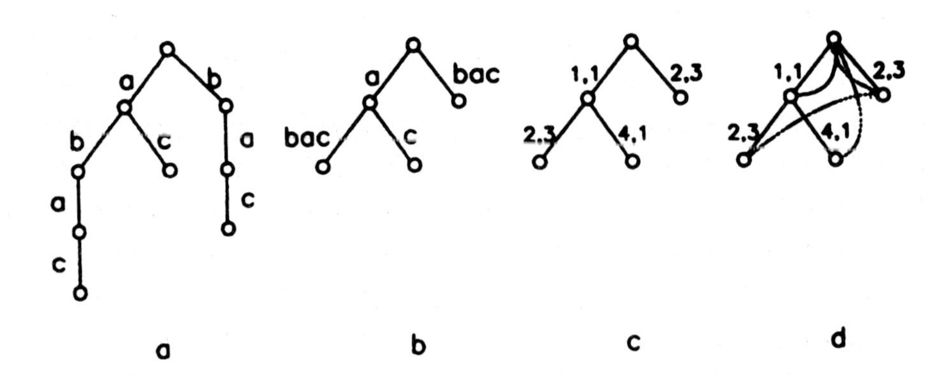

Figure 4. Various versions of the suffix tree of x = abac (suffix links appear as dotted lines)

While suffix trees have been defined and constructed for a single string, it is a straightforward generalization to build the suffix tree of a set of strings. In the sequel, let T_i be the suffix tree of the set $\{s_1, s_2 \ldots s_i\}$ (see Figure 5).

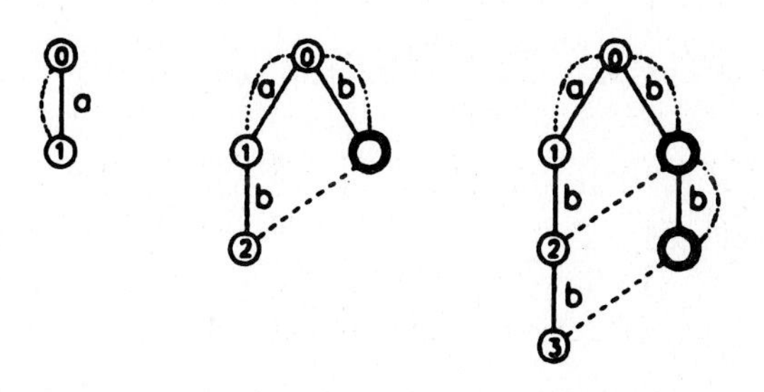

Figure 5. T_1, T_2 and T_3 of x=aababbabbabba

In Figure 5 we have overlayed the suffix trees T_1, T_2 and T_3 and Ziv and Lempel's trees (see Figure 3). Vertices which belong to Ziv and Lempel's tree are called *essential* and are drawn as single circles. The others are called *auxiliary vertices*, and are doubly circled.

An important property of T_i is that it is identical to its non-compact version. This follows from the special way in which the set $\{s_1, \ldots, s_i\}$ is constructed.

3.3 An Algorithm for Self Parsing

Let T_i be the suffix tree of the strings $s_1, s_2, \ldots, s_i$. Also, assume that the next characters of the input form a string $\bar{v}$ for some $v \in T_i$ and that *child(v,a)* is undefined for the character a which is about to be read. The algorithm below uses v as an input parameter. The interrelationships between the variables employed by the algorithm are presented pictorially in Figure 6.

The variable $\bar{w}$ may be conceived as a variable length ruler one end point of which moves to the right while the other one stays in place. For this reason the algorithm is linear.

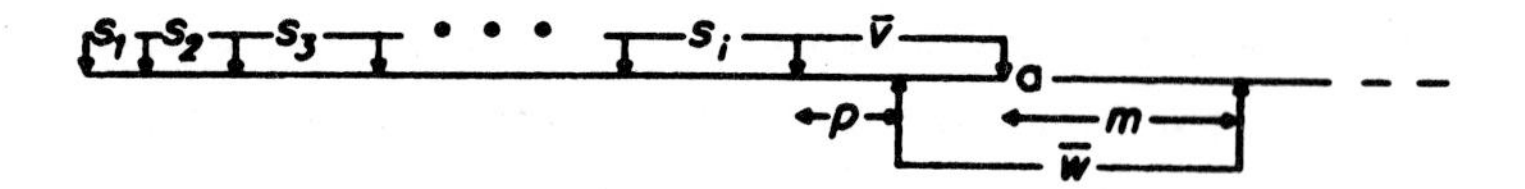

Figure 6. The interrelationships between the variables of Algorithm 2

Algorithm 2: (To find the optimal location of the next comma)

```
a:=input character;
(The assumption here is that the input is not empty, and
child(v,a)  is undefined)
m:=-1;
p:=0;
w:=v;
while |w̄| ≥ m do;
      if child(w,a) is essential then do;
         w:=child(w,a);
          if the input is empty then go to L;
          a:=input character;
          m:=m+1;
          p:=|v̄|+m-|w̄|+1;
          end;
      end do;
          w:=sl(w);
          while  w is an auxiliary vertex and |w̄| ≥ m do
                w:=sl(w);
          end;
      end;
L: output(p);
```

Example 10: Referring to Figure 5, assume that $s_1 = a$, $s_2 = ab$ and $s_3 = abb$. The suffix tree representing these substrings is the T_3 shown in Figure 5. The next characters of the input are *abbabba*. Among these, the first three lead to the vertex $v=3$ (see Figure 3). Thus, we have to process the suffix *abba* of the input string.

To start, the algorithm sets $m:=-1$, $p:=0$ and $w:=v$. Since $|\bar{w}| = 3 \geq -1$ and $child(w,a)$ is undefined we set $w:=sl(3)$. The new vertex w is an auxiliary vertex and we apply sl twice more before ending with an essential w, which in our case is equal to the root $(w=0)$. At this point

$|\bar{w}| = 0 \geq m = -1$ and *child(w,a)* is essential. Therefore, $w:=1$, $m:=0$ and $p:=3$. Since *child(w,b)* is essential, $w:=2$, $m:=1$ and $p:=3$. Again, *child(w,b)* is essential and therefore, $w:=3$, $m:=2$ and $p:=3$. Now *child(w,a)* is not defined and therefore *sl* is applied to w to yield the new value of w which is an auxiliary vertex. *sl* is applied again and causes $|\bar{w}| < m$ to hold. The algorithm terminates with $p = 3$. Therefore, s_4 has been found to be *abb* (contrary to Ziv and Lempel's value of s_4 which is *abba*).

Consecutive values of $\bar{w}$ are shown in Figure 7.

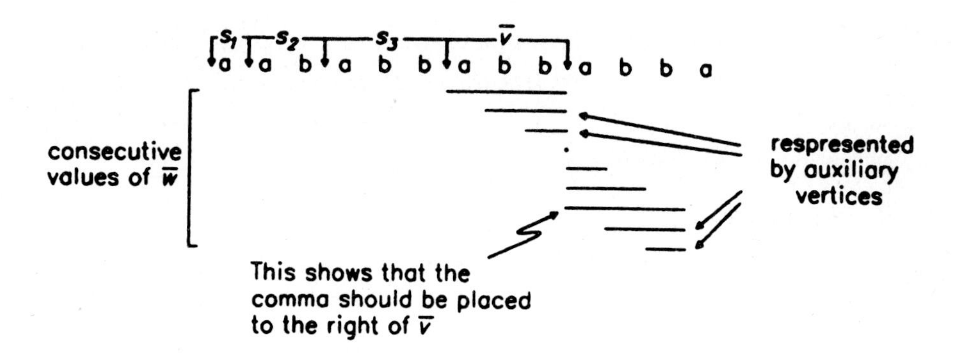

Figure 7. Consecutive values of $\bar{w}$

□

3.4 Supporting the Two Variants of Tree Build-Up

To support Variant 1, we use McCreight's algorithm to add the new vertex to T_i, and use Algorithm 2 to optimally position the comma.

To support Variant 2, we invoke McCreight's algorithm only when the optimal location of the comma is identical to that of Ziv and Lempel's algorithm.

3.5 Real Time Considerations

Algorithm 2 may be modified slightly to yield the value of v for the iteration. However, when the optimal position of the next comma has been found, we may want to emit the corresponding vertex of T_i — an operation which requires rescanning and buffering (see Section 2.3).

A more difficult problem is that of iterative application of the sl operator, without reading more characters. There may be as many as $|\bar{v}|$ operations and real time is not achieved.

4. CONCLUSIONS

An improvement to Ziv and Lempel's parsing technique has been suggested by way of a linear algorithm which is based on McCreight's algorithm. The suggested algorithm does not have a real time behavior which may be required in certain applications.

A real time algorithm has been suggested for data compression with respect to arbitrary trees. The preprocessing of the reference tree depends on the size of the alphabet. It is not clear to us whether the dependency on the alphabet size is essential.

REFERENCES

[LZ1] A. Lempel and J. Ziv, "On the Complexity of Finite Sequences", IEEE Trans. Inf. Theory, IT-22 (Jan. 1976), 75-81.

[LZ2] A. Lempel and J. Ziv, "Compression of Two Dimensional Data", submitted to IEEE Trans. Inf. Theory.

[M] E.M. McCreight, "A Space-Economical Suffix Tree Construction Algorithm", J. ACM 23, 2 (April 1976), 262-272.

[RPE] M. Rodeh, V.R. Pratt and S. Even, "Linear Algorithm, for Data Compression Via String Matching", J. ACM 28, 1 (Jan. 1981), 16-24.

[SS] J.A. Storer and T.G. Szymanski, "Data Compression Via Textual Substitution", J. ACM 29, 4 (October 1982), 928-951.

[Wa] R.A. Wagner, "Common Phrases and Minimum-Space Text Storage", CACM 16, 3 (March 1973), 148-152.

[We] P. Weiner, "Linear Pattern Matching Algorithm", IEEE 14th Annual Symposium on Switching and Automata Theory, (October 1973), 1-11.

[ZL1] J. Ziv and A. Lempel, "A Universal Algorithm for Sequential Data Compression," IEEE Trans. Inf. Theory, IT-23 (May 1977), 337-343.

[ZL2] J. Ziv and A. Lempel, "A Compression of Individual Sequences Via Variable-Rate Coding", IEEE Trans. Inf. Theory, IT-24 (Sept. 1978), 530-536.

Novel Compression of Sparse Bit-Strings — Preliminary Report [1]

Aviezri S. Fraenkel[2] and Shmuel T. Klein[3]

ABSTRACT

New methods for the compression of large sparse binary strings are presented. They are based on various new numeration systems in which the lengths of zero-block runs are represented. The basis elements of these systems, together with the non-zero blocks, are assigned Huffman codes. Experiments run on bit-maps of the Responsa Retrieval Project, and for comparison on randomly generated maps and on a digitized picture, yield compressions superior to previously known methods.

1. Introduction and Motivation

In some full-text retrieval systems, an inverted file, also called "concordance", is constructed that contains, for every different word W in the data base, a list $L(W)$ of all the documents in which W occurs. In order to find all the documents that contain the words A and B, one has to intersect $L(A)$ with $L(B)$. Usually, A and B stand for families of words A_i and B_j, each family consisting of terms which are considered synonymous for the given query. In this case one has to perform

$$\left(\bigcup_i L(A_i)\right) \cap \left(\bigcup_j L(B_j)\right)$$

which can seriously affect the response time in an online retrieval system when the number of involved sets and their sizes are large.

A different approach might be to replace the concordance of a system with l documents by a set of *bitmaps* of equal length l. For every different word W in the data base a bitmap $B(W)$ is constructed, the i-th bit of which is 1 if and only if W occurs in the i-th document. Processing queries reduces here to performing logical operations with bit-strings, which is easily done on most machines. It would be wasteful to store the bitmaps in their original form, since they are usually very sparse (the great majority of the words occur in very few documents). Beside the savings in secondary storage space, the processing time can be improved by compressing the maps, as more information can be read in a single input

[1] Supported in part by Canadian Research Grant NSERC 69-0695.

[2] Department of Mathematics and Statistics, The University of Calgary, Alberta, Canada. Permanent address: Department of Applied Mathematics, The Weizmann Institute of Science, Rehovot, Israel.

[3] Department of Applied Mathematics, The Weizmann Institute of Science, Rehovot, Israel.

NATO ASI Series, Vol. F12
Combinatorial Algorithms on Words
Edited by A. Apostolico and Z. Galil

operation, thus reducing the total number of I/O accesses. We are interested in a coding procedure which reduces the space needed to store long sparse binary strings without losing any information; in fact, there should be a simple algorithm which, given the compressed string, can reconstruct the original one.

In the next section, some known methods for bit-string compression are rewieved and new methods are proposed. The basic idea of the new methods is to use Huffman codes to represent items of different nature such as ***bit patterns*** of constant size on one hand and ***lengths*** of 0-bit-runs on the other. These lengths are chosen in a systematic way, using new exotic numeration systems. Some of the methods consistently yield compression factors* superior to the previously known ones. Section 3 then describes the experiments in which all these methods were applied to different files of bit-maps.

2. The Methods

The first three methods have been reported on before and are described here succinctly by way of introduction. The remaining methods are new. Each method is identified by a label which appears between brackets following the explanations, together with a number referring to the corresponding line in the appended Table 5, which displays the experimental results.

2.1 Methods Without 0-Run-Length Coding

(a) The following two methods were recently developed at the Responsa Retrieval Project (see for example ⟦**1**⟧ or ⟦**2**⟧): The original bit-vector v_1 of length n_1 (bits) is partitioned into k_1 blocks of equal length, and a second vector v_2 of length $n_2 = k_1$ is constructed, each bit in it being the ORing of the corresponding block in v_1; the 0-blocks of v_1 are then omitted. The procedure is recursively repeated until a level is reached, where the vector-length reduces to a few bytes [method TREE; line number 1 in Table 5].

(b) A refinement of the above approach consists of pruning some of the tree branches if they ultimately point to very few documents; the numbers of these documents are then just added to an appended list, which is compressed by the prefix-omission technique [TRECUT;2].

(c) A variant of Huffman coding was proposed in Jakobson ⟦**5**⟧. The bit-string is partitioned into k-bit blocks and is compressed according to the probabilities of occurrence of the 2^k possible block patterns [NORUN;3].

2.2 Incorporating 0-Run-Length Coding

As we are interested in sparse bit-strings, we can assume that the probability p of a block of k consecutive bits being zero is high. If $p \geq \frac{1}{2}$, method NORUN assigns to this 0-block a code of length one bit, so we can never expect a better compression factor than k. On the other hand, k cannot be too large since we must generate codes for 2^k different blocks.

---

*The *compression factor*, CF, is defined as the ratio of the size of the original file to the size of the compressed file.

In order to get a better compression, we extend the idea of method NORUN in the following way: there will be codes for the $2^k - 1$ non-zero blocks of length k, plus some additional codes representing runs of zero-blocks of different length. In the sequel, we use the term 'run' to designate a run of zero-blocks of k bits each.

The length (number of k-bit blocks) of a run can take any value up to n_1/k, so it is impractical to generate a code for each: as was just pointed out, k cannot be very large, but n_1 is large for applications of practical importance (see Sect. 3.1 below). On the other hand, using a fixed-length code for the run length would be wasteful since this code must suffice for the maximal length, while most of the runs are short. The following methods attempt to overcome these difficulties.

Definition of Classes of Run-Lenghts

Starting with a fixed-length code for the run-length, we like to get rid of the leading zeros in the binary representation $B(l)$ of run-length l, but we clearly cannot simply omit them, since this would lead to ambiguities. We *can* omit the leading zeros if we have additional information such as the position of the leftmost 1 in $B(l)$. Hence, partition the possible lengths into classes C_i, containing run-lengths l which satisfy $2^{i-1} \leq l < 2^i$, $i = 1, \ldots, \lfloor \log_2(n_1/k) \rfloor$. The $2^k - 1$ non-zero block-patterns and the classes C_i are assigned Huffman codes corresponding to the frequency of their occurrence in the file; a run of length l belonging to class C_i is encoded by the code for C_i, followed by $i - 1$ bits representing the number $l - 2^{i-1}$. For example, a run of 77 0-blocks is assigned the code for C_7 followed by the 6 bits 001101. Note that a run consisting of a single 0-block is encoded by the code for C_1, without being followed by any supplementary bits.

The Huffman decoding procedure has to be modified in the following way: The table contains for every code-word the corresponding class C_i as well as $i - 1$. Then, when the code which corresponds to class C_i is identified, the next $i - 1$ bits are considered as the binary representation of an integer m. The code for C_i followed by those $i - 1$ bits represent together a run of length $m + 2^{i-1}$; the decoding according to Huffman's procedure resumes at the i-th bit following the code for C_i [LLRUN;4].

Representing the Run-Length in some Numeration System

Method LLRUN seems to be efficient since the number of bits in the binary representation of integers is reduced to a minimum and the lengths of the codes are optimized by Huffman's algorithm. But encoding and decoding are admittedly complicated and thus time consuming. We therefore propose other methods for which the coded file will consist only of codewords, each representing a certain string of bits. Even if their compression factor is lower than LLRUN's, these methods are justified by their simpler processing.

To the $2^k - 1$ codes for non-zero blocks, a set S of t codes is adjoined representing $l_0, l_1, \ldots, l_{t-1}$ consecutive 0-blocks. Any run of zero-blocks will now be encoded by a suitable linear combination of some of these codes. The number t depends on the numeration system according to which we choose the l_i's and on the maximal run-length M, but should be low compared to 2^k. Thus in comparison with method NORUN, the table used for compressing and decoding should only slightly increase in size, but long runs are handled more efficiently. The encoding algorithm now becomes:

Step 1: Collect statistics on the distribution of run-lengths and on the set NZ of the $2^k - 1$ possible non-zero blocks. The total number of occurrences of these blocks is denoted by N_0 and is fixed for a given set of bit-maps.

Step 2: Decompose the integers representing the run-lengths in the numeration system with set S of "basis" elements; denote by TNO(S) the total number of occurrences of the elements of S.

Step 3: Evaluate the relative frequency of appearance of the $2^k - 1 + t$ elements of $\mathrm{NZ} \cup S$ and assign Huffman codes accordingly.

For any $x \in (\mathrm{NZ} \cup S)$, let $p(x)$ be the probability of the occurrence of x and $l(x)$ the length (in bits) of the code assigned to x by the Huffman algorithm. The weighted average length of a codeword is then given by $\mathrm{AL}(S) = \sum_{x \in (\mathrm{NZ} \cup S)} p(x)l(x)$ and the size of the compressed file is $\mathrm{AL}(S) \times (N_0 + \mathrm{TNO}(S))$. After fixing k so as to allow easy processing of k-bit blocks, the only parameter in the algorithm is the set S. In what follows, we propose several possible choices for the set $S = \{1 = l_0 < l_1 < \ldots < l_{t-1}\}$. To overcome coding problems, the l_i and the bounds on the associated digits a_i should be so that there is a unique representation of the form $L = \sum_i a_i l_i$ for every natural number L.

Given such a set S, the representation of an integer L is obtained by the following simple procedure:

```
for i ← t − 1 to 0 by − 1
    a_i ← ⌊L/l_i⌋
    L ← L − a_i × l_i
end
```

The digit a_i is the number of times the code for l_i is repeated. This algorithm produces a representation $L = \sum_{i=0}^{t-1} a_i l_i$ which satisfies

$$\sum_{i=0}^{j} a_i l_i < l_{j+1} \qquad \text{for} \quad j = 0, \ldots, t-1. \tag{1}$$

Condition (1) guarantees uniqueness of representation (see [[3]]).

Selection of the Set of Basis Elements

A natural choice for S is the standard binary system, $l_i = 2^i$, $i \geq 0$, or higher base numeration systems such as $l_i = n^i$, $i \geq 0$ for $n > 2$. If the run-length is L it will be expressed as $L = \sum_i a_i n^i$, with $0 \leq a_i < n$ and if $a_i > 0$, the code for n^i will be repeated a_i times. Higher base systems can be motivated by the following reason.

If p is the probability that a k-bit block consists only of zeros, then the probability of a run of m blocks is roughly $p^m(1-p)$, i.e. the run-lengths have approximately geometric distribution. The distribution is not exactly geometric since the involved events (some adjacent blocks contain only zeros, i.e. a certain word does not appear in some consecutive documents) are not independent. Nevertheless the experiments showed that the number of runs of a given length is an exponentially decreasing function of run-length (see Figure

2 below). Hence with increasing base of the numeration systems, the relative weight of the l_i for small i will rise, which yields a less uniform distribution for the elements of $\mathrm{NZ} \cup S$ calculated in step 3. This has a tendency to improve the compression obtained by the Huffman codes. Therefore passing to higher order numeration systems will reduce the value of AL(S).

On the other hand, when numeration systems to base n are used, TNO(S) is an increasing function of n. Define m by $n^m \leq M < n^{m+1}$ so that at most m n-ary digits are required to express a run-length. If the lengths are uniformly distributed, the average number of basis elements needed (counting multiplicities) is proportional to $(n-1)m = (n-1)\log_n M$, which is increasing for $n > 1$. For our nearly geometric distribution this is also the case as can be seen from the experiments. Thus from this point of view, lower base numeration systems are preferable. Numeration systems to base n were checked for $n = 2, \ldots, 10$ [POW2,..,POW10; 5–13]. For $n = 3$, we experimented with another variant: instead of using certain codes twice where a digit 2 is needed in the ternary representation, we added a special code indicating that the following code has to be doubled [POW3M; 14].

Trying to Reduce the Total Number of Codes

As an attempt to reduce TNO(S), we pass to numeration systems with special properties, such as systems based on Fibonacci numbers $F_0 = 0$, $F_1 = 1$, $F_i = F_{i-1} + F_{i-2}$ for $i \geq 2$.

(a) The binary Fibonacci numeration system: $l_i = F_{i+2}$ [FIBBIN; 15]. Any integer L can be expressed as $L = \sum_{i \geq 0} b_i F_{i+2}$ with $b_i = 0$ or 1, such that this binary representation of L consisting of the string of b_i's contains no adjacent 1's (see Knuth [[**6**]], Fraenkel [[**3**]]). This fact for a binary Fibonacci system is equivalent to condition (1), and reduces the number of codes we need to represent a specific run-length, even though the number of added codes is larger than for POW2 (instead of $t(\mathrm{POW2}) = \lfloor \log_2 M \rfloor$ we have $t(\mathrm{FIBBIN}) = \lfloor \log_\phi(\sqrt{5}M) \rfloor - 1$, where $\phi = (1+\sqrt{5})/2$ is the golden ratio). For example, when all the run-lengths are equally probable, the average number of codes per run is assymptotically (as $k \to \infty$) $\frac{1}{2}(1 - 1/\sqrt{5})t(\mathrm{FIBBIN})$ instead of $\frac{1}{2}t(\mathrm{POW2})$.

(b) A ternary Fibonacci numeration system: $l_i = F_{2(i+1)}$, i.e. we use only Fibonacci numbers with even indices. This system has the property that there is at least one 0 between any two 2's ([[**3**]]). This fact for a ternary Fibonacci system is again equivalent to (1) [FIBTER; 16].

(c) Like **(b)**, but with a special code for the digit 2 like in POW3M [FIBTERM; 17].

Trying to Reduce the Average Code Length

New systems with similar properties are obtained from generalizations of the Fibonacci systems to higher order. The idea is to lower AL(S) at the cost of TNO(S), while TNO(S) is kept smaller than for the POWn systems.

(a) Methods which are generalizations of method FIBBIN, based on the following sequence of integers which is defined recursively:

$$\begin{aligned} u_{-1}^{(n)} &= 1, \qquad u_0^{(n)} = 1 \\ u_i^{(n)} &= n u_{i-1}^{(n)} + u_{i-2}^{(n)}, \quad i \geq 1, \end{aligned} \tag{2}$$

(This is the recursion satisfied by the convergents of the continued fraction $[1,\dot{n}]$, where $\dot{n}$ represents the infinite concatenation of n with itself.) For $n = 1$ we get FIBBIN. The system based on the sequence $u_0^{(n)}, u_1^{(n)}, \ldots$ is an $(n+1)$-ary numeration system, i.e. there exists a unique representation of any integer L as $L = \sum_i a_i u_i^{(n)}$, such that $0 \leq a_i \leq n$, and such that if a_{i+1} reaches its maximal value n, then a_i is zero. These methods were checked for $n = 2, \ldots, 10$ [REC2,..,REC10; 18–26].

(b) Methods which are generalizations of method FIBTER, based on the sequence:

$$\begin{aligned} u_{-1}^{(ab)} &= 1 - \epsilon a, \qquad u_0^{(ab)} = 1 \\ u_i^{(ab)} &= (ab+2)\, u_{i-1}^{(ab)} - u_{i-2}^{(ab)}, \quad i \geq 1, \end{aligned} \tag{3}$$

where a,b and ϵ are parameters. This is the recursion satisfied by the convergents p_{2i}/q_{2i} of the continued fraction $[1, b, a, b, a, \ldots]$. Specifically, (3) is the recursion for the q_{2i} when $\epsilon = 0$ and for the p_{2i} when $\epsilon = 1$. The numeration system based on the q_{2i} is denoted by AaBbQ and that based on the p_{2i} by AaBbP. Any nonnegative integer L can be represented in these numeration systems in the form $L = \sum_{i=0}^{n} c_i u_i^{(ab)}$, where

$$0 \leq c_0 \leq a(b+\epsilon), \qquad 0 \leq c_i \leq ab+1 \quad (i > 0),$$

and the following condition holds: if for some $0 \leq i < m \leq n$, c_i and c_m assume their maximal values, then there exists an index j satisfying $i < j < m$, for which $c_j < ab$. In particular for A1B1P, which is our ternary system FIBTER, we have $c_j = 0$.

From (3) follows that the numeration system AaBbQ is equivalent to the system AcBdQ if $ab = cd$. In particular, AaBbQ is equivalent to AbBaQ. We have checked the following methods, which are listed in lexicographic increasing order of their basis elements: A1B2Q [27], which is ternary in c_0, but 4-ary in c_j for every $j > 0$; A1B2P [28], a 4-ary numeration system; A1B3Q [29], A2B1P [30], A1B3P [31], A2B2Q [32], A1B4P [33], A3B1P [34], A2B2P [35] and A4B1P [36].

Name	Arity	Sequence of basis elements									
FIBBIN	2	1	2	3	5	8	13	21	34	55	89
				144	233	377	610	987	1597	2584	4181
FIBTER	3	1	3	8	21	55	144	377	987	2584	
REC2	3	1	3	7	17	41	99	239	577	1393	3363
REC3	4	1	4	13	43	142	469	1549			
REC4	5	1	5	21	89	377	1597				
REC5	6	1	6	31	161	836	4341				
REC6	7	1	7	43	265	1633					
REC7	8	1	8	57	407	2906					
REC8	9	1	9	73	593	4817					
REC9	10	1	10	91	829						
REC10	11	1	11	111	1121						
A1B2Q	4,3	1	3	11	41	153	571	2131			
A1B2P	4	1	4	15	56	209	780	2911			
A1B3Q	5,4	1	4	19	91	436	2089				
A2B1P	4,5	1	5	19	71	265	989	3691			
A1B3P	5	1	5	24	115	551	2640				
A2B2Q	6,5	1	5	29	169	985					
A1B4P	6	1	6	35	204	1189					
A3B1P	5,7	1	7	34	163	781	3742				
A2B2P	6,7	1	7	41	239	1393					
A4B1P	6,9	1	9	53	309	1801					

Table 1: *Sequences defining the numeration systems*

Table 1 lists the sequences of the first few basis elements of the various numeration systems. A system that is l-ary in c_0 and n-ary in all other digits contains n, l in the column labeled "Arity".

2.3 Error Detection

One of the weak points of Huffman codes is their sensitivity to errors: a single wrong bit makes the code useless. In order to locally restrict the damage caused by errors, one usually adds some redundant bits so as to ensure that for some constant block-size m, the bits indexed $im + 1$, $i \geq 0$, start a new codeword. A single error can then affect at most m bits of the coded file, but for small m the loss of compression may be significant. We therefore compare the error-detecting capabilities of our methods supposing that no "synchronizing" bits are added and restrict ourselves to the case when a single error occurred.

For NORUN, suppose the error occurred in x, which was the i-th codeword of the compressed file. Now the i-th codeword is interpreted as some codeword y. If x and y are of the same length, the decoding from the $(i + 1)$-st codeword on is correct and the error will not be detected. However, only a single k-bit block of the decompressed file is garbeled. If x and y are not of the same length, chances are good to reveal the existence of an error at the end of the string either because the last bits do not form a codeword or because the size of the decompressed file is not as expected.

For LLRUN, error detection can be enhanced by checking that codes for classes C_i, representing runs of zero-blocks, do not appear consecutively. This may happen after a wrong bit which transforms the following into a sequence of independent codewords.

For all the methods using codes for the basis elements $\{l_0, l_1, \ldots, l_{t-1}\}$ of a numeration system, these codes should appear for every given run-length in monotone order, e.g. by decreasing basis elements. An error will quickly be discovered by checking that the decoded run-lenghts indeed appear in decreasing order. A wrong bit would tend to mix up the codewords and there are even better chances that an error will break this rule than for LLRUN.

3. Experimental Results

All the above mentioned methods were checked on bitmaps corresponding to different words of a full-text information retrieval system. By examining the statistics of the distribution of non-zero k-bit blocks, we found that although blocks with a fixed number s of 1-bits were nearly equiprobable, the frequency decayed rapidly as s increased. One exception should be noted: as s passed from $k-1$ to k, the frequency rose. This fact can be explained by the "clustering effect": adjacent bits represent usually documents written by the same author and there is a positive correlation for a word to appear in consecutive documents because of the specific style of the author or simply because such documents often treat the same subject.

In order to evaluate the influence of this clustering effect on the compression, a test was designed on randomly generated bitmaps, where each bit has a constant (low) probability to be 1, independently of each other.

A digitized picture can also be considered as a long bit-string, and for pictures of technical drawings for example, these maps are usually sparse. The problem of image compression is not quite the same as the one treated in this paper, since often several bits can be changed without losing the general impression of the picture. Thus methods can be applied which use noise removal, edge detection, void filling, etc. in both dimensions (see Ramachandran ⟦**8**⟧) and are thus not generalizable to information retrieval bit-maps, where every single bit must be kept. Nevertheless, we wanted to see how our methods perform on digitized pictures and to compare them with other methods used for image compression. The results are summarized below.

3.1 Bit-Maps for Information Retrieval Systems

The maps correspond to the 56588 different words of the current Responsa database ⟦**1**⟧ which occur more than 70 times. There are 42272 documents, so each map is of length 5284 bytes. These maps were divided into three files of different size and density. The first file corresponds to words of length 1 to 4 characters, which are the words with highest frequency of occurrence; the second file contains words of length 5–8 and the third of length 8–13 (lowest frequency); for technical reasons, the 8-letter words were split between the two last files. Finally, the three files were unified.

For the test, the following values were chosen for the various parameters: the size of each block in method **2.1(a)** was set to 2 bytes (16 bits); since $n_1 = 42272$, we had four levels, the last 5 bits of v_4 being always 0.

For all the methods using Huffman coding, the block-size was chosen as $k = 8$, i.e. one byte. Thus codes were generated for 255 non-zero bytes and for 3 to 18 run-lengths, depending on the numeration system chosen. For LLRUN, we had classes C_1 to C_{13}.

3.2 Random Bit-Maps

To allow comparison, the random maps were constructed with the same length (5284 bytes) and with a density-distribution similar to that of the Responsa maps. Toward this end the 56588 Responsa maps were partitioned into 101 classes corresponding to different ranges of the number of 1-bits in the maps and the number of generated maps was taken to be 10% of the number of the Responsa maps for each class. For each map a number K was chosen uniformly from the range, then K "random" bits were set to 1. In fact, the correct method of constructing the maps consists of deciding for every bit with a very low probability p if it should be set to 1, yielding an expected number of 1-bits equal to pn_1. We preferred the first mentioned method which is more economical to implement and introduced only a slight error, the probability of a given bit to be set to 1 being $1-(1-1/n_1)^K \simeq K/n_1 = p$, since n_1 is large and K is relatively small (there are very few maps with large K). The random numbers were generated following Knuth ⟦7⟧.

3.3 Image Compression

The picture we chose was used by several authors to test their compression methods (Hunter and Robinson ⟦4⟧, see Figure 1). After digitizing and after thresholding, a bit-map of $512 \times 512 = 262144$ bits was obtained. Figure 1 is the photocopy of this map as produced by the Versatec plotter on which this paper was originally printed. Other authors used a much higher resolution (1728×2376). Moreover even when different scanners operate at the same resolution and scan the same document, they can give significantly different statistics and compression factors (see ⟦4⟧). Since our equipment did not allow a higher resolution, we decided to compare our methods with the "Modified Huffman Codes" (MHC) proposed in ⟦4⟧, applying all the methods to our low-resolution picture.

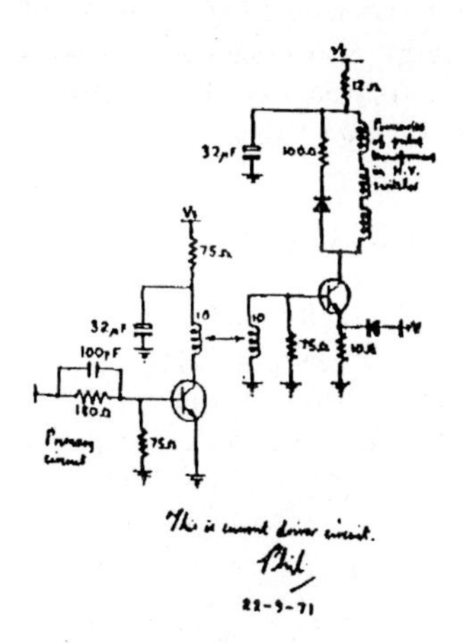

Figure 1: *Digitized picture of a technical drawing*

The latter method consists of coding runs of 0-bits. There are 64 so-called "Terminal codes" TC(i) representing a run of i zeros followed by a 1, $i = 0, \ldots, 63$, and some "Make-up" codes MU(j) standing for a run of $64j$ 0-bits, $j \geq 1$. A run of length l is encoded by TC(l) if $l < 64$, otherwise by MU(j) followed by TC(i) such that $64j + i = l$. The codes are generated using the Huffman algorithm. One could also add codes for runs of 1-bits, but for the sparse bit-strings we are interested in for our information retrieval applications, their influence will be negligible.

3.4 Results

Table 2 is a summary of statistical information on the three files of Responsa-maps, on the unified file, on the file of random maps and the picture.

Contents of the file: words of length	Number of maps	Number of non-zero bytes (N_0)	Nbr of runs	Density: avg nbr of 1-bits per map	Percentage of 1-bits per map	Average length of run (in byte)	Average number of runs per map
1–4	15378	6,091,250	3,214,770	720.15	1.70 %	23.38	209.05
5–8	36004	5,951,808	4,336,727	218.09	0.52 %	42.50	120.45
8–13	5206	518,919	427,947	117.69	0.28 %	63.08	82.20
unified (1-13)	56588	12,561,977	7,979,444	345.29	0.817%	35.90	141.01
random	5664	1,468,517	1,053,433	342.38	0.810%	27.02	185.99
picture	1	3135	1470	11398	4.35 %	20.16	1470

Table 2: *Statistics of the bit-maps*

Figure 2 displays the relative frequency (in %) of runs of a certain length as a function of run-length (number of k-bit blocks), for both the unified and the random files.

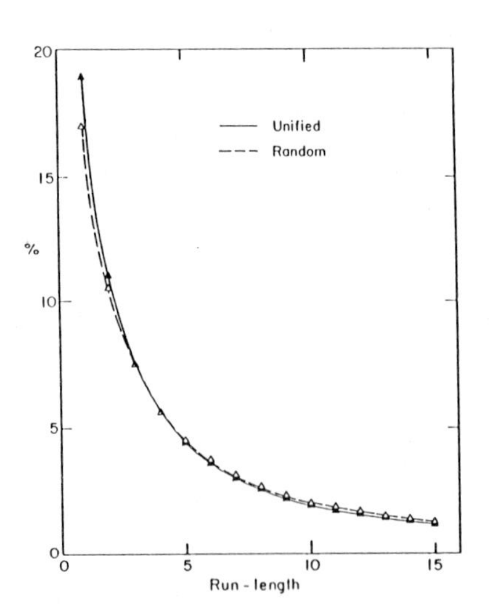

Figure 2: *Run-length distribution*

As we started to get results, we saw that the idea of adding a special code for the digit 2 in ternary systems (methods POW3M and FIBTERM) was not a good one, therefore we did not expand it to numeration systems with larger base.

As expected, LLRUN was the best method, but REC2 which was the second best on the Responsa and random files used only about 5–8% additional space. One of the striking results was that for several methods there are on the average less than 2 codes per run; if the run-lengths were uniformly distributed, the expected number for POW2 for example was 6. Table 3 is a sample for some of the good methods.

	POW2	FIBBIN	FIBTER	REC2	A1B2Q
unified	2.17	1.84	2.46	2.39	2.79
random	2.17	1.85	2.47	2.40	2.79
picture	2.22	1.70	2.29	2.20	2.38

Table 3: *Average number of codes needed per run*

Table 4 lists the first few codes (corresponding to the highest probabilities) of a method giving high compression (FIBTER), one giving a lower compression (NORUN), and a method from about the middle of the range (REC5). The probabilities correspond to the unified file of 56588 maps.

Encoded pattern	**Probability**	**Huffman code**
	Method FIBTER	
run of length 1	0.2405	10
run of length 3	0.1536	000
run of length 8	0.1025	110
run of length 21	0.0617	0111
00000001	0.0367	00100
10000000	0.0357	00101
01000000	0.0357	00110
00000010	0.0355	00111
	Method REC5	
run of length 1	0.4280	1
run of length 6	0.1859	000
run of length 31	0.0729	0011
00000001	0.0280	01001
10000000	0.0273	01010
01000000	0.0272	01011
00000010	0.0271	01100
	Method NORUN	
00000000	0.9580	0
00000001	0.0039	1100
10000000	0.00385	1101
01000000	0.00384	1110
00000010	0.00382	1111
00100000	0.00380	10000
00010000	0.00377	10001

Table 4: *Sample of generated codes*

The appended Table 5 gives the results of the experiments on the unified, the random and the picture files. The methods are listed by order of their appearance in the explanations. For LLRUN, the average length is calculated as follows: let $p(\mathrm{NZ}_i)$ denote the probability of occurrence of the i-th non-zero block and $l(\mathrm{NZ}_i)$ the length (in bits) of the corresponding Huffman code; define $p(C_i)$ and $l(C_i)$ similarly for the classes of zero-block runs; then

$$\mathrm{AL} = \sum_i p(\mathrm{NZ}_i)\, l(\mathrm{NZ}_i) + \sum_i p(C_i)\,(l(C_i) + i - 1)$$

As can be seen, TNO(S) is indeed increasing and AL(S) decreasing when passing to higher order numeration systems, with few exceptions on the picture file, which is a very small sample (the size of the picture file is only 0.1% of the size of the random file). The average length of a single code was low, about half a byte. This was due to the Huffman procedure, which takes advantage of the great differences in the frequencies of the non-zero blocks.

When the methods are sorted according to the compression they yield, one obtains a similar order on all the files. The compression factors for methods 5–36 vary only slightly for a given file; on the three files of Responsa maps, the CF varies from 10.03 to 10.61 for the words of length 1–4, from 20.68 to 22.28 for the words of length 5–8 and from 30.16 to 33.13 for the words of length 8–13, thus the CF is a decreasing function of the density of the file (see Table 2).

One can see that the results of the experiment on random maps are similar to the results obtained on the Responsa maps. Generally, the compression is lower in the random case so we conclude that the clustering effect in the Responsa maps improves the performance of our compression methods.

Though the digitized picture is less sparse than the other maps — the probability for a 1-bit is 4.35%, see Table 2 — the strong correlation between the bits yields a high compression factor of up to 9.2 approximately with the best method, which is again LLRUN. The MHC method of ⟦4⟧ gave a compression factor of only 8.29 (which is less than for some of our methods).

It has been proposed to drop the codes for the 255 non-zero blocks and to apply our methods to 0-*bit* runs as in the MHC-method, instead of 0-*byte* runs. A special code for a run of length 0 must be added. The advantage of this variant is that it considerably reduces the size of the Huffman translation tables. On the other hand, execution time for encoding and decoding will increase due to the bit-manipulations. On the random file, this method gave slightly inferior compression factors than the byte-oriented variants, for all the methods. On the Responsa maps, the difference was more evident, as well as for the picture, which was not surprising, since there we have relatively long runs of consecutive 1's which are associated with much waste in the bit-oriented method.

3.5 Concluding Speculations

Some of the new techniques of this paper gave higher compression factors on sparse bit-strings than other known methods, while being easy to implement and allowing for fast decoding. Nevertheless, the experiments showed that the "optimal" numeration system depends on the statistics of the bit-map file. Therefore, given a specific file, one should evaluate the CF for several systems and finally compress with the system giving the best results. It is even not necessary to stick to one of the above numeration systems, nor to any well-known system at all. As a matter of fact, *every* increasing sequence of integers $l_0 = 1, l_1, l_2, \ldots$ can be considered as the set of basis elements of some numeration system, and the representation of any integer in it will be unique if we use the algorithm of **2.3**. If we make only small perturbations in the sequences of Table 1, changing one of the elements slightly and continuing with the same recurrence formula to obtain the following, the desirable properties of the digits of the representation will be preserved at all places where the recurrence formula holds.

We thus suggest the following heuristic to improve the CF:

Step 1: Start the search with the numeration system giving the best CF among the methods 5–36; $i \leftarrow 1$.

Step 2: Create new sequences of basis elements by increasing or decreasing l_i; for $j > i$, l_j are obtained by the recurrence formula of the system found in Step 1; for each of the sequences evaluate its CF; continue until a first value of l_i is found which gives a local optimum.

Step 3: Repeat Step 2 for $i \leftarrow i + 1$ until the obtained improvement is smaller than a certain predetermined amount.

There is of course no certainty that the optimal sequence will be found or even approached. Perhaps the local optima in Step 2 are not global; perhaps starting the Step 3 iteration with i greater than 1 can lead to better results, but since long runs are rare, perturbations in l_i for larger i will have less influence on the CF.

This heuristic was applied to the unified file, which led to the following sequences:

	Sequence of basis elements	CF
REC2	1 3 7 17 41 99 ...	17.442
	1 3 7 18 43 104 ...	17.463
	1 3 7 18 44 106 ...	17.478
	1 3 7 18 44 107 ...	17.480

Note that the last few systems give a slightly better compression than method TRECUT.

ACKNOWLEDGMENT

We wish to express our gratitude to Professor Y. Choueka, Head of the Institute for Information Retrieval and Computational Linguistics (IRCOL) at Bar Ilan University, for kindly placing at our disposal the Responsa database.

	S	UNIFIED FILE TNO	UNIFIED FILE AL	UNIFIED FILE CF	RANDOM FILE TNO	RANDOM FILE AL	RANDOM FILE CF	PICTURE FILE TNO	PICTURE FILE AL	PICTURE FILE CF
1	TREE	.	.	10.533	.	.	8.451	.	.	5.560
2	TRECUT	.	.	17.451	.	.	15.906	.	.	5.809
3	NORUN	286449015	0.151	6.568	28460059	0.152	6.585	29633	0.189	5.284
4	LLRUN	7979444	0.776	18.768	1053433	0.730	16.251	1470	0.770	9.241
5	POW2	17314002	0.587	17.039	2289097	0.543	14.664	3259	0.615	8.332
6	POW3	21355385	0.517	17.047	2826247	0.471	14.798	3683	0.562	8.553
7	POW4	24769215	0.473	16.924	3277474	0.430	14.675	4865	0.501	8.182
8	POW5	27794543	0.445	16.642	3688763	0.403	14.414	5225	0.476	8.227
9	POW6	30650875	0.418	16.545	4058894	0.375	14.458	5078	0.473	8.428
10	POW7	33231411	0.396	16.507	4401817	0.353	14.456	5441	0.451	8.465
11	POW8	35581405	0.377	16.489	4717956	0.335	14.447	6764	0.407	8.139
12	POW9	37871111	0.361	16.443	5030555	0.320	14.382	6377	0.415	8.300
13	POW10	39961965	0.348	16.380	5310817	0.308	14.328	6773	0.398	8.317
14	POW3M	21355385	0.541	16.293	2826247	0.499	13.964	3683	0.586	8.202
15	FIBBIN	14716716	0.631	17.360	1944357	0.586	14.966	2504	0.663	8.766
16	FIBTER	19648611	0.533	17.410	2598588	0.489	15.040	3369	0.581	8.666
17	FIBTERM	19648611	0.558	16.637	2598588	0.518	14.208	3369	0.601	8.377
18	REC2	19089063	0.542	17.442	2523209	0.497	15.080	3231	0.588	8.750
19	REC3	23063838	0.494	17.001	3055201	0.446	14.825	3989	0.534	8.612
20	REC4	26531683	0.455	16.800	3515323	0.410	14.663	4921	0.491	8.292
21	REC5	29577710	0.423	16.766	3918559	0.380	14.617	4988	0.477	8.463
22	REC6	32385333	0.399	16.668	4294663	0.358	14.517	5411	0.451	8.500
23	REC7	34921144	0.380	16.559	4630351	0.338	14.500	6456	0.413	8.277
24	REC8	37284439	0.364	16.491	4957755	0.323	14.432	6353	0.415	8.329
25	REC9	39481131	0.350	16.427	5249356	0.310	14.365	6782	0.397	8.329
26	REC10	41572115	0.338	16.334	5528749	0.299	14.284	7273	0.380	8.287
27	A1B2Q	22284147	0.503	17.073	2943319	0.455	14.898	3497	0.565	8.740
28	A1B2P	23927839	0.481	17.037	3167224	0.436	14.794	4522	0.513	8.346
29	A1B3Q	25647807	0.463	16.916	3395956	0.420	14.643	4649	0.503	8.370
30	A2B1P	25811643	0.450	16.949	3423471	0.414	14.786	4883	0.492	8.314
31	A1B3P	27306536	0.449	16.719	3621359	0.405	14.502	5275	0.476	8.180
32	A2B2Q	28601027	0.438	16.588	3785631	0.397	14.346	5345	0.473	8.174
33	A1B4P	30315745	0.420	16.602	4016360	0.377	14.486	5132	0.471	8.411
34	A3B1P	31045059	0.404	16.962	4127455	0.361	14.807	5147	0.462	8.556
35	A2B2P	32098563	0.400	16.736	4261207	0.358	14.577	5033	0.464	8.644
36	A4B1P	35720479	0.368	16.817	4761363	0.328	14.641	5869	0.425	8.558

Table 5: *Results of the experiments on three files*

REFERENCES

[[1]] **Choueka Y.,** *The Responsa Project: What, How and Why 1976–1981,* Institute for Inf. Retr. and Comp. Linguistics, Bar-Ilan Univ. (1982).

[[2]] **Fraenkel A.S.,** *All about the Responsa Retrieval Project you always wanted to know but were afraid to ask, Expanded Summary,* Proc. Third Symp. on Legal Data Processing in Europe, Oslo (1975) 134–141 (Reprinted in Jurimetrics J. 16 (1976) 149–156 and in Informatica e Diritto II, No. 3 (1976) 362–370).

[[3]] **Fraenkel A.S.,** *Systems of Numeration,* Amer. Math. Monthly, to appear.

[[4]] **Hunter R. & Robinson A.H.,** *International Digital Facsimile Coding Standards,* Proc. IEEE, Vol 68 (1980) 854 – 867.

[[5]] **Jakobson M.,** *Huffman coding in bit-vector compression,* Information Processing Letters 7 (1978) 304–307.

[[6]] **Knuth D.E.,** *The Art of Computer Programming, Vol. I, Fundamental Algorithms,* Addison-Wesley, Reading, Mass. (1973).

[[7]] **Knuth D.E.,** *The Art of Computer Programming, Vol. II, Semi-numerical Algorithms,* Addison-Wesley, Reading, Mass. (1973).

[[8]] **Ramachandran K.,** *Coding Method for Vector Representation of Engineering Drawings,* Proc. IEEE, Vol 68 (1980) 813 – 817.

4 - COUNTING

The Use and Usefulness of Numeration Systems*

Aviezri S. Fraenkel†

Abstract.

The proper choice of a counting system may solve mathematical problems or lead to improved algorithms. This is illustrated by a problem in combinatorial group theory, compression of sparse binary strings, encoding of contiguous binary strings of unknown lengths, ranking of permutations and combinations, strategies of games and other examples. Two abstract counting systems are given from which the concrete ones used for the applications can be derived.

1. Introduction

The thesis of this article is that numeration systems are very useful. The choice of an appropriate ***numeration system***, that is, a set of integer basis elements such that every integer can be represented uniquely over the set using bounded integer digits, may be the mathematical key to solving a problem; or it may lead — much the same as the proper choice of a data structure — to more efficient algorithms.

As an example, consider the following problem: let $\sigma_{p^t} = \{0, 1, \ldots, p^t - 1\}$ be the cyclic group of order p^t under addition mod p^t, where p is prime and t a positive integer. Then for every integer $s \in [0, t]$, there is precisely one subgroup $\sigma_{p^s} = \{jp^{t-s} : 0 \leq j < p^s\}$ of order p^s. Unless $s = t$, the elements of σ_{p^s}, though equispaced on the interval $[0, p^t - 1]$, do not form an ***interval*** of consecutive lattice points on it. Prove either existence or nonexistence

---

*Work supported, in part, by Canadian Research Grant NSERC 69-0695.

†Queens College, Flushing, N.Y. Permanent Address: Department of Applied Mathematics, The Weizmann Institute of Science, Rehovot, Israel 76100. Part of this work was done during a visit at the Department of Mathematics and Statistics, The University of Calgary, Calgary, Alberta, Canada.

NATO ASI Series, Vol. F12
Combinatorial Algorithms on Words
Edited by A. Apostolico and Z. Galil

of a bijection $\phi : [0, p^t - 1] \rightarrow [0, p^t - 1]$ which maps *every* coset of every subgroup σ_{p^s} into an interval of consecutive lattice points with the additional requirement that every subgroup σ_{p^s} is mapped into a group $\{0, 1, \ldots, p^s - 1\}$ under addition mod p^s.

We will solve this problem in the final Section 7. In Section 2 we present two abstract systems of numeration from which all but one of the concrete ones of interest to us can be derived easily. The reader may wish to skip this part, at least in a first reading. All the subsequent sections are concerned with applications.

In Section 3 we show that a proper choice of a numeration system can lead to high compression of sparse binary strings. At the same time the codes used are more robust than, say, usual Huffman codes. In Section 4 we show how to use binary numeration systems based on Fibonacci numbers of orders $m \geq 2$ for encoding a sequence of contiguous binary integers of varying lengths. The codes used are considerably more robust and normally also shorter than those that have been used to date. The ranking of permutations of various kinds and of combinations is taken up in Section 5, in which also an application to the compression of dictionaries is indicated. Applications to 2-player games are presented in Section 6.

For additional numeration systems and applications, see Knuth [15].

2. Two Abstract Numeration Systems

Let $1 = u_0 < u_1 < u_2 \ldots$ be a finite or infinite sequence of integers. Let N be a nonnegative integer, and suppose that u_n is the largest number in the sequence not exceeding N. The algorithm:

```
for i = n to 0 by -1
        d_i ← ⌊N/u_i⌋
     N ← N - d_i u_i
end
```

produces the representation of N in the numeration system $S = \{u_0, u_1, \ldots\}$ with digits

d_i satisfying

$$\sum_{i=0}^{j} d_i u_i < u_{j+1} \quad (0 \leq j < n). \tag{1}$$

THEOREM 1. Let $1 = u_0 < u_1 < u_2 < \ldots$ be any finite or infinite sequence of integers. Any nonnegative integer N has precisely one representation in the numeration system $S = \{u_0, u_1, u_2 \ldots\}$ of the form $\sum_{i=0}^{n} d_i u_i$, where the digits d_i are nonnegative integers satisfying (1).

For the proofs of Theorems 1 and 2 see [9], where the derivation of many concrete numeration systems from these theorems is also carried out. We shall now derive another abstract numeration system.

For $m \geq 1$, let $b_1 = b_1^{(n)}, b_2, \ldots, b_m$ be integers satisfying

$$1 \leq b_m \leq \ldots \leq b_2 \leq b_1^{(n)}$$

for all $n \geq 1$, where $b_1 = b_1^{(n)}$ may depend on n. Let

$$\left.\begin{array}{ll} u_{-m+1}, u_{-m+2}, u_{-1} \text{ be fixed nonnegative integers, and let} & \\ u_0 = 1, u_n = b_1^{(n)} u_{n-1} + b_2 u_{n-2} + \cdots + b_m u_{n-m} & (n \geq 1) \end{array}\right\} \tag{2}$$

be an increasing sequence.

THEOREM 2. Let $S = \{u_i\}$ be a sequence of the form (2). Any nonnegative integer N has precisely one representation in S of the form $N = \sum_{i=0}^{n} d_i u_i$ if the digits d_i are nonnegative integers satisfying the following (two-fold) condition:

(i) Let $k \geq m-1$. For any j satisfying $0 \leq j \leq m-2$, if

$$(d_k, d_{k-1}, \ldots, d_{k-j+1}) = (b_1^{(k+1)}, b_2, \ldots, b_j), \tag{3}$$

then $d_{k-j} \leq b_{j+1}$; and if (3) holds with $j = m-1$, then $d_{k-m+1} < b_m$.

(ii) Let $0 \leq k < m-1$. If (3) holds for any j satisfying $0 \leq j \leq k-1$, then $d_{k-j} \leq b_{j+1}$; and if (3) holds with $j = k$, then $d_0 < \sum_{i=k+1}^{m} b_i \, u_{k+1-i}$.

It turns out that the condition of Theorem 2 is equivalent to (1) and therefore it yields unique representation. It is equivalent to

$$0 \leq d_i < b_1^{(i+1)} \quad (i \geq 0)$$

for $m = 1$ (recurrence of length 1). For $m > 1$, the condition of Theorem 2 implies

$$0 \leq d_i \leq b_1^{(i+1)} \; (i \geq 1), \quad 0 \leq d_0 < b_1^{(1)} + \sum_{j=2}^{m} b_j u_{1-j},$$

but is not implied by these inequalities. In fact, if the recurrence (2) has length $m > 1$, then uniqueness of representation requires a condition on *sequences* of digits rather than only on every digit alone, namely the condition of Theorem 2 is required.

3. Compression of Sparse Binary Strings

Given a sparse binary string. We consider it being partitioned into k-bit blocks, where k is fixed. For example, $k = 8$ gives a partition into bytes. Jakobson [12] proposed to compress the string according to the probabilities of occurrence of the 2^k possible block patterns which are assigned Huffman codes.

For sufficiently sparse blocks, the probability for a 0-block of length k is $\geq \frac{1}{2}$, so Jakobson's method assigns to every 0-block a code of length 1 bit. Hence we cannot expect a compression factor better than k from this method, where the compression factor is defined as the ratio of the size of the original file to the size of the compressed file.

To improve the compression, we adjoin to the $2^k - 1$ nonzero blocks a small number of basis elements $u_0, \ldots, u_{t-1}$ of a suitable numeration system, in which we will represent all the 0-run lengths. We then Huffman-code the $2^k - 1$ nonzero blocks together with $u_0, \ldots, u_{t-1}$. The compression achieved will depend on the type of numeration system selected.

It turns out that especially good compression is gotten when the recurrence (2) is

$$u_0 = 1, \quad u_1 = 3, \quad u_n = 2u_{n-1} + u_{n-2} \qquad (n \geq 2).$$

The condition for uniqueness eqivalent to the condition of Theorem 2 is that $0 \le d_i \le 2$ for every digit d_i, with the proviso that $d_{i+1} = 2$ implies $d_i = 0$ $(i \ge 0)$. (This ternary system is the special case $a_i = a = 2$ of the p-system of Section 6 below.)

Another ternary numeration system, whose basis elements are the even Fibonacci numbers $u_0 = 1$, $u_2 = 3$, $u_4 = 8$, $u_6 = 21, \ldots$ also gives good compression. Here the condition for uniqueness is $0 \le d_i \le 2$ $(i \ge 0)$ with the additional proviso that between any two consecutive digits 2 there is a digit 0. This system can be derived from Theorem 1.

The experiments were run on six files. The largest of these contained 56,588 strings, each string being partitioned into 5,284 bytes. (Each string represented $8 \times 5,284 = 42,272$ documents of an information retrieval system. Bit i in string j is 1 if and only if document i contains word w_j, $1 \le i \le 42,272$, $l \le j \le 56,588$.) The percentage of 1-bits per string was 0.817%. For the first ternary system mentioned above, the ten basis elements 1, 3, 7, 17, 41, 99, 239, 577, 1393, 3363 were adjoined to the 255 nonzero bytes, and then Huffman-coded. A compression of 17.44 was achieved, compared to 6.57 for Jakobson's method. For the second ternary system, a compression of 17.41 was obtained. For each of the two methods, the average number of codes per 0-run was less than 2.5.

The Huffman-codes of the basis elements for representing 0-run lengths are arranged in monotone order. This should help to make the very error-sensitive Huffman codes more robust, by checking that the decoded basis elements indeed appear in monotone order. For further details, see [10].

We remark that, more generally, codes based on Fibonacci numeration systems can be defined which are constant, that is, they do not have to be generated for every probability distribution and are therefore easier to use than Huffman codes. Although the latter are more economical, the former are more robust — less error sensitive — in a sense that can be made quite precise. The robustness enables to save on the error detection bits, which are indispensable when transmitting Huffman codes.

In the next section we will be concerned even more intimately with robustification.

The second ternary numeration system mentioned above was also used by Chung and Graham [3] to investigate irregularities of distribution of sequences.

4. Encoding Strings of Varying Length

Suppose we like to transmit a sequence of binary strings whose lengths lie in an unknown range. Since a comma is not a binary bit, we cannot separate the strings by commas. Even, Rodeh and Pratt [5, 21] proposed to overcome the difficulty by preceding the string by its length and repeating this recursively until a small length, say 3, is attained. Since all strings thus produced start with a most significant bit 1, the bit 0 can be used to signal the end of the logarithmic ramp and the beginning of the string itself. Thus if S is a string, $E_1(S)$ its representation in this coding scheme, then

$$E_1(520) = 100\ 1010\ 0\ 1000001000.$$

The major disadvantage of this representation, however, lies in its vulnerability to transmission errors. A single error occurring in the logarithmic ramp prefix plays complete havoc with the decoding, which normally cannot be resumed.

An alternative scheme, which is quite insensitive to errors, is one based on an m-order Fibonacci numeration system. If we denote by $E_2(S) = E_2^{(m)}(S)$ the m-order Fibonacci representation of S when a sequence of strings S is concatenated, then asymptotically, $|E_2^{(m)}(S)| > |E_1(S)|$ $(|S| \to \infty)$, where $|E_i(S)|$ and $|S|$ denote the lengths of $E_i(S)$ and S. But for very large initial values of $|S|$, depending on m, we have actually $|E_2^{(m)}(S)| < |E_1(S)|$ as we shall see below.

Fibonacci numbers of order m $(m \geq 2)$ are defined by

$$u_{-m+1} = u_{-m+2} = \cdots = u_{-2} = 0, \quad u_{-1} = u_0 = 1,$$
$$u_n = u_{n-1} + u_{n-2} + \cdots + u_{n-m} \quad (n \geq 1).$$

(For simplicity we write u_i instead of $u_i^{(m)}$.) This definition reduces to the ordinary Fibonacci numbers for $m = 2$.

Fibonacci numbers of order m have been used for polyphase merge and sort of data runs: We like to sort-merge data runs stored on $m + 1$ magnetic tape transports such that at each stage m tapes merge into one tape and so that the tapes run continuously to save time. This can be done if at each stage the number of runs on one of the tapes is an m-th order Fibonacci number, and a simple function thereof on the other tapes. See Knuth [16] and Lynch [18].

It follows directly from Theorem 2 that every nonnegative integer N has precisely one binary representation of the form

$$N = \sum_{i=0}^{n} d_i u_i \qquad (d_i \in \{0,1\}, \quad 0 \leq i \leq n)$$

such that there is no run of m consecutive 1's. This class of binary numeration system is denoted by $F^{(m)}$. For $m = 2$ it gives the ordinary binary Fibonacci numeration system (in which two adjacent 1-bits never occur). See Zeckendorf [27]. This system lies behind the Fibonacci search [16].

If the most significant bit of a string is 1 — in which case we say that S is an *integer* — then the representation of S of length n in the $F^{(m)}$ numeration system requires $|E_2(S)| = n+m$ bits in the contiguous string representation, which includes a 0-bit and $m-1$ postfixed 1-bits, which form a "synchronizer" or "comma" of m 1-bits between adjacent integers. A similar but somewhat more involved representation can be defined for strings that are not necessarily integers. Below we assume that S is an integer.

Let lg denote log to the base 2. Then asymptotically (for large n and large m),

$$D = |E_2(S)| - |E_1(S)| = m - 1 + \frac{n\delta}{2^m} lge - \lceil lg(k+1) \rceil - \lceil lg \lceil lg(k+1) \rceil + 1 \rceil - \ldots - 3,$$

where the last $lg\ lg \ldots$ term is 3, $\delta = \delta(m)$ is a suitable real number satisfying $\frac{1}{2} < \delta < 1$, and

$$k \approx n\left(1 - \left(\frac{\delta}{2^m} + \frac{\delta^2}{2^{2m+1}}\right) lge\right).$$

This is derived in [1]. We see that for any fixed m we have $D > 0$ if n is sufficiently large, so ultimately the method of [5] gives shorter representations. This kind of asymptotic behavior also holds for other Fibonacci representations.

For integers of practical length, however, the methods of [1] give in fact shorter representations, in addition to providing robustness. The following computational results refer to a Fibonacci representation somewhat more economical than the one alluded to above.

For $m = 2$, $|E_1(n)| = 4$ bits and $|E_2(n)| = 5$ bits, where $n = 5, 6, 7$. But $|E_2(n)| \leq |E_1(n)|$ for all positive integers $n \leq u_{27} - 1 = 514,228$ ($n \neq 5, 6, 7$). Beyond this point, the representation $E_2(n)$ becomes slowly longer than $E_1(n)$.

For $m = 3$, $|E_1(n)| < |E_2(n)|$ for $3 \leq n \leq 7$. But $|E_2(n)| \leq |E_1(n)|$ for all

$$8 \leq n \leq \frac{1}{2}(u_{63}^{(3)} + u_{61}^{(3)} - 1) = 34,696,689,675,849,696 \simeq 3.470 \times 10^{16}.$$

for larger n, $|E_2(n)|$ becomes slowly larger than $|E_1(n)|$. Thus for $n = \frac{1}{2}(u_{80}^{(3)} + u_{78}^{(3)} - 1) \simeq 1.095 \times 10^{21}$, we have $|E_2(n)| - |E_1(n)| = 1$, and this difference is 5, for example at $n = u_{146} + u_{144} \simeq 3.208 \times 10^{38}$. Incidentally, the difference does not increase monotonically: it decreases at points n where $E_1(n)$ picks up a new *lg lg* term on its logarithmic ramp.

For $m = 4$, $|E_1(n)| < |E_2(n)|$ for $2 \leq n \leq 7$, $|E_2(n)| \leq |E_1(n)|$ for $8 \leq n \leq 116$, and $|E_2(n)| - |E_1(n)| = 1$ for $117 \leq n \leq 127$. But $|E_2(n)| \leq |E_1(n)|$ for all

$$128 \leq n \leq \frac{1}{3}(u_{231}^{(4)} + 2u_{229}^{(4)} + u_{228}^{(4)} - 1 \simeq 4.194 \times 10^{65}.$$

Beyond this point, the representation $E_2(n)$ becomes very slowly longer than $E_1(n)$.

The above computations are based on a list of higher-order Fibonacci numbers which Gerald Bergum has kindly prepared for us.

5. Ranking Permutations and Combinations

Permutations and combinations are fundamental to many combinatorial problems. The *rank* of a permutation or combination is the serial number of the permutation or combination in some linear ordering of all of them. In many applications the rank of a given permutation or combination has to be found, or conversely, the permutation or combination corresponding to a given rank has to be determined. One case where a ranked list of permutations is required is that of generating random permutations for Monte Carlo procedures. Ranked lists of combinations, permutations, permutations with repetitions and Cayley-permutations have been used for data compression. This application is indicated briefly below.

Ranking algorithms are normally based on the following systems of numeration.

(i) *Mixed radix.* Let $1 = a_0, a_1, a_2, \ldots$ be any sequence of integers satisfying $a_i > 1$ $(i \geq 1)$, and let $u_n = a_0 a_1 \cdots a_n$, that is, $u_{n+1} = a_{n+1} u_n$ $(n \geq 0)$. The representation $N = \sum_{i=0}^{n} d_i u_i$ is the *mixed radix* representation of N. Theorem 2 implies immediately that the representation is unique if and only if $0 \leq d_i < a_{i+1}$ $(i \geq 0)$. The mixed radix representation has also been used for a constructive proof of the generalized Chinese Remainder Theorem (see [6] and [15]).

(ii) *Factorial representation.* This is the special case of the mixed radix represention where $a_n = n + 1$, leading to $u_n = (n+1)!$ $(n \geq 0)$. Thus the representation $N = \sum_{i=1}^{n} d_i (i+1)!$ is unique if and only if $d_i \leq i + 1$ $(i \geq 0)$.

(iii) *Reflected factorial representation.* To represent a nonnegative integer N, select h with $h! > N$, and let $u_n = h!/(h-n)!$, that is, $u_{n+1} = u_n(h-n)$ $(n \geq 0)$. Since again the recurrence has length 1 only, the representation $N = \sum_{i=0}^{h-2} d_i h!\ (h-i)!$ is seen to be unique if and only if $0 \leq d_i < h - i$ $(0 \leq i \leq h-2)$.

(iv) *Combinatorial representation.* For every integer $k \geq 1$, there is a unique representation of any nonnegative integer N in the form

$$N = \binom{a_1}{1} + \binom{a_2}{2} + \cdots + \binom{a_k}{k}, \quad 0 \leq a_1 < a_2 < \ldots < a_k.$$

This numeration system does not fit into the general framework of Theorems 1 and 2. For a fast algorithm for computing the a_i, see [11, Sect. 2].

A simple algorithm to generate permutations in an orderly way is based on having one element "sweep" over the other elements, using a single transposition at each step. This leads in a natural way to a ranking of the generated permutations in the numeration systems (ii) or (iii). Not all the transpositions involve adjacent elements in the method of Wells [23], in which the permutations are ranked using the system (ii). If all the transpositions are between adjacent elements, the system (iii) replaces (ii). This is the method of Johnson [13]. See Even [4] and Lehmer [17] for general descriptions of these methods and Pleszczynski [20] for algorithms realizing transformations between permutations and their ranks.

Combinations are ranked using the combinatorial representation. This ranking and the mixed radix representation have been used in [19] for ranking permutations with repetitions and Cayley permutations. A ***Cayley permutation*** of length m on $S = \{1, 2, \ldots, n\}$ is a permutation p of length m on S with possible repetitions, such that if $j \in p$ then also $i \in p$ for all $i < j$. The ranking of Cayley permutations has been used in [11] for compressing and partitioning large dictionaries. The main idea is to replace each dictionary word w by a pair (L, I), where the ***lexform*** L is an ordered string of the ***distinct*** letters of w, and I is an ***index*** which permits transforming L back into w. If the letters of L are numbered $1, 2, \ldots, |L|$, then I is a Cayley permutation of length $|W|$ on $1, 2, \ldots, |L|$, specifying the order of the letters of W.

The main variation investigated was that when the lexforms reside in fast memory and the indexes are relegated to disk. The information contained in the lexforms seems to be almost the same as that of the w's: experiments indicate that the entropy increase in transforming the latter to the former is very small.

Further compression can be achieved by replacing each lexform of length ℓ over an alphabet of size n by its serial number in some linear ordering of all $\binom{n}{\ell}$ combinations. Similarly, every index can be replaced by its serial number in some linear ordering of the Cayley permutations.

The overall savings of fast memory is very high, and may enable storage of a large dictionary in form of its lexforms in fast memory, which otherwise could not be kept in it because of lack of space. Typical applications are for information retrieval, where: (1) Most accesses to the dictionary are unsuccessful, that is, the word sought is not in the dictionary. (2) Many accesses are successful, but additional Boolean or metrical constraints — verifiable without consulting the disk — reject the search. In both of these cases there are many accesses to fast memory and few to disk, whose access time is typically 10^4 times slower than that of fast memory.

6. Strategies for Games

I. *Wythoff games.* Let a be a positive integer. Given two piles of tokens, two players move alternately in a generalized Wythoff game. The moves are of two types: a player may remove any positive number of tokens from a single pile, or he may take from both piles, say k (> 0) and ℓ (> 0) from the other, provided that $|k - \ell| < a$. In *normal* play, the player first unable to move is the loser, his opponent the winner. In *misère* play, the outcome is reversed: the player first unable to move is the winner, his opponent the loser.

Wythoff game positions are denoted by (x, y) with $x \leq y$, where x and y denote the number of tokens in the two piles. Positions from which the Previous player can win whatever move his opponent will make, are called *P-positions* and those from which the Next player can win whatever move his opponent will make are called *N-positions*. We also denote by P (N) the set of all P (N)-positions. For any set S of nonnegative integers, if $\overline{S}$ denotes the complement of S with respect to the set Z of nonnegative integers, we define mex S (minimum excluded value of S) by

$$\text{mex}\, S = \min \overline{S} = \text{least nonnegative integer not in } S.$$

Thus mex $\emptyset = 0$.

We restrict attention to normal play below. The following result holds:

THEOREM 3. $P = \cup_{i=0}^{\infty} \{(A_i, B_i)\}$, where

$$A_n = \text{mex}\{A_i, B_i : 0 \leq i < n\}, \text{ and } B_n = A_n + an \quad (n \geq 0).$$

The first few P-positions for $a = 2$ are depicted in Table 1. It is not difficult to see that $(\cup_{n=0}^{\infty} A_n) \cap (\cup_{n=0}^{\infty} B_n) = \{0\}$, and $(\cup_{n=0}^{\infty} A_n) \cup (\cup_{n=0}^{\infty} B_n) = Z$.

Table 1. The first few P-positions for $a = 2$.

n	A_n	B_n
0	0	0
1	1	3
2	2	6
3	4	10
4	5	13
5	7	17
6	8	20
7	9	23
8	11	27
9	12	30
10	14	34

Given a position (x, y) for which we wish to compute a winning strategy, if it exists. We may assume $y \le x + ax$, since otherwise it is easy to see that $(x, y) \in N$. Since $A_n \le 2n$ for all a (follows from $B_n - B_{n-1} \ge 2$), the computation based on Theorem 3 requires computing $O(x)$ table entries (A_n, B_n) and $O(x)$ words of memory space. This is an exponential algorithm, since the input length is only $= O(lg\ x + lg\ y) = O(lg\ x)$. Below we describe a polynomial strategy, based on the following two exotic numeration systems, and a connection between them.

Let $\alpha = [1, a, a, \ldots]$ be the simple continued fraction expansion

$$\alpha = 1 + \cfrac{1}{a + \cfrac{1}{a + \cfrac{1}{a + \ldots}}}$$

whose convergents $p_n/q_n = [1, a, \ldots, a]$ (n terms a) are defined recursively by

$$p_0 = 1, \quad p_1 = a_1 + 1, \qquad p_n = a_n p_{n-1} + p_{n-2} \qquad (n \ge 2),$$
$$q_0 = 1, \quad q_1 = a_1, \qquad q_n = a_n q_{n-1} + q_{n-2} \qquad (n \ge 2).$$

Theorem 2 implies directly that every nonnegative integer has precisely one representation of the form

$$N = \sum_{l=0}^{k} s_i p_i, \quad 0 \le s_i \le a_{i+1}; \quad s_{i+1} = a_{i+2} \Rightarrow s_i = 0 \quad (i \ge 0) \quad (p - system)$$

and also precisely one representation of the form

$$N = \sum_{l=0}^{l} t_i q_i, \quad 0 \leq t_0 < a_1, \quad 0 \leq t_i \leq a_{i+1}; \ t_i = a_{i+1} \Rightarrow t_{i-1} = 0 \ (i \geq 1) \ (q-system).$$

In Table 2 the first few P-positions for $a = 2$ appear in the left columns (copied from Table 1), followed by $R_p(A_n)$ and $R_p(B_n)$, the represention of A_n and B_n in the p-system, and $R_q(n)$, the representation of n in the q-system. The following properties hold for every a. They can be verified in Table 2 for $a = 2$ and small n.

Table 2. The representation of the first few P-positions (A_n, B_n) in the p-system and n in the q-system for $\alpha = [1, 2, 2, 2, \ldots]$.

			$R_p(A_n)$	$R_p(B_n)$	$R_q(n)$
n	A_n	B_n	7 3 1	17 7 3 1	5 2 1
0	0	0	0	0	0
1	1	3	1	1 0	1
2	2	6	2	2 0	1 0
3	4	10	1 1	1 1 0	1 1
4	5	13	1 2	1 2 0	2 0
5	7	17	1 0 0	1 0 0 0	1 0 0
6	8	20	1 0 1	1 0 1 0	1 0 1
7	9	23	1 0 2	1 0 2 0	1 1 0
8	11	27	1 1 1	1 1 1 0	1 1 1
9	12	30	1 1 2	1 1 2 0	1 2 0
10	14	34	2 0 0	2 0 0 0	2 0 0

Property 1. The set of numbers A_n $(n \geq 1)$ is identical with the set of numbers which end in an even number of zeros in the p-system; and the set of numbers $B_n = A_n + an$ $(n \geq 1)$ is identical with the set of numbers which end in an odd number of zeros in the p-system. (Thus in Table 2, $R_p(4) = 11$ ends in an even number (zero) of zeros, and so does $R_p(7) = 100$ (ending in two zeros). Both 4 and 7 appear in the A_n-column.)

Property 2. For every $n \geq 1$, the representation$R_p(B_n)$ is a "left shift" of the representation $R_p(A_n)$. (Thus $(A_n, B_n) = (1, 3)$ and (5,13) have representations (1,10) and (12,120) respectively in the p-system.)

Property 3. Let n be any positive integer. If $R_q(n)$ ends in an even number of zeros, then $R_q(n) = R_p(A_n)$. (Thus $R_q(5) = R_p(A_5) = 100$.) If $R_q(n)$ ends in an odd number of zeros, then $R_q(n) = R_p(A_n + 1)$. (Thus $R_q(9) = R_p(A_9 + 1) = R_p(13) = 120$.)

These properties imply the existence of a polynomial strategy for this class of games. For the details of normal play see [7]; misère play [8]. Normal play with $a = 1$ is the classical Wythoff game. See Wythoff [25] and Yaglom and Yaglom [26].

II. ***Single pile games.*** The analysis of the following 2-player game of Whinihan [24, 14] is based on the Fibonacci system of numeration. See also Schwenk [22].

Given a pile of n tokens. The first player removes any (positive) number of tokens less than n. The players alternate their moves, each person removing one or more tokens, but not more than twice as many tokens as the opponent has taken in the preceding move. The player first unable to move is the loser, his opponent the winner.

III. ***Multiple pile games.*** The analyses of Nim, Moore's Nim, N-person Nim and N-person Moore's Nim also depend on systems of numeration. Nim, as well as other ***disjunctive sums*** of games, depend only on the conventional binary numeration system, whereas the other three games depend on a combination of binary and additional numeration systems.

7. The Combinatorial Group Theory Problem

The requirements that the bijection ϕ is to fulfil are too stringent for the subgroups of σ_{p^t} alone, so a fortiori for all the cosets. So it seems. However, a bijection ϕ with the desired properties ***does*** exist. Here it is.

Every $N \in [0, p^t - 1]$ has a unique representation in the p-ary numeration system of the form

$$N = f(p) = \sum_{i=0}^{t-1} a_i p^i \quad (0 \le a_i < p).$$

We define

$$\phi(N) = p^{t-1} f\left(\frac{1}{p}\right) = \sum_{i=0}^{t-1} a_i p^{t-1-i}.$$

The intuition why this works is that the elements of the subgroup $\sigma_{p^s} = \{jp^{t-s} : 0 \leq j < p^s\}$ are represented by the high-order digits in the above representation of N. They transform into the low-order digits in $\phi(N)$, thus producing an interval of lattice points.

Example. Let G be the cyclic additive group whose order is the number of references of this article. Then $\sigma_3 = \{0, 9, 18\}$, and $\phi(\sigma_3) = \{0, 1, 2\}$. Also $\sigma_9 = \{0, 3, 6, 9, 12, 15, 18, 21, 24\}$, and $\phi(\sigma_9) = \{0, 3, 6, 1, 4, 7, 2, 5, 8\}$.

The proof that this bijection has the required properties appears in [2].

Acknowledgment. I wish to thank Joseph Kahane and Yehoshua Zanger and their families for the warmest and finest of hospitalities they extended to my family and myself during our recent visit in New York.

References

1. A. Apostolico and A.S. Fraenkel, Robust complete universal Fibonacci representations, and transmission of unbounded strings, Technical Report, December 1983.

2. M. Berger, A. Felzenbaum and A.S. Fraenkel, An improvement to the Newman-Znám result for disjoint covering systems, Technical Report, Department of Mathematics, The Weizmann Institute of Science, September 1984.

3. F.R.K. Chung and R.L. Graham, On irregularities of distribution of real sequences, *Proc. Natl. Acad. Sci.* USA **78** (1981) 4001.

4. S. Even, Algorithmic Combinatorics, Macmillan, New York, N.Y., 1973.

5. S. Even and M. Rodeh, Economical encoding of commas between strings, *CACM* **21** (1978) 315–317.

6. A.S. Fraenkel, New proof of the generalized Chinese Remainder Theorem, *Proc. Amer. Math. Soc.* **14** (1963) 790–791.

7. A.S. Fraenkel, How to beat your Wythoff games' opponent on three fronts, *Amer. Math. Monthly* **89** (1982) 353–361.

8. A.S. Fraenkel, Wythoff games, continued fractions, cedar trees and Fibonacci searches, *Theoretical Computer Science* **29** (1984) 49–73.

9. A.S. Fraenkel, Systems of numeration, *Amer. Math. Monthly,* to appear (Jan. 1985).

10. A.S. Fraenkel and T. Klein, Novel compression of sparse bit-strings — preliminary report, These Proceedings.

11. A.S. Fraenkel and M. Mor, Combinatorial compression and partitioning of large dictionaries, *The Computer J.* **26** (1983) 336–343.

12. M. Jakobson, Huffman coding in bit-vector compression, *Inform. Process. Letters* **7** (1978) 304–307.

13. S.M. Johnson, Generation of permutations, *Math. Comp.* **17** (1963) 282–285.

14. D.E. Knuth, The Art of Computer Programming, Vol. 1: Fundamental Algorithms, Second Printing, Addison-Wesley, Reading, MA, 1973.

15. D.E. Knuth, The Art of Computer Programming, Vol. 2: Seminumerical Algorithms, 2nd Edition, Addison-Wesley, Reading, MA, 1981.

16. D.E. Knuth, The Art of Computer Programming, Vol. 3: Sorting and Searching, Second Printing, Addison-Wesley, Reading, MA, 1975.

17. D.E. Lehmer, The machine tools of combinatorics, in Applied Combinatorial Mathematics (E.F. Beckenbach, Ed.), Wiley, New York, NY, pp. 5–31, 1964.

18. W.C. Lynch, The t-Fibonacci numbers and polyphase sorting, *The Fibonacci Quarterly* **8** (1970) 6–22.

19. M. Mor and A.S. Fraenkel, Cayley permutations, *Discrete Mathematics* **48** (1984) 101–112.

20. S. Pleszcynski, On the generation of permutations, *Inform. Process. Letters* **3** (1975) 180–183.

21. M. Rodeh, V.R. Pratt and S. Even, Linear algorithm for data compression via string

matching, *J. ACM* **28** (1981) 16–24.

22. A.J. Schwenk, Take-away games, *The Fibonacci Quarterly* **8** (1970) 225-234.

23. M.B. Wells, Generation of permutations by transpositions, *Math. Comp.* **15** (1961) 192–195.

24. M.J. Whinihan, Fibonacci Nim, *The Fibonacci Quarterly* **1 No. 4** (1963) 9–12.

25. W. Wythoff, A modification of the game of Nim, *Nieuw Arch. Wisk.* **7** (1907) 199–202.

26. A.M. Yaglom and I.M. Yaglom, Challenging Mathematical Problems with Elementary Solutions, translated by J. McCawley, Jr., revised and edited by B. Gordon, Vol. II, Holden-Day, San Francisco, 1967.

27. E. Zeckendorf, Représentation des nombres naturels par une somme de nombres de Fibonacci ou de nombres de Lucas, *Bull. Soc. Royale Sci. Liège* **41** (1972) 179–182.

ENUMERATION OF STRINGS

A. M. Odlyzko
AT&T Bell Laboratories
Murray Hill, New Jersey 07974
USA

ABSTRACT

A survey is presented of some methods and results on counting words that satisfy various restrictions on subwords (i.e., blocks of consecutive symbols). Various applications to comma-free codes, games, pattern matching, and other subjects are indicated. The emphasis is on the unified treatment of those topics through the use of generating functions.

1. Introduction

Enumeration of words that satisfy restrictions on their subwords is a very satisfying subject to deal with. Not only are there many applications for results in this area, but in addition one can use a wide range of mathematical techniques which lead to a variety of elegant results. These methods are usually not applicable when instead of subwords (i.e., blocks of consecutive symbols) one studies general subsequences.

We will consider strings (words and patterns will be used synonymously) over a fixed finite alphabet Ω of size $q \geqslant 2$. The basic results, from which most of the others are derived, enumerate strings of a given length that contain no elements of some fixed set of strings as subwords. The enumeration does not yield simple explicit formulas for the number of such strings, as no such formulas exist. Instead, formulas are obtained for the generating functions of these quantities. These generating functions are rational, and their special form leads to numerous results, both in asymptotic enumeration and in more algebraic and combinatorial applications.

In Section 2 we briefly review how rational generating functions can be used in both asymptotic enumeration and other settings. In Section 3, the important notion of a *correlation* of two strings is defined. Using that notion, Section 4 then presents the main formulas for the generating functions that arise in the enumeration of strings. Section 5 discusses some applications of these generating functions to the study of worst-case behavior of pattern matching algorithms, recognizing unavoidable sets of words, and

NATO ASI Series, Vol. F12
Combinatorial Algorithms on Words
Edited by A. Apostolico and Z. Galil

nontransitive games. Section 6 discusses some of the main applications of these generating functions to asymptotic enumeration, especially to the study of prefix-synchronized codes and occurrences of long repetitive patterns in strings.

This paper is based largely on the author's joint work with L. J. Guibas [11-15] and B. Richmond [21]. Only a small selection of results are presented, and they are chosen so as to emphasize how generating functions can be used as a unifying theme in a variety of contexts. Many of those results can also be obtained by other methods.

2. Rational generating functions

If $f(0), f(1), \ldots$ is any sequence, we define its generating function $F(z)$ to be

$$F(z) = \sum_{n=0}^{\infty} f(n) z^{-n}, \tag{2.1}$$

where z is an indeterminate. We use the form (2.1), instead of the more common form $F(z^{-1})$, which is a sum of terms of the form $f(n)z^n$, since (2.1) yields somewhat more elegant expressions for the generating functions arising in string enumeration.

In essentially all of the cases we will be concerned with, the generating function $F(z)$ will be rational, so that

$$F(z) = \frac{P(z)}{Q(z)}, \tag{2.2}$$

where $P(z)$ and $Q(z)$ are polynomials, which we may assume satisfy $\gcd(P(z), Q(z)) = 1$, $Q(z) \not\equiv 0$. Several conclusions follow when $F(z)$ is rational. First of all, the $f(n)$ satisfy a linear recurrence with constant coefficients; if we write

$$Q(z) = \sum_{j=0}^{d} a_j \, z^{q-j}, \tag{2.3}$$

where d is the degree of $Q(z)$, so that $a_0 \not\equiv 0$, then (2.1) and (2.2) show that

$$f(n) = -a_0^{-1} \sum_{j=1}^{d} a_j f(n-j) \quad \text{for } n > d \,. \tag{2.4}$$

This linear recurrence can already be used to prove nontrivial results. For example, when coupled with some information about the coefficients a_j, it can be used to obtain lower bounds on the worst case performance of string searching algorithms (see Section 5).

Perhaps the most important consequence of the rationality of a generating function is that it yields significant information about the asymptotic behavior of the sequence being enumerated. If $F(z)$ has the form (2.2), and in addition $Q(z)$ has the form

$$Q(z) = a_0 \prod_{j=1}^{d} (z-\rho_j) \,, \tag{2.4}$$

where the ρ_j are distinct, then the partial fraction expansion of $F(z)$ is

$$F(z) = c_0 + \sum_{j=1}^{d} \frac{c_j}{z-\rho_j} \,, \tag{2.5}$$

where c_0 is some constant (there is no polynomial part to this expansion, since deg $P(z) \leqslant$ deg $Q(z)$ by (2.1)), and

$$c_j = \frac{P(\rho_j)}{Q'(\rho_j)} \,, \quad 1 \leqslant j \leqslant d \,. \tag{2.6}$$

If the ρ_j are not all distinct, then (2.5) has to be replaced by a slightly more complicated expression, but we will not need to use it. The partial fraction expansion (2.5) shows that

$$F(z) = c_0 + \sum_{j=1}^{d} \frac{c_j z^{-1}}{1-\rho_j z^{-1}}$$

$$= c_0 + \sum_{j=1}^{d} c_j z^{-1} \sum_{n=0}^{\infty} \rho_j^n z^{-n} \tag{2.7}$$

$$= c_0 + \sum_{n=1}^{\infty} z^{-n} \sum_{j=1}^{d} c_j \rho_j^{n-1} .$$

This shows that

$$f(n) = \sum_{j=1}^{d} c_j \rho_j^{n-1} , \quad n \geqslant 1 . \tag{2.8}$$

In particular, the expansion (2.8) shows that asymptotically, the growth rate of $f(n)$ is determined by the largest of the ρ_j;

$$|f(n)| \leqslant c \max_{1 \leqslant j \leqslant d} |\rho_j|^n \tag{2.9}$$

for some constant c. For finer details of the asymptotic behavior of $f(n)$, Eq. (2.8) by itself is often insufficient. The large zeros (those for which $|\rho_j|$ is maximal) can combine so as to make $f(n)$ small for some values of n, as happens, for example, for $F(z) = z^2/(z^2-1)$, where $f(n) = 0$ if n is odd, and $f(n) = 1$ if n is even. Some very sophisticated methods have been developed to deal with problems of this nature, as can be seen in [18] and the references cited there. For our purposes, though, such methods are not suitable.

Another source of difficulty with the use of expansions of the form (2.8) is that even when there is only one ρ_j with maximal absolute value, the other ρ_j can be important for small n. To obtain uniform estimates from (2.8), it becomes necessary to estimate the sizes of the c_j as well as of the ρ_j.

Fortunately for us, in string enumeration we are able to use a simple approach that lets us bypass the difficulties that arise in using the full partial fraction expansion. We have the following well-known general result.

Lemma 2.1. Suppose that $F(z)$ is analytic in $|z| \geqslant r > 0$ with the exception of $z = \rho$, $|\rho| > r$, where it has a first order pole with residue α, and that

$$|F(z)| \leqslant C \quad \text{for} \quad |z| = r \,. \tag{2.10}$$

Then, if

$$F(z) = \sum_{n=0}^{\infty} f_n \, z^{-n} \,,$$

we have

$$|f_n - \alpha\rho^{n-1}| \leqslant r^n (C + |\alpha| (|\rho| - r)^{-1}) \quad \text{for} \quad n \geqslant 1 \,. \tag{2.11}$$

Proof. We use Cauchy's theorem. Since

$$F(z) - \frac{\alpha}{z-\rho} = f_0 + \sum_{n=1}^{\infty} (f_n - \alpha\rho^{n-1}) z^{-n}$$

and is analytic in $|z| \geqslant r$, we have for $n \geqslant 1$

$$f_n - \alpha\rho^{n-1} = \frac{1}{2\pi i} \int_{|z|=r} z^{n-1} \left\{ F(z) - \frac{\alpha}{z-\rho} \right\} dz,$$

and so

$$|f_n - \alpha\rho^{n-1}| \leqslant r^n \max_{|z|=r} \left| F(z) - \frac{\alpha}{z-\rho} \right|$$

$$\leqslant r^n \left(C + \frac{|\alpha|}{|\rho| - r} \right),$$

which was to be proved. □

In the applications to string enumeration, the rational functions we will deal with satisfy the conditions of the lemma above. They have a single pole (which is real) and no other singularities nearby. Lemma 2.1 enables us to obtain very sharp asymptotic estimates without having to worry about existence of higher order poles or sizes of the residues other than at the dominant singularity. Thus the basic method of asymptotic estimation is very simple, especially when compared to some of the other asymptotic enumeration methods that are used in other situations (cf. [4,20]). The main difficulty is to obtain the generating functions for the quantities that are of interest and show that they satisfy the conditions of Lemma 2.1. As the following sections show, this is possible in a surprising variety of cases, but not always.

3. Correlations of strings

In order to describe the generating functions that arise in string enumeration, we introduce notation (from [13]) that expresses the way that two strings can overlap each other. If X and Y are two words, possibly of different lengths, over some alphabet, the *correlation* of X and Y, denoted by XY, is a binary string of the same length as X. The i-th bit (from the left) of XY is determined by placing Y under X so that the leftmost character of Y is under the i-th character (from the left) of X, and checking whether the characters in the overlapping segments of X and Y are identical. If they are identical, the i-th bit of XY is set equal to 1, and otherwise it is set equal to 0. For example, if $X = cabcabc$ and $Y = abcabcde$, then $XY = 0100100$, as depicted below:

```
X: c a b c a b c
Y: a b c a b c d e              0
     a b c a b c d e            1
       a b c a b c d e          0
         a b c a b c d e        0
           a b c a b c d e      1
             a b c a b c d e    0
               a b c a b c d e  0
```

Note that in general $YX \neq XY$ (they can even be of different lengths), and in the example above YX = 00000000. The *autocorrelation* of a word X is just XX, the correlation of X with itself. In the example above, XX = 1001001.

It is often convenient to interpret a correlation XY as a number in some base t, or else as a polynomial in the variable t, with the most significant bits being the ones on the left. Thus in the example above,

$$XY_t = t^5 + t^2, \quad XY_2 = 36 .$$

String correlations are important because they provide very compact descriptions of how two words can overlap. They are also fascinating subjects of study in their own right, and characterizations of strings that are correlations and enumeration of strings that have a given correlation are contained in [14] (see also [10]).

4. Basic generating functions

Now that correlations of strings have been defined, we can present the basic generating function results. Suppose that A is a word over some finite alphabet Ω of $q \geqslant 2$ letters, and let $f(n)$ be the number of strings of length n over that alphabet that contain no occurrences of A as a subword. We set $f(0) = 1$, and define, as in (2.1),

$$F(z) = \sum_{n=0}^{\infty} f(n) z^{-n} . \tag{4.1}$$

Then the basic result is that

$$F(z) = \frac{z\, AA_z}{1+(z-q)AA_z} . \tag{4.2}$$

This elegant formula, which apparently was first published in a somewhat different form by Solov'ev [26], is quite easy to prove, as we will now demonstrate. First define $f_A(n)$ to be the number of strings over Ω of length n that end with A (at the right end) but contain no other occurrence of A as a subword anyplace else. Set

$$F_A(z) = \sum_{n=0}^{\infty} f_A(n) z^{-n} . \tag{4.3}$$

To prove the formula (4.2) for $F(z)$, we will show that

$$(z-q)\ F(z) + zF_A(z) = z\,, \tag{4.4a}$$

$$F(z) = zAA_z F_A(z)\,, \tag{4.4b}$$

from which (4.2) and

$$F_A(z) = \frac{1}{1+(z-q)AA_z} \tag{4.5}$$

follow immediately. To prove (4.4a), consider any of the $f(n)$ strings of length n that do not contain A as a subword, and adjoin to it (at the right end) any one of the letters from Ω. In this way we obtain $qf(n)$ distinct strings of length $n+1$. Any one of them either contains an appearance of A or not, but if it does, then that appearance has to be in the rightmost position. Therefore we obtain

$$qf(n) = f(n+1) + f_A(n+1)\,, \tag{4.6}$$

and this is valid for all $n \geqslant 0$. Multiplying (4.6) by z^{-n} and summing on $n \geqslant 0$, we easily obtain (4.4a).

To obtain (4.4b), consider any one of the $f(n)$ strings of length n that do not contain A, and append A at its right end. Suppose that the first (from the left) appearance of A in this string has its rightmost character in position $n+r$. Then $0 < r \leqslant |A|$. Furthermore, since A also appears in the last $|A|$ positions, the prefix of A of length r has to be equal to the suffix of A of length r, and so the r-th entry in AA (counted from the right) is 1, which we write as $r \in AA$. The string of length $n+r$ is then counted by $f_A(n+r)$. Conversely, given any string counted by $f_A(n+r)$, $r \in AA$, there is a unique way to extend it to a string of length $n+|A|$ in which the last $|A|$ characters equal A. Therefore for $n \geqslant 0$ we have

$$f(n) = \sum_{r \in AA} f_A(n+r) \, . \tag{4.7}$$

Multiplying (4.7) by z^{-n} and summing on n yields (4.4b).

The main result about string enumeration is proved by methods that are only slightly more elaborate than those shown above, and so we will only state it. First we introduce the notion of a *reduced* set of words. The set $\{A,B, \ldots, T\}$ is called reduced if no word in this set contains any other as a subword. Since we are primarily interested in strings that contain none of a set $\{A, \ldots, T\}$ of words, we can reduce to the case of a reduced set of words by eliminating from that set any words that contain any others as subwords, since if G is a subword of Q, say, strings not containing G also fail to contain Q. Given a reduced set $\{A,B, \ldots, T\}$ of words, we let $f(n)$ denote the number of strings of length n over the alphabet Ω that contain none of $A,B, \ldots, T$, with $f(0) = 1$. We also let $f_H(n)$ denote the number of strings of length n over Ω that contain none of $A,B, \ldots, T$, except for a single appearance of H at the right end of the string. We define the generating functions $F(z)$ of $f(n)$ and $F_H(z)$ of $f_H(n)$ by (4.1) and (4.3).

Theorem 4.1 ([13,25]). If $\{A, \ldots, T\}$ is a reduced set of patterns, the generating functions $F(z)$, $F_A(z), \ldots, F_T(z)$ satisfy the following system of linear equations:

$$\begin{aligned}
&(z-q)F(z) + zF_A(z) + zF_B(z) + \ldots + zF_T(z) = z \\
&F(z) - zAA_zF_A(z) - zBA_zF_B(z) - \ldots - zTA_zF_T(z) = 0 \\
&\ldots \\
&F(z) - zAT_zF_A(z) - zBT_zF_B(z) - \ldots - zTT_zF_T(z) = 0 \, .
\end{aligned} \tag{4.8}$$

It is easy to show ([13]) that the system (4.8) is nonsingular, and therefore determines $F(z)$, $F_A(z), \ldots, F_T(z)$. Moreover, Cramer's rule shows that these generating functions are all rational. When there is only a single excluded word, A, the system (4.8) reduces to (4.4). When more than a single word is excluded, the expressions become much more complicated than the simple forms of (4.2) and (4.5), a topic we will return to later.

Of the many other rational generating functions that arise in string enumeration, we will discuss just one more. Suppose that A is a word over some alphabet Ω of q letters, and let $g(n)$ denote the number of strings of length n over Ω that contain two distinct appearances of A at the beginning and at the end of the string, but no other occurrences of A as a subword anyplace else. (In particular, $g(n) = 0$ for $n \leqslant |A|$.) Then the generating function

$$G(z) = \sum_{n=0}^{\infty} g(n) z^{-n} \tag{4.9}$$

satisfies

$$G(z) = z^{-|A|} + \frac{(q-z)z^{-1}}{1+(z-q)AA_z} . \tag{4.10}$$

The proof of this result, which will be used in the discussion of prefix-synchronized codes in Section 6, is very similar to the proof of the other results mentioned above and can be found in [11]. Estimates for $g(n)$ are also central to the work described in [1,6].

There are many other rational generating functions that occur in string enumeration. In fact, essentially all generating functions that enumerate strings which satisfy restrictions on the number of appearances of particular subwords are rational, even when those generating functions are multivariate (cf. [9]). We will not discuss them any further, however, because they usually cannot be utilized profitably in asymptotic enumeration. Even Theorem 4.1 above has its limitations in this regard, as we will show.

5. Elementary applications of generating functions

In this section we discuss some applications of the generating functions presented in Section 4 to subjects other than asymptotic enumeration.

5.1 Coin-tossing games

The first application is to a rather amusing coin-tossing game. Suppose that A is some sequence of heads or tails, and that we toss a fair coin until A appears as a sequence of consecutive coin tosses (subword). What is the expected number of coin tosses until this happens? We can answer that question very easily using the generating functions of Section 4 (cf. [7,19]). The probability that A does not appear in the first n coin tosses is just $f(n)2^{-n}$, where $f(n)$ denotes the number of strings of heads or tails of length n that do not contain A as a subword. Hence the expected number of coin tosses until A appears is

$$\sum_{n=0}^{\infty} f(n)2^{-n} = F(2),$$

and by (4.2) this is $2AA_2$. In particular, this waiting time is strongly dependent on the periods of A.

W. Penney [22] has invented a penny-tossing game that exhibits the somewhat paradoxical feature of having a nontransitive dominance relation. In Penney's game, two players agree on some integer $k \geqslant 2$ at the start of the game. Player I then selects a pattern A of heads and tails of length k. Player II is shown A, and then proceeds to select another pattern B of k heads and tails. A fair coin is then tossed until either A or B appears as a subword of the sequence of outcomes. If A appears first, Player I wins, and if B appears first, Player II wins.

There are several possible reactions to Penney's game. One is to say that clearly the game is fair as everything is open. Another is to note that if the two players made their choices of A and B independently of each other (with the event $A = B$ resulting in a draw) the game would obviously be fair because of the symmetry of the players' positions, and so Penney's game ought to be favorable to Player II since that player receives additional information, namely the choice of Player I. A third reaction (usually restricted to those who know the waiting time result discussed above) is to say that Player I should have the edge in Penney's game, since he can choose A to have very small waiting time.

The most interesting feature of Penney's game is that if $k \geqslant 3$, then no matter what A is, Player II can choose B so as to obtain probability of winning strictly greater than 1/2. (If $k = 2$ and $A = HT$, say, then Player II can only draw.) Thus this game is nontransitive in that for any A, one can find B that beats it, but then one can find C that beats B, etc. We do not have space to discuss the theory of Penney's game to any great extent (see [13]), but we indicate how the generating functions of Section 4 can be used in its analysis.

Let $\{A,B\}$ be the set of excluded patterns. Then solving the linear system of Theorem 4.1 gives

$$F_A(z) = \frac{BB_z - BA_z}{(z-2)(AA_z \cdot BB_z - AB_z \cdot BA_z) + AA_z + BB_z - AB_z - BA_z} . \tag{5.1}$$

Now $f_A(n)2^{-n}$ is the probability that Player I wins exactly on the n-th throw. Hence Player I wins with probability

$$\sum_{n=0}^{\infty} f_A(n)2^{-n} = F_A(2) = \frac{BB_2 - BA_2}{AA_2 + BB_2 - AB_2 - BA_2} . \tag{5.2}$$

Similarly the probability that Player II wins is given by a formula like (5.2) but with A and B interchanged. Hence the odds of Player II winning are given by

$$\frac{AA_2 - AB_2}{BB_2 - BA_2} . \tag{5.3}$$

This elegant formula, first proved by Conway by a much more complicated method, can be used to analyze Penney's game, and to show that Player II always has a winning strategy. Together with some information about possible sets of periods of strings, it can also be used to determine the optimal choice for B. Details can be found in [13].

5.2 Worst-case behavior of pattern-matching algorithms

The next application of the generating function of Section 4 is to the study of pattern-matching algorithms. We consider the problem of deciding whether a given word A of length k occurs in a string S of length n, with the cost of the algorithm being measured by the number of characters of S that the algorithm looks at. Ever since the appearance of [17], it has been known that one can decide whether A appears in S in $O(n)$ character comparisons, and a lot of work has been done on deriving improved algorithms. Furthermore, starting with the Boyer-Moore algorithm [3], methods have been known that on average look at only αn of the characters of S for some $\alpha < 1$ (which depends on the alphabet and on the algorithm). This situation gave rise to the question whether there is any pattern-matching algorithm that is sublinear in the worst case; i.e., which never looks at more than βn characters of S for $n \geqslant n_0$, where $\beta < 1$. This question was answered in the negative by Rivest [24]. We present a simplified version of Rivest's proof which relies on the explicit form of the generating function (4.2) [13]. The problem is not trivial, since in many cases it is possible to decide whether A appears in S without looking at all the characters of S. For example, if $A = 00$ and $n = 4$, so $S = s_1 s_2 s_3 s_4$, we can query whether $s_2 = 0$. If the answer is that $s_2 \neq 0$, then there will be no need to even look at s_1. Hence we may assume that the answer is that $s_2 = 0$. But then we can query whether $s_3 = 0$. If the answer is that $s_3 = 0$, then $A = s_2\, s_3$, and we do not need to look at s_1 and s_4. If the answer is that $s_3 \neq 0$, though, we do not need to look at s_4. Thus we can always decide whether A is a subword of S by looking at $\leqslant 3$ characters of S. Note that we have phrased the question in terms of deciding whether A occurs as a subword of S, instead of asking for the determination of all such occurrences of A. In the latter situation the problem is indeed trivial (at least in general), since if $A = 0...0$, and the answer to a query about any character is that it's 0, any algorithm will clearly have to look at all characters of S to determine all the $n-k+1$ occurrences of A.

To prove that there is no sublinear worst-case algorithm for determining whether A is a subword of S, suppose that there is an algorithm which, given any string $S = s_1 \cdots s_n$ of length n, can determine whether A is a subword of S by looking at $\leqslant n-1$ characters of S. Suppose that for a particular string S, the algorithm looks only at the characters in positions $i_1, \ldots, i_r$. Then, for any of the q^{n-r} strings $S' = s'_1 \cdots s'_n$ which have $s'_{i_j} = s_{i_j}$, $1 \leqslant j \leqslant r$, the algorithm will examine the same characters as it does in examining S, and will come to the same conclusion as to whether A is a subword of S or not. Since $r \leqslant n-1$, $q \mid q^{n-r}$, and so the total number of strings that do not contain A as a subword is divisible by q. This can of course happen for some values of n. The generating function (4.2) shows that this cannot happen for too many values of n.

Let $f(n)$ be the number of strings of length n that do not contain A as a subword, with $f(0) = 1$. Then (4.2) implies that if

$$1+(z-q)\, AA_z = z^k - \sum_{j=0}^{k-1} h_j z^j, \tag{5.4}$$

then

$$f(n) = \sum_{j=0}^{k-1} h_j f(n-k+j) \quad \text{for } n \geqslant k .$$

But by (5.4), $h_0 \equiv 1 \pmod q$, so

$$f(n-k) \equiv f(n) - \sum_{j=1}^{k-1} h_j f(n-k+j) \pmod q \tag{5.5}$$

for $n \geqslant k$. If for some n we had $f(n) \equiv f(n-1) \equiv \cdots \equiv f(n-k+1) \equiv 0 \pmod q$, then by (5.5) we would have $f(n-k) \equiv 0 \pmod q$, and by induction $f(m) \equiv 0 \pmod q$ for all m, $0 \leqslant m \leqslant n$. But that contradicts $f(0) = 1$.

The conclusion to be drawn from the above discussion is that among any k consecutive values of n, there is at least one for which $f(n)$ is not divisible by q. For such a value of n, any algorithm will have to examine all n characters of some string S to determine whether A is a subword of S or not. Since the number of characters that have to be examined in the worst case is a monotone function of n, this also shows that for any n, at least $n-k+1$ characters have to be examined in some string S, and so there is no algorithm that is sublinear in the worst case.

5.3 Unavoidable sets of words

The final application to be discussed in this section is to the problem of determining when a set of words is *unavoidable*; i.e., any sufficiently long string over the given alphabet contains at least one of these words as a subword. As an example, over the binary alphabet, the set {000, 1111, 01} is unavoidable. How can we recognize when a set is unavoidable? One way is to use Theorem 4.1. If $\{A, B, \ldots, T\}$ is the set in question, we can use the system (4.8) to determine $F(z)$, the generating function for the number of strings that contain none of $\{A, \ldots, T\}$ as subwords. This set of words is then unavoidable if and only if $F(z)$ is a polynomial in z^{-1}.

Another way to determine unavoidability of a set of words is through the construction of a finite-state automaton recognizing the words $A, \ldots, T$. The two approaches seem to be of comparable complexity, but the one presented above might be somewhat simpler to implement.

The main reason for discussing the problem of determining unavoidability here is that it points out some of the limitations of the use of generating functions in enumeration. Unavoidable sets $\{A, \ldots, T\}$ lead to the collapse of the generating functions $F(z)$ to polynomials in z^{-1}. Yet almost all the basic asymptotic enumeration results, such as those of Section 6, depend on the use of Lemma 2.1, which depends on the generating function having a single dominant singularity away from 0. Thus we cannot expect to be able to apply the simple method of Lemma 2.1 even to all the generating functions determined by Theorem 4.1.

6. Asymptotic enumeration of strings

The end of the preceding section concluded with remarks about the limitations on the use of generating functions in asymptotic string enumeration. This section demonstrates that in spite of these limitations, numerous interesting asymptotic estimates can be obtained. The basic method relies on nothing more sophisticated than Lemma 2.1 and Rouche's theorem. However, usually substantial work is needed to apply these basic tools.

Perhaps the simplest case to consider is that of counting the number of strings, $f(n)$, of length n over an alphabet of size 2 that do not contain some fixed pattern A as a subword. (The restriction to $q=2$ is imposed here for the sake of clarity, to make it easier to compare various quantities.) By (4.2), the generating function of $f(n)$ has the form

$$F(z) = \frac{zAA_z}{1+(z-2)AA_z} . \tag{6.1}$$

To apply Lemma 2.1, we need to show that $F(z)$ has a single dominant pole. To do this, we need to study the zeros of the polynomial in the denominator on the right side of (6.1). As it turns out, the only feature of the denominator polynomial that we will use is that AA_z has all its coefficients 0 and 1. It is easy to obtain the following result (cf. Lemma 2 of [11]).

Lemma 6.1. There is a constant $c_1 > 0$ such that if $u(z)$ is a polynomial,

$$u(z) = z^{k-1} + \sum_{j=2}^{k} u_j z^{k-j} \ , \ u_j = 0 \text{ or } 1, \tag{6.2}$$

then for $|z| \geqslant 1.7$,

$$|u(z)| > c_1 \, (1.7)^k \ . \tag{6.3}$$

Using Lemma 6.1, it is easy to show that $F(z)$ has a single dominant singularity (cf. Lemma 3 of [11]).

Lemma 6.2. There is a constant $c_2 > 0$ *such that if* $u(z)$ *satisfies the conditions of Lemma 6.1 and* $k \geqslant c_2$, *the* $1+(z-2)\,u(z)$ *has exactly one zero* ρ *(of multiplicity one) with* $|\rho| \geqslant 1.7$.

Proof. On the circle $|z| = 1.7$,

$$|z-2|\,|u(z)| > 0.3c_1(1.7)^k > 1 \ \text{ for } \ k \geqslant c_2 \, .$$

Hence by Rouche's theorem, $(z-2)\,u(z)$ and $1+(z-2)\,u(z)$ have exactly the same number of zeros in $|z| \geqslant 1.7$, namely one. □

It is also possible to localize the dominant zero ρ quite precisely. In a neighborhood of $z=2$, $u(z)$ is essentially $u(2)$, so we expect the zero ρ to be close to the solution of $1+(z-2)\,u(2) = 0$; i.e., $\rho \approx 2 - u(2)^{-1}$. This can be shown to be true, and we obtain the following result. (A more precise estimate is contained in Lemma 4 of [11].)

Lemma 6.3. If $u(z)$ *satisfies the conditions of Lemma 6.1, and* $k \geqslant c_2$, *then the single zero* ρ *of* $1+(z-2)\,u(z)$ *in* $|z| \geqslant 1.7$ *satisfies*

$$\rho = 2 - \frac{1}{u(2)} - \frac{u'(2)}{u(2)^3} + O(k^2 2^{-3k}) \quad \text{as} \quad k \to \infty \, .$$

On the circle $|z| = 1.7$, the generating function $F(z)$ of (6.1) is $O(1)$ by Lemma 6.1. Hence Lemma 2.1 applies directly, and we find that (cf. Eq. (7.3) of [13])

$$f(n) = \frac{\rho^n}{1-(2-\rho)^2 u'(\rho)} + O((1.7)^n) \quad \text{as } n \to \infty, \tag{6.4}$$

where $u(z) = AA_z$.

Estimates of the form (6.4) can be used in a variety of applications. We will discuss only the case of prefix-synchronized codes. A code $\mathcal{C}$ of block length n (over the binary alphabet, say) is said to be *comma-free* if for every pair $(a_1, \ldots, a_n)$, $(b_1, \ldots, b_n)$ of codewords in $\mathcal{C}$, no subword of length n of their concatenation $(a_1, \ldots, a_n, b_1, \ldots, b_n)$, other than the first and last blocks of length n, is a codeword of $\mathcal{C}$. When such a code is used in a communication system, the receiver can always recover synchronization by examining the transmitted signal until a codeword is encountered. It is known [5,16] that at least for odd n, comma-free codes of length n can be found which are of size

$$n^{-1} 2^n (1+o(1)) \quad \text{as } n \to \infty . \tag{6.5}$$

One disadvantage of general comma-free codes, though, is that it is usually hard to decide whether a word is in the code. In order to overcome this disadvantage, E. N. Gilbert invented *prefix-synchronized codes*, a special class of comma-free codes. Such a code is characterized by a prefix A of length $k < n$. Every codeword has A as a prefix, and the remaining $n-k$ characters of the codewords are chosen so that in any concatenation of codewords, A occurs only at the beginnings of codewords. Synchronization is then very easy, since all that is needed is to scan the incoming stream of characters until the prefix A is encountered, and that can be done very efficiently using string searching algorithms. The question then arises as to the efficiency of prefix-synchronized codes. Gilbert [8] showed that one can construct such codes of size

$$n^{-1} 2^{n-1} (1+o(1)) \log 2 \sim (0.346\ldots) n^{-1} 2^n \quad \text{as } n \to \infty,$$

which is not much less than the bound (6.5) attainable (at least for odd n) by maximal comma-free codes. Gilbert also conjectured that optimal prefixes can be chosen to be of the form $A = 11...110$ for an appropriate length k. This conjecture has been shown to be true, and a more precise estimate of the size of the largest prefix-synchronized code has been obtained [11].

Theorem 6.1. Given a positive integer n, choose the unique integer $k = k(n)$ so that $\beta = n2^{-k}$ satisfies

$$\log 2 \leqslant \beta < 2 \log 2 .$$

Then the maximal prefix-synchronized code of length n has cardinality

$$n^{-1} 2^n \beta e^{-\beta} (1+o(1)) \quad \text{as} \quad n \to \infty .$$

Moreover, if n is large enough, maximal prefix-synchronized codes can be constructed with prefixes $A = 11...10$ of length $k-1$, k, or $k+1$.

The results of Theorem 6.1 can be generalized (see [11]) to non-binary alphabets. However, if the alphabet is of size $q \geqslant 5$, then it is not always possible to construct maximal codes using prefixes of the form $A = 11...10$. For all alphabets, the phenomenon of asymptotic oscillation holds, so that the size of the maximal prefix-synchronized codes is not asymptotic to a constant multiple of $n^{-1} q^n$. For example, Theorem 6.1 shows that for $q=2$, the ratio of the maximal code size to $n^{-1} 2^n$ oscillates between $(\log 2)/2 = 0.3465...$ and $e^{-1} = 0.3678....$

The proof of Theorem 6.1 relies on the generating function formula (4.10) and the stronger forms of lemmas 6.1-6.3 that are presented in [11]. Note first of all that given a prefix A, the allowable codewords are precisely those strings of length n which, when A is adjoined to them at the right end, have A as a prefix and a suffix, but do not contain A as a subword elsewhere. Hence, in the notation of Section 4, the maximal size of a code of length n with prefix A is just $g(n+|A|)$, for which we have the generating function (4.10). In particular, this formula shows that the size of the maximal code determined by a prefix A depends only on the autocorrelation of A, and not on A itself. The proof of Theorem 6.1 is quite involved, but in principle quite straightforward, involving very careful analysis.

So far we have discussed asymptotic enumeration results only in cases where we were basically excluding a single word A, and so the generating functions ((4.2) and (4.10)) had denominators of the simple form $1+(z-q)AA_z$. When we count strings that exclude larger numbers of words, we cannot expect our denominator to be this nice and easy to deal with. The unavoidable sets of words discussed at the end of Section 5 show that in fact quite pathological behavior can occur. However, in some cases it is possible to obtain sharp asymptotic estimates from Theorem 4.1. That is the case with repetitive runs, which were studied extensively in [15]. For simplicity we will again consider only the binary alphabet. Let $B = b_1 \cdots b_n$ be a word that is nonperiodic; i.e., B cannot be written as $B = CC \cdots C$ for any word C shorter than B. (As an example, 010 is nonperiodic while 0101 is not.) A *B-run* of length k is a word $A = BB \cdots BB'$ of length k, where B' is a prefix of B. For example, if $B = 010$, then 01001 is a B-run of length 5. We will discuss the appearance of long B-runs in random sequences, when B is held fixed. In many situations, a much more interesting question to study is that of appearance of B^*-runs. A *B^*-run* of length k is a word $A = B''BB \cdots BB'$, where B' is a prefix of B and B'' is a suffix of B. Thus, if $B = 010$, then 001001 is a B^*-run of length 6.

The theory of B-runs follows easily from the results we have already discussed. If $A = BB \cdots BB'$, $|A| = k$, and $f(n)$ denotes the number of binary strings of length n that do not contain A, then the generating function for $f(n)$ is given by (4.2). Moreover, since B is nonperiodic, it is easy to show that

$$AA_z = z^{k-1} + z^{k-m-1} + z^{k-2m-1} + \ldots + z^{k-rm-1} + p(z),$$

where $r = [k/m]-1$ and $\deg p(z) < m$, so we obtain the following result (Theorem 2.2 of [15]).

Theorem 6.2. If B is a nonperiodic pattern of length m, and A is a B-run of length k, then the generating function of $f(n)$, the number of strings of length n that do not contain A, is of the form

$$F(z) = \frac{zQ(z)}{1+(z-2)Q(z)},$$

where

$$Q(z) = \frac{z^{k+m-1}}{z^m-1} + \frac{q(z)}{z^m-1},$$

and $q(z)$ is a polynomial of degree $<2m$ with coefficients bounded by 2 in absolute value.

The form of $F(z)$ given by Theorem 6.2 is sufficiently explicit to obtain very good estimates for the dominant singularity ρ and the residue at ρ so that

$$f(n) \approx 2^n \exp(-n(1-2^{-m})2^{-k}) \tag{6.6}$$

is a very good approximation. (See [15] for more precise estimates.)

The theory of B^*-runs is considerably more complicated. The number of strings of length n that do not contain a B^*-run of length k is the number $f(n)$ of Theorem 4.1 when we exclude the m words $A(1), \ldots, A(m)$ of length k given by

$$A(r) = b_{m-r+1}\, b_{m-r+2} \cdots b_m\, BB \cdots BB',$$

where $B' = B'(r)$ is a prefix of B of the appropriate length. The denominator of the generating function $F(z)$ of $f(n)$ is then the determinant of an $m+1$ by $m+1$ matrix. The entries of that matrix are very strongly constrained; for example, if $1 \leqslant r \leqslant s \leqslant m$, then it can be shown that

$$A(r)A(s)_z = \frac{z^{k-1-s+r} + g_{s,r}(z)}{z^m-1},$$

where $g_{s,r}(z)$ is a polynomial of degree $<2m$ with coefficients that are $\leqslant 2$ in absolute value. Therefore by going through some very involved algebra, one can obtain a relatively explicit form for $F(z)$ (Theorem 2.3 of [15]).

Theorem 6.3. If B is a nonperiodic word of length m, then the generating function $F(z)$ of $f(n)$, the number of binary strings of length n that contain no B^*-run of length $k \geqslant 3m$ has the form

$$F(z) = \frac{z^{(k+1)m+1}\,(z^m-1)^{m-1} + g_1(z)}{(z-2)z^{(k+1)m}\,(z^m-1)^{m-1} + g_2(x)}\ .$$

Here $g_1(z)$ and $g_2(z)$ are polynomials of degrees $\leqslant k(m-1)+c_1$ with coefficients that are $\leqslant c_2$ in absolute value, and c_1 and c_2 are constants that depend only on m, but not on k. Furthermore,

$$g_1(2) = O(2^{k(m-1)}),$$

$$g_2(2) = m2^{km-k+m}(2^m-1)^{m-1} + O(2^{k(m-2)}),$$

where the constants implied by the O-notation depend only on m.

Once Theorem 6.3 is available, it is not too difficult to show that $F(z)$ has a dominant singularity and to estimate its location and residue. One finds that

$$f(n) \approx 2^n \exp(-nm2^{-k-1}) \tag{6.7}$$

is a very good approximation in this case (cf. (6.6), which corresponds to B-runs).

Once estimates like (6.6) and (6.7) (or rather, the more precise estimates established in [15]) are obtained, one can obtain a variety of results from them. For example, of $E_B(n)$ denotes the expected length of a B-run in a random string of length n, and $E_B^*(n)$ the expected length of a B^*-run, then [15]

$$E_B(n) = \lg n - 1/2 + \gamma(\log 2)^{-1} + \lg(1-2^{-m})$$

$$+ \nu(\lg n + \lg(1-2^{-m})) + o(1) \quad \text{as} \quad n \to \infty,$$

$$E_B^*(n) = \lg n - 3/2 + \gamma(\log 2)^{-1} + \lg m$$

$$+ \nu(\lg n + \lg m) + o(1) \quad \text{as} \quad n \to \infty,$$

where lg x is the logarithm of x to base 2, γ is Euler's constant (0.577...), and $\nu(x)$ is a certain nonconstant continuous periodic function of period 1 and mean 0. (There is a mistake in [15] in the statement of the result about $E_B(n)$, with -1/2 incorrectly replaced by -5/2.) When B is a single letter, such results were obtained first by Boyd [2].

References

1. A. Apostolico and A. S. Fraenkel, Fibonacci representations of strings of varying length using binary separators, to be published.

2. D. W. Boyd, Losing runs in Bernoulli trials, unpublished manuscript (1972).

3. R. S. Boyer and J. S. Moore, A fast string searching algorithm, *Comm. ACM 20* (1977), 762-772.

4. N. G. de Bruijn, *Asymptotic Methods in Analysis*, North-Holland 1958.

5. W. L. Eastman, On the construction of comma-free codes, *IEEE Trans. Information Theory IT-11* (1965), 263-266.

6. A. S. Fraenkel, The use and usefulness of numeration systems, to be published.

7. H. U. Gerber and S. R. Li, The occurrence of sequence patterns in repeated experiments and hitting times in a Markov chain, *Stoch. Proc. Appl. 11* (1981), 101-108.

8. E. N. Gilbert, Synchronization of binary messages, *IRE Trans. Inform. Theory IT-6* (1960), 470-477.

9. I. P. Goulden and D. M. Jackson, *Combinatorial Enumeration*, Wiley 1983.

10. L. J. Guibas, Periodicities in strings, to be published in this volume.

11. L. J. Guibas and A. M. Odlyzko, Maximal prefix-synchronized codes, *SIAM J. Appl. Math. 35* (1978), 401-418.

12. L. J. Guibas and A. M. Odlyzko, A new proof of the linearity of the Boyer-Moore string searching algorithm, pp. 189-195 in Proc. 18th IEEE Found. Comp. Sci. Symposium, 1977. Revised version in *SIAM J. Comput. 9* (1980), 672-682.

13. L. J. Guibas and A. M. Odlyzko, String overlaps, pattern matching, and nontransitive games, *J. Comb. Theory (A) 30* (1981), 183-208.

14. L. J. Guibas and A. M. Odlyzko, Periods in strings, *J. Comb. Theory (A) 30* (1981), 19-42.

15. L. J. Guibas and A. M. Odlyzko, Long repetitive patterns in random sequences, *Z. Wahrscheinlichkeits. v. Geb. 53* (1980), 241-262.

16. B. H. Jiggs, Recent results in comma-free codes, *Canad. J. Math. 15* (1963), 178-187.

17. D. E. Knuth, J. H. Morris, and V. R. Pratt, Fast pattern matching in strings, *SIAM J. Comput. 6* (1977), 323-350.

18. K. K. Kubota, On a conjecture of Morgan Ward, I and II, *Acta Arith. 33* (1977), 27-48.

19. P. T. Nielsen, On the expected duration of a search for a fixed pattern in random data, *IEEE Trans. Inform Theory IT-19* (1973), 702-704.

20. A. M. Odlyzko, Some new methods and results in tree enumeration, *Congressus Numerantium 42* (1984), 27-52.

21. A. Odlyzko and B. Richmond, On the compositions of an integer, pp. 199-210 in *Combinatorial Mathematics VII*, Proc. Seventh Austr. Conf. Comb. Math., R. W. Robinson, G. W. Southern, and W. D. Wallis, eds., Lecture Notes in Mathematics #829, Springer 1980.

22. W. Penney, Problem: penney-ante, *J. Recreational Math. 2* (1969), 241.

23. P. Révész, Strong theorems on coin tossing, pp. 749-754 in *Proc. 1978 Int'l. Congress of Mathematicians*, Helsinki, 1980.

24. R. L. Rivest, On the worst-case behavior of string-searching algorithms, *SIAM J. Comput. 6* (1977), 669-674.

25. S. W. Roberts, On the first occurrence of any of a selected set of outcome patterns in a sequence of repeated trials, unpublished Bell Laboratories report (1963).

26. A. D. Solov'ev, A combinatorial identity and its application to the problem concerning the first occurrences of a rare event, *Theory Prob. Appl. 11* (1966), 276-282.

Two counting problems solved via string encodings

Andrei Broder[1]
Department of Computer Science
Stanford University
Stanford, CA 94305

1. Introduction

The common feature of the two problems presented in this paper is the use of bijective string encodings to solve counting questions. The first problem deals with a "natural" object – the size of the transitive closure of a random mapping. In the second problem certain contrived combinatorial objects are counted in two different ways in order to devise a new class of Abelian identities.

2. Preliminaries

The section presents some mathematical entities that will be used later.

2.1. The r-Stirling numbers

In this paper the Stirling numbers of the second kind are denoted $\left\{{n \atop m}\right\}$; they are defined combinatorially as the number of partitions of the set $\{1,\ldots,n\}$ into m non-empty disjoint unlabelled sets. Good expositions of their properties can be found, for example, in [Comtet74], [Riordan58], or [Jordan74].

The r-Stirling numbers, $\left\{{n \atop m}\right\}_r$, count certain restricted partitions and are defined, for all integer $r \geq 0$, as the number of partitions of the set $\{1,\ldots,n\}$ into m non-empty disjoint subsets, such that the numbers $1, 2, \ldots, r$ are in distinct subsets.

The properties of the r-Stirling numbers are discussed in [Broder84]. They were also studied under different names and notations in [Nielsen23], [Carlitz80a], [Carlitz80b], [Neuman82], and [Koutras82]. Their asymptotics were studied in [IM65]. Here we shall need only the fact that

$$\left\{{n+m \atop n}\right\}_r = \frac{n^{2m}}{(2m)!!} + O(n^{2m-1}), \tag{1}$$

as $n \to \infty$, for fixed m and r. (The notation $x!!$ means $x(x-2)(x-4)\ldots$.)

---

[1] Current affiliation: DEC Systems Research Center, 130 Lytton Ave., Palo Alto, California 94301. This work was supported in part by the National Science Foundation under grant number MCS-83-00984

NATO ASI Series, Vol. F12
Combinatorial Algorithms on Words
Edited by A. Apostolico and Z. Galil

2.2. The Q-series

Knuth defined the Q-series in [Knuth83] as

$$Q_n(a_1, a_2, \ldots) = \sum_{k \geq 1} \binom{n}{k} k! n^{-k} a_k. \tag{2}$$

For a certain fixed sequence $a_1, a_2, \ldots$, this function depends only on n. In particular, $Q_n(1,1,1,\ldots)$ is denoted $Q(n)$. The asymptotic behaviour of $Q(n)$ is well understood [Ramanujan12],

$$Q(n) = \sqrt{\frac{\pi n}{2}} - \frac{1}{3} + \frac{1}{12}\sqrt{\frac{\pi}{2n}} + \cdots, \tag{3}$$

and in fact in many cases $Q(n)$ behaves like a discrete analog of $\sqrt{n}$.

The Q-series are relevant to many problems in the analysis of algorithms [Knuth83], for instance representation of equivalence relations [KS78], hashing [Knuth73, §6.4], interleaved memory [KR75], labelled trees counting [Moon71], optimal cacheing [Knuth83], permutations *in situ* [Stanf81], and random mappings [Knuth81, §3.1].

There is an interesting relation between Q-series and regular and r-Stirling numbers; For $h \geq 1$

$$Q_n\left(\left\{ {h \atop 1} \right\}_r, 2\left\{ {h+1 \atop 2} \right\}_r, \ldots\right) = \sum_{k \geq 1} \frac{n^{\underline{k}}}{n^k} k \left\{ {h+k-1 \atop k} \right\}_r = n^h \frac{n^{\underline{r}}}{n^r}, \tag{4}$$

and in particular for $r = 0$ (i.e. regular Stirling numbers) we have

$$Q_n\left(\left\{ {h \atop 1} \right\}, 2\left\{ {h+1 \atop 2} \right\}, \ldots\right) = n^h. \tag{5}$$

The half-integer Stirling numbers [Knuth83] were defined to satisfy a "half-integer" analog of the above equation, with $Q(n)$ playing the role of $\sqrt{n}$. That is, the half-integer Stirling numbers, satisfy

$$Q_n\left(\left\{ {h+1/2 \atop 1} \right\}, 2\left\{ {h+3/2 \atop 2} \right\}, \ldots\right) = n^h Q(n). \tag{6}$$

It can be shown that

$$\left\{ {n+1/2 \atop m} \right\} = \sum_{r \geq 1} \left\{ {n \atop m} \right\}_r, \tag{7}$$

which is in agreement with equations 4 and 6.

We shall later make use of a generalization of equations 3; for s fixed

$$Q(1^{2s}, 2^{2s}, \ldots) = \sum_{k \geq 1} \frac{n^{\underline{k}}}{n^k} k^{2s} = (2s-1)!!\, n^s Q(n) + O(n^s), \tag{8}$$

and

$$Q(1^{2s+1}, 2^{2s+1}, \ldots) = \sum_{k \geq 1} \frac{n^{\underline{k}}}{n^k} k^{2s+1} = (2s)!!\, n^{s+1} + O(n^{s+1/2}). \tag{9}$$

3. The first problem – The transitive closure of a random mapping

Assume that all the n^n mappings $f : \{1, \ldots, n\} \to \{1, \ldots, n\}$ are equally probable. Denote f^* the transitive closure of f. Let $A \subset \{1, \ldots, n\}$ be a fixed set of size s. Then the size of the image of A under f^* is a random variable. What is its probability distribution?

The motivation for this problem is given, among others, by the study of the crypt-analytic time – memory tradeoff [Hellman80]. It can also be seen as an epidemic process on a certain type of random graphs, which is a question of interest in statistics. (see e.g. [Gertsbach77]) Another solution to this problem based on a quite different approach appears in [Pittel82].

The first step is to reduce our question to a somewhat simpler instance. Denote $F_n(A, k)$ to be the family of functions f such that $|f^*(A)| = k$ for a fixed set $A \subset \{1, \ldots, n\}$ of size s. To construct an arbitrary function $f \in F_n(A, k)$ proceed as follows:

1. Choose $k - s$ points not in A to be in $f^*(A)$. (This can be done in $\binom{n-s}{k-s}$ ways)
2. The restriction of f to the k points in $f^*(A)$ must be some function in $F_k(A, k)$. ($|F_k(A, k)|$ ways)
3. At each of the $n - k$ points not in $f^*(A)$ give f an arbitrary value. (n^{n-k} ways)

From this contruction, it is clear that

$$|F_n(A, k)| = \binom{n-s}{k-s} |F_k(A, k)|\, n^{n-k}, \tag{10}$$

and hence, in order to determine $|F_n(A, k)|$, it suffices to compute $|F_k(A, k)|$, which is the size of the family of functions $f : \{1, \ldots, k\} \to \{1, \ldots, k\}$ such that for a specific set $A \subset \{1, \ldots, k\}$ of size s, the transitive closure of f satisfies $f^*(A) = \{1, \ldots, k\}$.

Let $A = a_1, a_2, \ldots, a_s$. Encode any function $f \in F_k\ (A, k)$ as a string of the form

$$
\begin{aligned}
f \leftrightarrow a_1 a_2 \ldots a_s \\
& f(a_1) f^2(a_1) \ldots f^{i_1}(a_1) \\
& \quad f(a_2) f^2(a_2) \ldots f^{i_2}(a_2) \\
& \quad\quad \ldots f(a_s) f^2(a_s) \ldots f^{i_s}(a_s),
\end{aligned}
$$

where $f^{i_j}(a_j)$ is the first "already seen" element that is encountered starting from a_j.

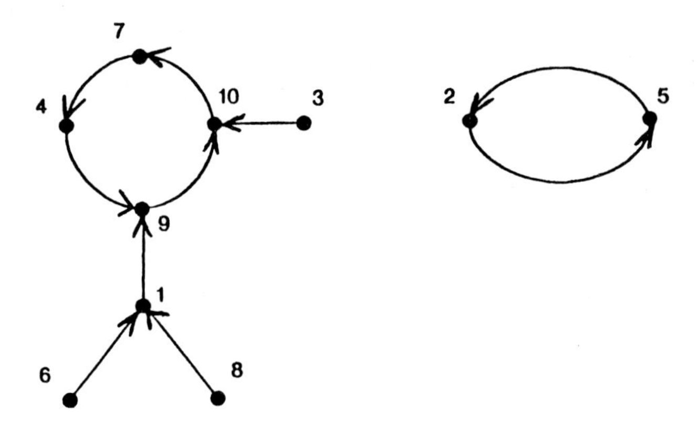

Figure 1 An encoding example

For example for the mapping in Figure 1, we have

$$f^*(\{2,3,6,8,9\}) = \{1,\ldots,10\},$$

and the encoding of f is

$$f \leftrightarrow 2 \quad 3 \quad 6 \quad 8 \quad 9 \quad 5 \quad 2 \quad 10 \quad 7 \quad 4 \quad 9 \quad 1 \quad 9 \quad 1 \quad 10.$$

It is obvious that the encoding is 1–1. Hence it suffices to count how many possible legal words exists. First remark that

1. The length of each word is exactly $s + k$.
2. Within the first $s + k - 1$ positions each of the k letters appears at least once, and the first s positions contain the letters $a_1, \ldots, a_s$ in that order.

   To construct a legal word:

- Partition the first $k + s - 1$ positions into k non-empty subsets such that positions $1, \ldots, s$ are in different subsets. (Each subset corresponds to a certain letter.) This can be done in $\left\{ {k+s-1 \atop k} \right\}_s$ ways.
- Associate to the subsets containing the positions $1, \ldots, s$ the letters $a_1, a_2, \ldots, a_s$ in this order. To each of the remaining $k - s$ subsets associate one of the remaining $k - s$ letters. ($(k - s)!$ ways)
- Choose any letter for the last position. (k possibilities)

  From this construction it follows that

$$|F_k(A,k)| = (k-s)!\,k \left\{ {k+s-1 \atop k} \right\}_s, \tag{11}$$

and therefore, by equation 10,

$$|F_n(A,k)| = \binom{n-s}{k-s}(k-s)!\,n^{n-k}k\left\{ {k+s-1 \atop k} \right\}_s, \tag{12}$$

and finally

$$\mathbf{Pr}\left(|f^*(A)| = k\right) = p_n(A,k) = \frac{|F_n(A,k)|}{n^n} = \frac{1}{n^{\underline{s}}}\frac{n^{\underline{k}}}{n^k}k\left\{ {k+s-1 \atop k} \right\}_s, \tag{13}$$

with $p_n(A,k)$ a notation.

As a quick check, note that by equation 4, we indeed have

$$\sum_k p_n(A,k) = \frac{1}{n^{\underline{s}}}\sum_k \frac{n^{\underline{k}}}{n^k}k\left\{ {k+s-1 \atop k} \right\}_s = 1. \tag{14}$$

Of more interest is the *average* size of the transitive closure, that is the sum

$$\sum_k k p_n(A,k) = \frac{1}{n^{\underline{s}}}\sum_k \frac{n^{\underline{k}}}{n^k}k^2\left\{ {k+s-1 \atop k} \right\}_s. \tag{15}$$

For fixed s, combining equations 1, 8, and 9, we obtain that

$$\begin{aligned}\frac{1}{n^{\underline{s}}}\sum_k \frac{n^{\underline{k}}}{n^k}k^2\left\{ {k+s-1 \atop k} \right\}_s &= \frac{1}{n^{\underline{s}}}\sum_k \frac{n^{\underline{k}}}{n^k}k^2\left(\frac{1}{(2s-2)!!}k^{2s-2} + O(k^{2s-3})\right)\\ &= \frac{n^s}{n^{\underline{s}}}\frac{(2s-1)!!}{(2s-2)!!}Q(n) + O(1) = \frac{(2s-1)!!}{(2s-2)!!}\sqrt{\frac{\pi n}{2}} + O(1).\end{aligned} \tag{16}$$

For $s = o(n)$ it can be shown [Pittel82] that

$$\frac{1}{n^{\underline{s}}}\sum_k \frac{n^{\underline{k}}}{n^k}k^2\left\{ {k+s-1 \atop k} \right\}_s \approx \sqrt{2sn}. \tag{17}$$

4. The second problem – "Half-integer" Abelian identities

This section presents a generalization and simplification of certain results first presented elsewhere [Broder83]. For sake of completeness the background material is repeated here.

4.1. Abelian identities

The sums of the type

$$A_n(x,y;p,q) = \sum_k \binom{n}{k}(x+k)^{k+p}(y+n-k)^{n-k+q}, \qquad p,q,n \text{ integers,}$$

were called "Abelian binomial sums" by Riordan [Riordan68],[Riordan69].

With this notation, the famous Abel identity [Abel1826] becomes

$$A_n(x,y;-1,0) = x^{-1}(x+y+n)^n, \tag{18}$$

that is

$$\sum_k \binom{n}{k}(x+k)^{k-1}\big(y+(n-k)\big)^{n-k} = x^{-1}(x+y+n)^n. \tag{19}$$

Another well known example of an identity involving Abelian sums is the Cauchy formula [Cauchy1826]

$$A_n(x,y;0,0) = \sum_k \binom{n}{k} k!(x+y+n)^{n-k}, \tag{20}$$

which for $x = y = 0$ gives

$$\sum_k \binom{n}{k} k^k(n-k)^{n-k} = n^n\big(Q(n)+1\big). \tag{21}$$

Riordan found a recurrence and a symmetry formula for A_n and used them to prove these identities and also to iteratively derive similar "Abelian identities" for p and q between -3 and 3.

Another proof method is due to Françon [Françon74], based on an encoding developed by Foata and Fuchs [FF70], which associates to a mapping $f:\{1,\ldots,n\}\to\{1,\ldots,n\}$, a word ω of length n over the alphabet $\Sigma = \{\overline{1},\ldots,\overline{n}\}$. Françon applied this encoding to a suitably chosen family of mappings in order to prove the Abel identity and the Cauchy formula by counting arguments. In a previous paper [Broder83] the author obtained a general explicit expression for the Abelian identities, for all $p, q \geq 0$, using similar word counting arguments. For $x = y = 0$ the general identity is

$$\sum_k \binom{n}{k} k^{k+p}(n-k)^{n-k+q} = n^{n-1}\sum_{k,l\geq 0}\frac{n^{\underline{k+l}}}{n^{k+l}}\left\{ {k+p \atop k} \right\}\left\{ {l+q \atop l} \right\}(k+l). \tag{22}$$

4.2. The Foata-Fuchs encoding

In this section we shall consider encodings that are 1-1 correspondences between the n^n distinct mappings from $\{1,2,\ldots,n\}$ into $\{1,2,\ldots,n\}$ and the n^n distinct words of length n over the alphabet $\Sigma = \{\overline{1},\overline{2},\ldots,\overline{n}\}$. Let f be such a mapping. The trivial encoding of f is given by

$$f \leftrightarrow \overline{f(1)}\,\overline{f(2)}\ldots\overline{f(n)},$$

that is the word associated to f is obtained by concatenating the letters corresponding to the values taken by f in $1,2,\ldots,n$. The trivial encoding of the mapping in Figure 2 is

$$\overline{4}\;\overline{3}\;\overline{15}\;\overline{15}\;\overline{6}\;\overline{3}\;\overline{16}\;\overline{13}\;\overline{9}\;\overline{4}\;\overline{10}\;\overline{7}\;\overline{16}\;\overline{9}\;\overline{5}\;\overline{7}\,.$$

The trivial encoding of a mapping f is denoted $C_0(f)$.

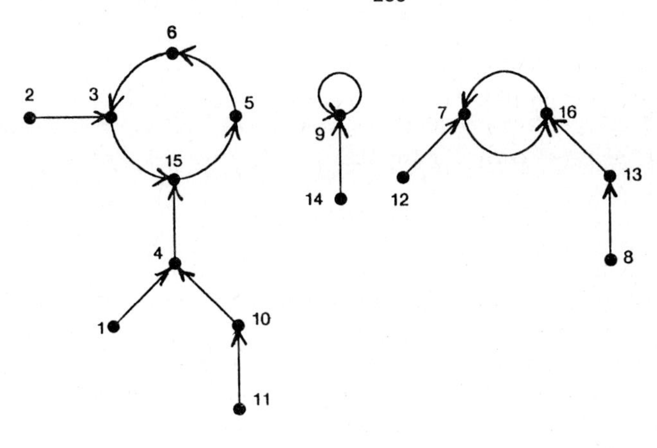

Figure 2 A mapping on 16 elements.

The Foata-Fuchs encoding (FF-encoding) of a mapping f is the concatenation of $l+1$ words, $\omega_0, \omega_1, \ldots, \omega_l$ where l is the number of leaves (nodes with indegree 0) in the graph of f. The word ω_0 describes the permutation induced by the recurrent (an element x is called recurrent if there exists an $i > 0$ such that $f^i(x) = x$.) elements and is generated by the following algorithm. (After each step, the result of this algorithm, applied to the mapping in Figure 2, is shown in square brackets.)

Algorithm A.

1. Write the permutation as a product of cycles.

   $[(3,15,5,6)(9)(7,16)]$

2. Reverse each cycle.

   $[(6,5,15,3)(9)(16,7)]$

3. Rotate each cycle such that the maximum element in each cycle is in the first position. The element in the first position of a cycle is called the cycle leader.

   $[(15,3,6,5)(9)(16,7)]$

4. Put the cycles in increasing order of their cycle leader.

   $[(9)(15,3,6,5)(16,7)]$

5. Remove parentheses and replace each number by its corresponding letter.

   $[\overline{9}\ \overline{15}\ \overline{3}\ \overline{6}\ \overline{5}\ \overline{16}\ \overline{7}]$

□

The result is actually another permutation of the same numbers. It is clear that the transformation is 1-1 because the algorithm can be reversed. The end of a cycle is "signalled" by a new left-to-right maximum. This transformation is implied in [Riordan58; chap. 8] and formalized and generalized in [Foata65]. It was later referred to as "the first fundamental transformation for permutations" [Foata83].

Now let $a_1, a_2, \ldots, a_l$, $a_1 < a_2 < \cdots < a_l$ be the leaves in the graph of f. The word ω_i consists of the labels of the nodes on the path from a_i to the first node already appearing in $\omega_0, \ldots, \omega_{i-1}$, including this node and excluding a_i, in reverse order. For the mapping in Figure 2 we have $l = 6$, $a_1 = 1$, $a_2 = 2$, $a_3 = 8$, $a_4 = 11$, $a_5 = 12$, $a_6 = 14$, and $\omega_1 = \overline{15}\ \overline{4}$, $\omega_2 = \overline{3}$, $\omega_3 = \overline{16}\ \overline{13}$, $\omega_4 = \overline{4}\ \overline{10}$, $\omega_5 = \overline{7}$, $\omega_6 = \overline{9}$. Therefore the complete encoding of the function in Figure 2 is

$$\overline{9}\ \overline{15}\ \overline{3}\ \overline{6}\ \overline{5}\ \overline{16}\ \overline{7}\ \overline{15}\ \overline{4}\ \overline{3}\ \overline{16}\ \overline{13}\ \overline{4}\ \overline{10}\ \overline{7}\ \overline{9}\ .$$

It is clear from its definition that the FF-encoding is a permutation of the trivial encoding. It is a 1-1 correspondence because it can be inversed. Given a word ω of length n the letters corresponding to $a_1, a_2, \ldots, a_l$ are exactly those letters among $\overline{1}, \ldots, \overline{n}$ that do not appear in ω, sorted in increasing order. The subword ω_0 ends before the first repeated letter, or $\omega_0 = \omega$ if no letter is repeated; the beginning of each subword describing a path is "signalled" by a repeated letter; and so on.

Given a mapping f its FF-encoding will be denoted $C_1(f)$. A set of mappings F encodes into a set of words, denoted $C_1(F)$.

Examples

1. Let F be the family of functions $f : \{1, \ldots, n\} \to \{1, \ldots, n\}$ such that the graph of f is connected and n is recurrent. Then

$$|F| = n^{n-1}, \tag{23}$$

   because any $f \in F$ has exactly one cycle, and n must be in it. Therefore $C_1(f)$ starts with $\overline{n}$ followed by an arbitrary word of length $n - 1$.

2. Let F be the family of functions $f : \{1, \ldots, n\} \to \{1, \ldots, n\}$ such that n is the unique recurrent element. (The graph of every f is a tree rooted at n.) Then

$$|F| = n^{n-2}, \tag{24}$$

   because for any $f \in F$, $C_1(f)$ starts with $\overline{n}\,\overline{n}$ followed by an arbitrary word of length $n - 2$.

   This means that the number of undirected trees on n labelled vertices is also n^{n-2}, which is the well known Cayley's theorem.

4.3. Half integer Abelian identities

Let A be the set $\{1, 2, \ldots, n + p + 2\}$. Define F_1 to be the family of functions $f : A \to A$ such that:

1. The graph of f has exactly two components, K_1 and K_2.
2. In this graph $n+p+1$ and $n+p+2$ are in different components, K_1 and respectively K_2, are both recurrent, and have no branches attached to them. (i.e. they have indegree one.)
3. The points $1, 2, \ldots, p$ are all in K_1 are all leaves.
4. The component K_1 has at least i recurrent elements, besides $n + p + 1$.

Our goal is to compute $|F_1|$ via two *different* encodings, and thus to obtain a generalization of equation 22.

The first way is to use the FF-encoding on each component separately.

Assume that there are k elements in the component K_1, *besides* $1,\ldots,p$, and $n+p+1$ that are there by definition. That means that the size of K_1 is $k+p+1$ and the size of K_2 is $n-k+1$. Let these k elements be the set $A=\{a_1,\ldots,a_k\}$. A word that encodes K_1 is constructed as follows:

- The first letter is $\overline{n+p+1}$, because $n+p+1$ is recurrent and maximal.
- The next i letters must be in A, and must be distinct, because there are at least i recurrent elements besides $n+p+1$. ($k^{\underline{i}}$ possibilities)
- The next $k+p-i$ letters must be all in A, because $1,\ldots,p$ are all leaves and hence do not appear in the encoding, and $n+p+1$ has no branches attached to it. (k^{k+p-i} possibilities)

A word that encodes K_2 has the following structure:

- The first letter is $\overline{n+p+1}$.
- The next $n-k$ letters are among the letters assigned to K_2 , but not $\overline{n+p+2}$, because there are no branches attached to $n+p+2$. (This gives $(n-k)^{n-k}$ possibilities.)

This construction proves that

$$|F_1|=\sum_k \binom{n}{k} k^{\underline{i}} k^{k+p-i}(n-k)^{n-k}. \tag{25}$$

For the second construction, we first encode the connected component H generated by $1,\ldots,p$, (i.e. the loop containing $n+p+1$, branch 1, branch 2, ... , branch p) and then encode the rest of the graph (i.e. the loop containing $n+p+2$, alias y, and all the remaining branches) using the usual FF-encoding but considering all the elements in H, "already seen."

Assume that there are k elements in H different from $1,\ldots,p$ and from $n+p+1$. Again let these k elements be the set $A=\{a_1,\ldots,a_k\}$. After choosing these k elements, the FF-encoding of H can be built as follows:

- The length of the word is $k+p+1$.
- The first letter must be $\overline{n+p+1}$.
- The next $k+p$ positions contain exactly k distinct letters, namely $\overline{a_1},\ldots,\overline{a_k}$, with the restriction that positions $2,\ldots,i+1$ contain distinct letters. Therefore these positions must be partioned into $\left\{{k+p \atop k}\right\}_i$ subsets, each subset correponding to a different letter. The letters in A can be associated to these subsets in $k!$ ways.

The word encoding the rest of the graph has length $n-k+1$ and has the following structure:

- First letter is $\overline{n+p+2}$.

- The next $n-k$ letters are arbitrarily chosen from the set $\{p+1,\ldots,p+n\}$, because now branches can be either in K_1 or in K_2. (n^{n-k} possibilities)

  This construction implies that

$$|F_1| = \sum_k \binom{n}{k} k! \left\{ {k+p \atop k} \right\}_i n^{n-k}, \tag{26}$$

which together with equation 25 gives

$$\sum_k \binom{n}{k} k^{k+p} \frac{k^{\underline{i}}}{k^i} (n-k)^{n-k} = n^n \sum_k \frac{n^{\underline{k}}}{n^k} \left\{ {k+p \atop k} \right\}_i, \qquad i > 0. \tag{27}$$

Summing on all $i > 0$ and using equation 7 we obtain that

$$\sum_k \binom{n}{k} k^{k+p} Q(k)(n-k)^{n-k} = n^n \sum_k \frac{n^{\underline{k}}}{n^k} \left\{ {k+p+1/2 \atop k} \right\} \tag{28}$$

If we modify the definition of the family of mappings F_1 to require that the component K_2 also has at least j recurrent elements besides $n+p+2$ and apply similar techniques we obtain that, for $i, j > 0$

$$\begin{aligned} \sum_k \binom{n}{k} k^{k+p} \frac{k^{\underline{i}}}{k^i} (n-k)^{n-k+q} \frac{(n-k)^{\underline{j}}}{(n-k)^j} \\ = n^{n-1} \sum_{k,l} \frac{n^{\underline{k+l}}}{n^{k+l}} \left\{ {k+p \atop k} \right\}_i \left\{ {l+q \atop l} \right\}_j (k+l), \end{aligned} \tag{29}$$

which after summing on $i > 0$ and $j > 0$ results in

$$\begin{aligned} \sum_k \binom{n}{k} k^{k+p} Q(k)(n-k)^{n-k+q} Q(n-k) \\ = n^{n-1} \sum_{k,l} \frac{n^{\underline{k+l}}}{n^{k+l}} \left\{ {k+p+1/2 \atop k} \right\} \left\{ {l+q+1/2 \atop l} \right\} (k+l). \end{aligned} \tag{30}$$

Comparing equation 30 with equation 22, one can see that if we make the syntactic convention that $x^{1/2} \equiv Q(x)$, then we can write

$$\sum_k \binom{n}{k} k^{k+p/2} (n-k)^{n-k+q/2} = n^{n-1} \sum_{k,l} \frac{n^{\underline{k+l}}}{n^{k+l}} \left\{ {k+p/2 \atop k} \right\} \left\{ {l+q/2 \atop l} \right\} (k+l), \tag{31}$$

which is valid for both even and odd p and q, and justifies the name "half-integer Stirling numbers" as well as the title of this section.

It is worth noting that there exists polynomial identities in two variables corresponding to equations 28 and 30; they can be derived by applying some variations of the counting methods presented here, as done in [Broder83] for the integer Abelian identities. The results though, are rather intricate.

References

[Abel1826] N. H. Abel, "Beweis eines Ausdruckes, von welchem die Binomial-Formel ein einzelner Fall ist," *Crelle*, **1**(1826), 159–160.

[Broder83] A. Z. Broder, "A general expression for Abelian identities," in: L. J. Cummings (ed.), *Combinatorics on Words*, Academic Press, 1983.

[Broder84] A. Z. Broder, "The r-Stirling numbers," *Discrete Mathematics*, **49**(1984), 241–259.

[Carlitz80a] L. Carlitz, "Weighted Stirling numbers of the first and second kind – I," *The Fibonacci Quarterly*, **18**(1980), 147–162.

[Carlitz80b] L. Carlitz, "Weighted Stirling numbers of the first and second kind – II," *The Fibonacci Quarterly*, **18**(1980), 242–257.

[Cauchy1826] A. Cauchy, *Exercises de Mathématiques*, Paris, 1826.

[Comtet74] L. Comtet, *Advanced Combinatorics*, Reidel, Dordrecht/Boston, 1974.

[FF70] D. Foata and A. Fuchs, "Réarrangements des fonctions et dénombrement," *Journal of Combinatorial Theory*, **8**(1970), 361–375.

[Foata65] D. Foata, "Etude algébrique de certain problèmes d'analyse combinatoire et du calcul des probabilités," *Publ. Inst. Statist. Univ. Paris*, **14**(1965), 81–241.

[Foata83] D. Foata, "Rearrangement of words," in: M. Lothaire, *Combinatorics on Words*, Encyclopedia of Mathematics and its Applications, vol. 17, Addison-Wesley, Reading, Mass., 1983.

[Françon74] J. Françon, "Preuves combinatoires des identités d'Abel," *Discrete Mathematics*, **8**(1974), 331-343.

[Gertsbakh77] I. B. Gertsbakh, "Epidemic processes on a random graph: some preliminary results," *J. Appl. Probability*, **14**(1977), 427–438.

[Hellman80] M. E. Hellman, "A cryptanalitic time-memory tradeoff," *IEEE Trans. Inform. Theory*, **IT-26**(1980), 401–406.

[IM65] G. I. Ivchenko and Yu. I. Medvedev, "Asymptotic representations of finite differences of a power function at an arbitrary point," *Theory of Probability and its Applications*, **10**(1965), 139–144.

[Jordan74] C. Jordan, *Calculus of Finite Difference*, Chelsea, New York, 1947.

[Knuth73] D. E. Knuth, *The Art of Computer Programming*, Vol. 3, Addison-Wesley, Reading, Mass., 1973.

[Knuth81] D. E. Knuth, *The Art of Computer Programming*, Vol. 2, Second edition, Addison-Wesley, Reading, Mass., 1981.

[Knuth83] D. E. Knuth, "The analysis of optimum cacheing," to appear in *Journal of Algorithms*.

[Koutras82] M. Koutras, "Non-central Stirling numbers and some applications," *Discrete Mathematics*, **42**(1982), 73–89.

[KR75] D. E. Knuth and G. S. Rao, "Activity in an interleaved memory," *IEEE Transactions on Computers*, **C-24**(1975), 943–944.

[KS78] D. E. Knuth and A. Schönhage, "The expected linearity of a simple equivalence algorithm," *Theoretical Computer Science*, **6**(1978), 281–315.

[Moon71] J. W. Moon, *Counting Labelled Trees*, Canadian Mathematical Monographs, 1971.

[Neuman82] E. Neuman, "Moments and Fourier transform of B-splines," *J. Comput. Appl. Math.*, **7**(1981), 51–62.

[Nielsen23] N. Nielsen, *Traité Élémentaire des Nombres de Bernoulli*, Gauthier-Villars, Paris, 1923.

[Pittel82] B. Pittel, "On distributions related to the transitive closures of random finite mappings," *Annals of Probability*, **11**(1983), 428–441.

[Ramanujan12] S. Ramanujan, "Questions for solution, number 294," *J. Indian Math. Soc.*, **4**(1912), 151–152.

[Riordan58] J. Riordan, *An Introduction to Combinatorial Analysis*, Wiley, New York, 1958.

[Riordan68] J. Riordan, *Combinatorial Identities*, Wiley, New York, 1968.

[Riordan69] J. Riordan, "Abel identities and inverse relations," in: R.C. Bose and T.A. Dowling, eds., *Combinatorial Mathematics and its Application*, Univ. of North Carolina Press, Chapel Hill, 1969.

[Stanf81] Stanford Computer Science Department, Qualifying examination in the analysis of algorithms, April 1981.

SOME USES OF THE MELLIN INTEGRAL TRANSFORM IN THE ANALYSIS OF ALGORITHMS

Philippe Flajolet and *Mireille Regnier*
I.N.R.I.A
Rocquencourt
78150-Le Chesnay (France)

Robert Sedgewick
Department of Computer Science
Brown University
Providence R.I. 02912 (U.S.A.)

ABSTRACT

We informally survey some uses of the Mellin integral transform in the context of the asymptotic evaluation of combinatorial sums arising in the analysis of algorithms.

1. Introduction

The *Mellin transform* is an integral transform that is part of the working kit of analytic number theory where its use is to be traced to Riemann's famous memoir on the distribution of primes. Its usefulness in asymptotic analysis comes from the fact that it relates *asymptotic properties* of a function around 0 and ∞ to the *singularities* of the transformed function.

We propose here to informally explore the uses of the Mellin transform in the context of the average case analysis of algorithms.

Average case analysis of algorithms starts with a counting of certain combinatorial configurations like permutations, words, trees, finite functions, distributions *etc*.... One then computes certain weighted averages of the counting results to determine the expected cost of algorithms. Depending on the particular problem under consideration, the counting results are either obtained explicitly or indirectly accessible through some generating function. Mellin transform techniques are especially valuable when number-theoretic functions (divisor function, functions related to binary representation of integers) and/or periodicities appear.

The Mellin transform of a real valued function $F(x)$ defined over $[0;+\infty[$ is the complex function $F^*(s)$ of the complex variable s given by:

$$F^*(s) = \int_0^\infty F(x)\, x^{s-1}\, dx \tag{1}$$

also written sometimes:

NATO ASI Series, Vol. F12
Combinatorial Algorithms on Words
Edited by A. Apostolico and Z. Galil

	$F(x)$	$F^*(x)$	
T1.	e^{-x}	$\Gamma(s)$	$0<\mathrm{Re}(s)$
	$e^{-x}-1$	$\Gamma(s)$	$-1<\mathrm{Re}(s)<0$
	$e^{-x}-1+x$	$\Gamma(s)$	$-2<\mathrm{Re}(s)<-1$
	$e^{+x}-1-x-\frac{x^2}{2}$	$\Gamma(s)$	$-3<\mathrm{Re}(s)<-2$
T2.	$\frac{1}{1+x}$	$\pi\operatorname{cosec}\pi s$	$0<\mathrm{Re}(s)<1$
T3.	$\log(1+x)$	$\frac{\pi}{s\sin\pi s}$	$-1<\mathrm{Re}(s)<0$
T4.	$\delta(x)$	$\frac{1}{s}$	$0<\mathrm{Re}(s)$
T5.	$\log x\,\delta(x)$	$\frac{1}{s^2}$	$0<\mathrm{Re}(s)$
T6.	$(1-x)^{m-1}\delta(x)$	$\frac{\Gamma(s)\Gamma(m)}{\Gamma(s+m)}$	$0<\mathrm{Re}(s);0<m$

Table 1: *Some basic transforms. ($\delta(x)$ is defined by $\delta(x)=1$ if $x\le 1$ and* 0 *otherwise.)*

$$\mathbf{M}[F(x);s] \quad \text{or} \quad \mathbf{M}[F] .$$

The interest of the Mellin transform comes from the combination of two types of properties:

P1 *Asymptotic properties:* under fairly general conditions, one obtains an asymptotic expansion of $F(x)$ from the singularities of its transform:

$$F(x) \sim \pm \sum_{\alpha\in H} Res\,(F^*(s)\,x^{-s}\ ;\ s=\alpha) \tag{2}$$

where H is either a left ($x\to 0$) or right ($x\to\infty$) half-plane. Expansion (2) *is* an asymptotic expansion.

P2 *Functional properties:* somewhat intricate sums often have simple transforms. An important paradigm is that of *harmonic sums*. A harmonic sum is of the form:

$$F(x) = \sum_k \lambda_k f\,(\mu_k x) \tag{3}$$

and its transform has, at least formally, the factored form

$$F^*(s) = (\sum_k \lambda_k \mu_k^{-s})\, f^*(s) \tag{4}$$

which is simply the product of a (generalized) *Dirichlet series* and the transform of the base function f.

The combination of (2) and (3) allows for derivation of very many asymptotic expansions. The generality of expansions (2) permits in particular to attack expressions whose expansions involve non trivial *periodicity phenomena* which correspond to poles of the transform function in expansion (2) that have a non-zero imaginary part.

2. Basic properties of the Mellin transform

Let $F(x)$ be piecewise continuous on $[0;\infty]$ and assume F satisfies:

$$F(x) = O(x^{\alpha}) \quad (x \to 0) \quad ; \quad F(x) = O(x^{\beta}) \quad (x \to \infty) \quad .$$

Then the Mellin transform of $F(x)$ is defined in the strip:

$$-\alpha < \mathrm{Re}(s) < -\beta \ .$$

This strip is called the ***fundamental strip*** of F^* and is sometimes denoted by $<\alpha;\beta>$. It is empty unless $\beta<\alpha$. Thus the Mellin transform of a function is only defined if its *order at infinity* is smaller than its *order at zero*.

Table 1 summarizes some basic transforms while Table 2 describes the main functional properties of the Mellin transform

	$F(x)$	$F^*(s)$	
P1	$F(x)$	$\int_0^\infty F(x) x^{s-1} dx$	definition
P2	$\frac{1}{2i\pi} \int_{c-i\infty}^{c+i\infty} F^*(s) s\, ds$	$F^*(s)$	inversion th.
P3	$F(ax)$	$a^{-s} F^*(s)$	$a>0$
P4	$x^{\nu} F(x^{\mu})$	$\frac{1}{\beta} F^*(\frac{s+\nu}{\mu})$	
P5	$\frac{d}{dx} F(x)$	$-(s-1)F^*(s-1)$	
P6	$F(x) \log x$	$\frac{d}{ds} F^*(s)$	
P7	$\sum \lambda_k f(\mu_k x)$	$(\sum \lambda_k \mu_k^{-s}) f^*(s)$	harmonic sum

Table 2: *Basic functional properties of the Mellin transform.*

2.1. Direct properties

The characteristic property of Mellin transforms is: *local properties* at 0 and ∞ are reflected by the *singularities* of the transform function.

Theorem 1: *Assume $F(x)$ has the following expansions:*

$$F(x) \sim \sum_{k>0} c_k x^{\alpha_k} \qquad (x \to 0)$$

$$F(x) \sim \sum_{k>0} d_k x^{\beta_k} \qquad (x \to \infty)$$

where the α_k form an increasing sequence that tends to $+\infty$ and the β_k form a decreasing sequence that tends to $-\infty$.

Then the transform $F^*(s)$ of $F(x)$ is meromorphic in the whole complex plane with simple poles at points $-\alpha_k$ and $-\beta_k$ with:

$$F^*(s) \sim \frac{c_k}{s+\alpha_k} \quad (s \to -\alpha_k) \ ,$$

$$F^*(s) \sim -\frac{d_k}{s+\beta_k} \quad (s \to -\beta_k) \ .$$

Proof: (Sketch) To translate the expansion at 0, consider the transform:

$$F_o^*(s) = \int_0^1 F(x)\,dx \ .$$

Formally applying this partial transform to the asymptotic expansion leads to (see the transform of the δ function in Table 1)

$$F_o^*(s) \sim \sum_k \frac{c_k}{s+\alpha_k} \ .$$

Theorem 1 admits a number of extensions: the α_k only need to be complex numbers whose sequence of real parts tends to $\pm\infty$. Also partial asymptotic expansions may be similarly translated: then the transform function is meromorphic in an extended strip (not necessarily the whole of the complex plane). Finally generalized expansions of the form $c_k(\log x)$ for a sequence of polynomials $c_k(t)$ may be allowed. The meromorphy result still holds but then $F^*(s)$ will have a pole of order $deg(c_k)-1$ at $-\alpha_k$ (obvious analogues are true for expansions at ∞).

In other words the asymptotic behaviour of F at 0 (resp. ∞) is reflected by the singularities of F^* in a *left half-plane* (resp. *right half-plane*) w.r.t. the fundamental strip.

Another important property of Mellin transforms is their smallness towards $i\infty$. From the Riemann-Lebesgue lemma, one has:

Theorem 2: *If $F(x)$ is infinitely differentiable over $[0;+\infty[$, then for s in the fundamental strip, one has for all $m>0$:*

$$\lim_{|s|\to\infty} s^m F^*(s) = 0 \ .$$

2.2. Inverse properties

One can come back from F^* to F by means of the following important inversion theorem:

Theorem 3: *For any real c inside the fundamental strip, one has:*

$$F(x) = \frac{1}{2i\pi} \int_{c-i\infty}^{c+i\infty} F^*(s) x^{-s}\, ds \ .$$

This theorem is crucial for proving a converse of Theorem 1, namely:

Theorem 4: *Let $<-\alpha;-\beta>$ be the fundamental strip of F^*, and assume that F^* is small towards $i\infty$. If $F^*(s)$ is meromorphic for $-L\le \mathrm{Re}(s)\le -\alpha$ with finitely many simple poles at $-\alpha_1 > -\alpha_2 > \cdots > -L$, with residue c_k at $-\alpha_k$, then $F(x)$ admits the asymptotic expansion:*

$$F(x) \sim \sum_{k\ge 1} c_k x^{\alpha_k} + O(x^L) \ .$$

If $F^*(s)$ is meromorphic for $-\beta \le \mathrm{Re}(s) \le -M$ with finitely many simple poles at $-\beta_1 < -\beta_2 < \cdots < -M$, with residue d_k at $-\beta_k$, then $F(x)$ admits the asymptotic expansion:

$$F(x) \sim -\sum_{k \ge 1} c_k x^{\beta_k} + O(x^M) .$$

Proof: Start with the inversion theorem to express $F(x)$ and evaluate the integral by shifting the line of integration to the left until the line $\mathrm{Re}(s) = -L$ only taking residues into account.

In this way, one obtains a form:

$$F(x) = \sum_k Res\,(F^*(s)x^{-s}; s=-\alpha_k) + \frac{1}{2i\pi} \int_{-L-i\infty}^{-L+i\infty} F^*(s)x^{-s}\,ds .$$

The other result about the expansion of F at $+\infty$ is proved similarly by moving the line of integration to the right.

Observation: The same process can be applied to multiple poles and be extended to cases where F^* has infinitely many poles in a finite width strip, provided F^* remains small along some horizontal lines towards $i\infty$. One observes that:

1. A simple pole of the form $s = \sigma + i\tau$ will give a term in the asymptotic expansion of F of order

$$x^{-s} = x^{-\sigma}x^{-i\tau} = x^{-\sigma}e^{-i\tau \log x}$$

Poles of F^* farther west contribute smaller terms to the expansion of F at 0. Poles of F^* farther east contribute smaller terms to the expansion of F at ∞.

Poles with non-zero imaginary parts correspond to asymptotic fluctuations.

2. A pole of F^* of order k at s contributes a term of the form (P_{k-1} being a polynomial of degree $k-1$):

$$P_{k-1}(\log x)x^{-s} .$$

Or:

Multiple poles introduce factors that are powers of $\log x$ *in asymptotic expansions.*

3. Full asymptotic expansions may or may not be convergent (for sums of residues in either a left or a right half plane). In most of the cases dealt with here, they are *not convergent*, thus only asymptotic.

3. Harmonic sums

Let f be a smooth enough function. Then a sum of the form:

$$F(x) = \sum_k \lambda_k f(\mu_k x) \tag{7}$$

is called a *harmonic sum*. Function f is called the *base function*; the coefficients λ_k are the *amplitudes* and the μ_k are the *frequencies*. Applying the basic functional property of the Mellin transform to (7), we find formally:

$$F^{*}(s) = \omega(s) f^{*}(s) \tag{8}$$

where ω is the (generalized) Dirichlet series:

$$\omega(s) = \sum_{k} \lambda_k \mu_k^{-s} . \tag{9}$$

The transformed equation (8) is valid inside the intersection of the fundamental domain of f^{*} and of the domain of absolute convergence of ω.
A pair of sequences $\{\lambda_k\}$, $\{\mu_k\}$ is said to be an ***arithmetic pair*** iff the series (8) is meromorphic in the whole complex plane.

Mellin transform techniques are well suited to the treatment of harmonic sums associated to arithmetic pairs amplitudes-frequencies when the base function is smooth (the exponential function in many applications).

From the preceding sections, we see that provided $\omega(s)$ is not too ill behaved at $i\infty$, the asymptotics of harmonic sums can be easily determined.

Example 1: *Harmonic numbers.*
Harmonic numbers are particular harmonic sums. The n-th harmonic number is classically defined by

$$H_n = \sum_{k=1}^{n} \frac{1}{k} .$$

It can be extended to a smooth real function by introducing:

$$h(x) = \sum_{k \geq 0} \left[\frac{1}{k} - \frac{1}{x+k} \right]$$

so that $h(n) = H_n$. The fundamental strip of h^{*} is $<-1;0>$ and rewriting $h(x)$ under the form:

$$h(x) = \sum_{n} \frac{1}{n} \left(\frac{x}{n}\right) \left(1 + \frac{x}{n}\right)^{-1}$$

we see that it is a harmonic sum. The Dirichlet series associated to amplitudes and frequencies is:

$$\omega(s) = \sum_{n \geq 1} \frac{1}{n} \left(\frac{1}{n}\right)^{-s}$$

that is $\omega(s) = \zeta(1-s)$ where ζ is the classical Riemann zeta function. The transform of the base function is obtained from Table 1, so that:

$$h^{*}(s) = \frac{\pi}{\sin \pi s} \zeta(1-s) \tag{10}$$

for $-1 < Re(s) < 0$. Thus we can choose $c = 1/2$ in the inversion theorem and get:

$$h(x) = \frac{1}{2i\pi} \int_{-1/2-i\infty}^{-1/2+i\infty} \frac{\pi}{\sin \pi s} \zeta(1-s)\, ds$$

The asymptotic behaviour of h towards $+\infty$ is obtained by moving the line of integration to the right. One first encounters a double pole at $s=0$ corresponding to a logarithmic dominant term and then poles at some of the integers; whence the classical expansion (the B_k are the Bernoulli numbers):

$$h(x) \sim \log x + \gamma + \frac{1}{2n} + \sum_{k \geq 2} \frac{(-1)^k B_k}{k\, n^k} \tag{11}$$

Example 2: *Special Euler-Maclaurin summations.*
Assume $f(x)$ is exponentially small towards ∞, and admits around 0 an expansion of the form:

$$f(x) \sim \sum_{k\geq 0} c_k x^{\alpha_k} \tag{12}$$

where $\alpha_0 = 0 < \alpha_1 < \cdots$. Consider the sum:

$$F(x) = \sum_{n\geq 1} f(nx) . \tag{13}$$

for which an expansion as $x \to 0$ is sought. The transform of F is:

$$F^*(s) = f^*(s)\zeta(s)$$

for $\mathrm{Re}(s)>1$. From Section 2.1, the poles of f^* are at the $-\alpha_k$; $\zeta(s)$ has a unique pole at $s=1$ with residue 1 so that:

$$F^*(s) \sim \frac{1}{s-1} f^*(1) \quad (s \to 1) \tag{14}$$

thus putting everything together, we find an Euler-Maclaurin-Barnes expansion:

$$F(x) \sim \frac{1}{x}\int_0^\infty f(t)\,dt + \frac{1}{2}f(0) + \sum_{k\geq 1} c_k \zeta(-\alpha_k) x^{\alpha_k} . \tag{15}$$

Cases where the expansion of f also involves logarithmic terms can be dealt with similarly (multiple poles appear that introduce the derivatives of the zeta function at $-\alpha_k$), yielding formulae of Gonnet.

Example 3: *Powers-of-two sums and periodicities.*
Assume here for f the same conditions as in the previous example and consider the sum:

$$F(x) = \sum_{k\geq 0} f(2^k x) \tag{16}$$

for which an asymptotic expansion is sought as $x \to 0$. The transform of F is for $\mathrm{Re}(s)>0$:

$$F^*(s) = \frac{f^*(s)}{1-2^{-s}} .$$

It has a double pole at $s=0$ if $f(0) \neq 0$ so that we need two terms in the expansion of f^* at 0. Writing:

$$f^*(s) = \frac{f(0)}{s} + \int_0^1 (f(x)-f(0))\, x^{s-1}\,dx + \int_1^\infty f(x) x^{s-1}\,dx$$

we see that around 0:

$$f^*(s) = \frac{f(0)}{s} + \gamma_f + O(s) ,$$

where:

$$\gamma_f = \int_0^1 (f(x)-f(0)) \frac{dx}{x} + \int_1^\infty f(x) \frac{dx}{x}. \tag{17}$$

Function F^* also has simple poles at $s = \chi_k$, $k \in \mathbb{Z}/\{0\}$ with $\chi_k = \dfrac{2ik\pi}{\log 2}$, as well as at the points $-\alpha_k$, whence the general expansion as $x \to 0$:

$$F(x) \sim -f(0)\log_2 x + \frac{1}{2}f(0) + \frac{\gamma_f}{\log 2} + P(\log_2 x) + \sum_{k\ge1} \frac{c_k}{1-2^{\alpha_k}} x^{\alpha_k},$$

with:

$$P(u) = \frac{1}{\log 2} \sum_{k \in \mathbf{Z}/\{0\}} f^*\left(\frac{2ik\pi}{\log 2}\right) e^{-2ik\pi u/\log 2}. \quad (18)$$

Example 4: *Perron's and other number-theoretic formulae*

The framework of harmonic sum is also a natural setting for formulae relating ordinary generating functions to Dirichlet series as well as for Perron's formula expressing partial sums of coefficients of Dirichlet series. This last formula follows from the application of our methods to sums of the form:

$$F(x) = \sum_{k\ge1} \lambda_k \delta(kx)$$

where function delta is defined in Table 1.

4. Harmonic power sums

We only give brief indications here on the idea underlying the asymptotic treatment of ***harmonic power sums*** referring the reader to M. Regnier's thesis for a precise statement of theorems, and validity conditions.

A harmonic power sum is a sum of the form:

$$F(x) = \sum_{k\ge1} \lambda_k f(\mu_k x)^{\gamma_k}. \quad (19)$$

We assume here that f and g are smooth enough functions. From (19), taking transforms, we see that one has at least formally:

$$F^*(s) = \sum_{k\ge1} \lambda_k \mu_k^{-s} f^*_{\gamma_k}(s) \quad (20)$$

where f^*_γ is the transform:

$$f^*_\gamma(s) = \int_0^\infty f^\gamma(x)\, x^{s-1}\, dx. \quad (21)$$

Our treatment concerns cases where the $\gamma_k \to \infty$. In that case, it is natural to expect from (21) that the behaviour of the sum F is determined by the behaviour of (21) as γ gets large.

Rewritting (21) in an exponential form, and assuming without loss of generality that $f(0)=1$, we should expect

$$f^*_\gamma(s) = \int_0^\infty e^{\gamma \log f(x)} x^{s-1}\, dx,$$

to be approximable for large γ by the *Laplace method for integrals*. If $\log f(x)$ admits around 0 an asymptotic expansion of the form $-c\, x^\alpha + O(x^\beta)$, then one can prove under certain conditions that F^* is "approximated" by a function

$$\omega_1(s) = \frac{1}{\alpha}\Gamma\left(\frac{s}{\alpha}\right) c^{-s/\alpha} \sum_{k\ge1} \lambda_k \mu_k^{-s} \gamma_k^{-s/\alpha}$$

in the sense that $F^{*}(s)-\omega_1(s)$ is analytic in a larger strip than $F^{*}(s)$. Whence, for F an asymptotic expansion of the form:

$$F(x) \sim \sum_{s_0 \in \mathbf{S}} Res(\omega_1(s)x^{-s}\ ;\ s=s_0) \tag{22}$$

where the sum is extended to poles s_0 in some strip $\mathbf{S}$ of the complex s-plane.

With these techniques one can estimate sums appearing in the analysis of the *Extendible Hashing* algorithm used for storing large files on disk while maintaining direct access in a dynamic context. For instance, under a Poisson model, the expected size of an Extendible Hashing directory formed with n records is $D(n)$ where:

$$D(x) = \sum_{k \geq 0} (1 - f_b(x2^{-k})^{2^k}) .$$

and $f_b(x)$ is $e^{-x} \sum_{k=0}^{b} \frac{x^k}{k!}$. Function $D(x)$ is reducible to a harmonic power sum.

5. Other methods

We only mention briefly here two other uses of Mellin techniques:

A. Complex Mellin inversion and singularity analysis of generating functions.

B. The so-called "Rice" formula, actually a classical formula from the calculus of finite differences.

A. It is known in many cases that the behaviour of a function around its (dominant) *singularities* determines the asymptotic behaviour of its coefficients. However in most cases (except for some scarce cases of application of Tauberian theorems) an asymptotic expansion of the function around its singularity *in the complex plane* is required. For instance it is known that the generating function of the quantity E_n representing the total number of registers to evaluate optimally all binary trees of size n is:

$$E(z) = \frac{1-u^2}{u} \sum_{p \geq 1} \frac{u^{2^p}}{1-u^{2^p}} \tag{23}$$

where:

$$u = u(z) = \frac{1-\sqrt{1-4z}}{1+\sqrt{1-4z}} .$$

The singularity of $E(z)$ is at $z=1/4$, where $u \to 1$ as $z \to 1/4$. The asymptotic behaviour of (23) when z is in a neighbourhood of $1/4$ in the complex plane (u in a neighbourhood of 1) determines the asymptotics of the E_n. Setting $u=e^{-x}$, the sum in (23) becomes:

$$F(x) = \sum_{p \geq 1} \frac{e^{-x2^p}}{1-e^{-x2^p}} , \tag{24}$$

typically a harmonic sum. whose transform is :

$$\frac{\Gamma(s)\zeta(s)}{2^s-1} .$$

The asymptotics of F for x in a complex neighbourhood of 0 can still be determined by appealing to *complex Mellin inversion* formulae, from which the asymptotics of E_n is found. (See a forthcoming paper of Flajolet and Prodinger for details).

The method also applies to the analysis of the expected height of general planted plane trees and of odd-even merge sorting networks. It allows derivation of full asymptotic expansions rather easily.

B. Let a_n be a sequence of numbers; their *Poisson generating function* is defined as:

$$a(x) = e^{-x} \sum_{n\geq 0} a_n \frac{x^n}{n!} . \tag{25}$$

Many applications require determining asymptotic properties of the a_n from their Poisson generating function.

One way of proceeding is to consider the *Mellin transform* of (25) which is easily found to involve a *Newton series:*

$$a^{*}(s) = \Gamma(s) \sum_{n\geq 0} a_n \frac{s(s+1)(s+2) \cdots (s+n-1)}{n!} . \tag{26}$$

Thus if everything goes well, a_n is nothing but the n-th difference of $\alpha(s) \equiv a^{*}(s)/\Gamma(s)$:

$$a_0 = \alpha(0) \ ; \ a_1 = \alpha(0)-\alpha(-1) \ ; \ a_2 = \alpha(0)-2\alpha(-1)+\alpha(-2) \ \cdots . \tag{27a}$$

From there, the a_n can be recovered by a classical formula from the calculus of finite differences and the theory of Newton series:

$$a_n = \frac{n!}{2i\pi} \int_{\Lambda} \frac{a^{*}(s)}{\Gamma(s)} \frac{ds}{s(s+1)(s+2) \cdots (s+n)} \tag{27b}$$

for a small contour around points $0,-1,-2,-3, \cdots$. This method was used by S.O. Rice as pointed out by Knuth. We call it the *Poisson-Mellin-Newton cycle*. A similar method is used to analyze digital search trees (see a forthcoming paper of Flajolet and Sedgewick). Notice that Knuth's use of the method is restricted to cases where an explicit form for the a_n has been obtained (by expanding generating functions) while our approach is more general since it accounts directly for the shape of the integrand in (27b). (See below an example of application to "trie sums").

6. Sample applications

We shall restrict ourselves here to presenting a few applications of the above methods: to carry propagation, to interpolation search, and to trie sums.

6.1. Interpolation search

Interpolation search is a way of searching an ordered table using the value of the record to be found in order to calculate an address at which it is likely to be found. It has been shown by several authors that the expected cost of such a search in a file of size n is $\log_2 \log_2 n + O(1)$. We propose here to reexamine some of the steps of a derivation of that result by Gonnet.

By a sequence of analytic and probabilistic arguments, Gonnet proves that the probability that more than k probes are required is majorized by the quantity:

$$p_k(t) = \prod_{i=1}^{k} (1-\frac{1}{2}e^{-t2^{-i}})$$

where $e^{-t} = \frac{8}{\pi n}$ $(t = \log \pi n / 8)$. Thus the expected number of probes is itself majorized by the function:

$$F(t) = \sum_{k=0}^{\infty} p_k(t) .$$

To evaluate F asymptotically for small t, introduce the infinite product:

$$Q(t) = \prod_{i=1}^{\infty} (1-\frac{1}{2}e^{-t2^{-i}})$$

so that $F(t)$ becomes:

$$F(t) = \frac{1}{Q(t)} \sum_{k=0}^{\infty} Q(t2^{-k}) ,$$

and the sum is a typical harmonic sum to which methods of Section 2 apply. One obtains in this way an asymptotic expansion of the form:

$$F(t) = K \log_2 t + P(\log_2 t) + \cdots \qquad (t \to \infty) \tag{28}$$

where P is a periodic function whose Fourier coefficients have simple expressions in terms of the values of Q^* at points $\chi_k = \frac{2ik\pi}{\log 2}$.

Equation (28) leads to the loglog result for interpolation search since t increases as $\log n$.

6.2. Carry propagation

The following problem arises in a work by Knuth relative to certain binary adders. What is the expected length of the longest sequence of (consecutive) *ones* in a random 0–1 string of length n?

Let $p_{n,k}$ denote the probability that this longest sequence has length k. Then it is known (see Feller's book) that:

$$p_k(x) = \sum_{n \geq 0} p_{n,k} x^n = \frac{1-2^{-k-1}x^{k+1}}{1-x+2^{-k-2}x^{k+2}} .$$

It is easily seen that $p_k(x)$ has a unique pole of smallest modulus that satisfies:

$$\rho_k = 1 + 2^{-k-2} + O(k 2^{-2k}) .$$

Numerically tight approximations for the probabilities that also prove sufficient for asymptotic analysis are obtained by retaining only the contribution in $p_{n,k}$ coming from the dominant pole. This leads to the approximations:

$$\begin{aligned} p_{n,k} &\approx \rho_k^{-n-1} \\ &\approx (1+2^{-k-2})^{-n-1} \\ &\approx e^{-n2^{-k-2}} . \end{aligned}$$

Using the above approximation in the expression $R_n = \sum_k 1-p_{n,k}$ representing the expected length of the longest run of ones in a random binary string of length n (this use can be justified rigorously), one is led to the approximation $R_n \sim F(n)$ where $F(x)$ is a familiar harmonic sum:

$$F(x) = \sum_k 1-e^{-x2^{-k-2}}$$

to which previously developed methods apply.

6.3. Trie sums

We refer the reader to vol. 3 of Knuth's *The Art of Computer Programming* for the origin of the problem (see the analysis of radix exchange sort, pp. 131 *et sq*. and the analysis of trie searching).

The question is to determine the asymptotic behaviour of a sequence of numbers $\{a_n\}$ whose exponential generating function $\sum_{n\geq 0} a_n x^n / n!$ satisfies the functional equation:

$$a(x) = 2e^{x/2} a(\frac{x}{2}) + x(e^x - 1) . \tag{30}$$

Given the initial conditions $a_0 = a_1 = 0$, one can solve (30) by iteration and take coefficients of the solution from which there results:

$$a_n = n \sum_{k\geq 0} [1-(1-\frac{1}{2^k})^{n-1}] . \tag{31}$$

One can also notice from (30), that the Poisson generating function $b(x)$ of the a_n satisfies the simpler equation:

$$b(x) = 2b(\frac{x}{2}) + x(1-e^{-x}) , \tag{32}$$

from which an alternative expression of the a_n follows by identifying coefficients.

We propose now a brief exposition of three different approaches to the asymptotic evaluation of the a_n.

Classical approach based on exponential approximations

This is Knuth's approach following suggestions by De Bruijn. it starts by using repeatedly the exponential approximation:

$$(1-a)^n \approx e^{-an} \tag{33}$$

which in the context of (31) leads to (after somewhat unpleasant real analysis):

$$a_n = n\,F(n) + o(n) \quad ; \quad F(x) = \sum_{k\geq 0} [1-e^{-x2^k}] . \tag{34}$$

Function $F(x)$ being clearly a harmonic sum, an asymptotic expansion involving periodicities appears, as usual:

$$F(x) = \log_2 x + P(\log_2 x) + \cdots \tag{35}$$

This method does not seem to allow for derivation of full asymptotic expansions, due to the limitations of real analysis methods involved in the derivation of approximation (34)

Direct approach

This consists in observing that the the a_n *are* simply expressible in terms of a harmonic sum; one has $a_{n+1}/n+1 = 1 + G(n)$, where G is given by:

$$G(x) = \sum_{k\geq 1} [1-\exp(-\mu_k x)] \quad ; \quad \mu_k = \log(1-\frac{1}{2^k})^{-1} . \tag{36}$$

Thus $G^*(s) = \mu(s)\Gamma(s)$, for $-1<\mathrm{Re}(s)<0$, with:

$$\mu(s) = \sum_{k\geq 1} \mu_k^{-s} . \tag{37}$$

To obtain an analytic continuation of $\mu(s)$, notice that:

$$(\log(1-u)^{-1})^{-s} = u^{-s}\,(1+\frac{u}{2}+\frac{u^2}{3}+\frac{u^3}{4}+\cdots)^{-s}$$
$$= u^{-s} \sum_{j\geq 0} c_j(s)u^j .$$

Thus using this expansion in the definition of μ, we find the meromorphic approximation of μ:

$$\mu(s) \approx \sum_{j=0}^{\infty} c_j(s)\,\frac{2^{s-j}}{1-2^{s-j}} \tag{38}$$

(convergence is not implied by this equation; only taking k terms of the sum will give a function that differs from $\mu(s)$ by a function analytic for $\mathrm{Re}(s)<k$).

Thus we see that $\mu(s)$ is meromorphic in the whole of the complex plane, with singularities that obtain from (38). Hence $G(x)$ admits a full asymptotic expansion of the form:

$$G(x) = \log_2 x + \sum_{j\geq 0} P_j(\log_2 x)\, x^{-j} \tag{39}$$

where the P_j are periodic with period 1 and with Fourier coefficients expressible in terms of values of the gamma function and the polynomials c_j.

The Poisson-Mellin-Newton cycle

One starts there from Equation (32) giving a functional equation for the Poisson generating function of the a_n; the Mellin transform of $b(x)$ is defined for $-2<\mathrm{Re}(s)<-1$ and satisfies the the transform equation:

$$b^*(s) = 2^{1+s}b^*(s) - s\Gamma(s) , \tag{40}$$

so that solving, we get:

$$b^*(s) = \frac{s\Gamma(s)}{1-2^{1+s}} . \tag{41}$$

From the "cycle" explained above, we see that:

$$a_n = \frac{n!}{2i\pi}\int_{\Lambda}(1-2^{1+s})^{-1}\,\frac{ds}{(s+1)(s+2)..(s+n)} \tag{42}$$

the integrand being exactly $b^*(s)/\Gamma(s)$ multiplied by the standard weight function $[s(s+1)\cdots s(n)]^{-1}$; there Λ is a skinny curve encircling the points $-2,\cdots,-n$. Moving the contour of integration and taking residues into account leads to closed form expressions and asymptotic expansions of the a_n that have a lot of similarities with Mellin inversion expansions.

7. Conclusion

We have presented here some applications of powerful Mellin transform techniques, trying to show that there are a few well recognizable cases of applications.

Applications to analysis of algorithms can be found in the following domains:

Tree parameters: height of trees; register allocation; height of random walks and analysis of dynamic data structures (stack "histories").

Trie parameters: radix exchange sort; trie searching; digital search trees; dynamic and extendible hashing; communication protocols.

Carry propagation and binary adders; odd-even merge sorting networks; interpolation search; longest probe sequence in hashing; probabilistic counting; approximate counting *etc*...

We do not have space here for a complete exposition or an even partial bibliography, for which the reader is referred to a forthcoming paper [1].

Bibliography

[1] P. Flajolet, M. Regnier, R. Sedgewick: "Mellin transform techniques and the analysis of algorithms", in preparation (1984).

5 - PERIODS AND OTHER REGULARITIES

PERIODICITIES IN STRINGS

Leo J. Guibas

Xerox Palo Alto Research Center
Palo Alto, California 94304

Abstract

In this talk we summarize what is known about the periodicities of strings. A period of a string is a shift that causes a string to match itself. We define z compact represention of the set of all periods of a string, called the (auto) correlation of the string. Two sets of necessary and sufficient conditions are obtained characterizing the correlations of strings. It is shown that the number of distinct correlations of strings of length n is of the order $n^{\log n}$. By using generating function methods we enumerate the strings having a given correlation, and investigate certain related questions. A more expanded version of these results appears in an article [0] by the author and A.M. Odlyzko.

1. INTRODUCTION

We consider strings over some finite alphabet Σ of size q. For a given string X of length n we will number its element from left to right by 0 through $n-1$, and write $X[i]$ for the i-th element. A non-negative integer p, $p < n$, will be called a *period* of X if we have $X[i]=X[i+p]$, for i in $[0,n-p)$. In other words, p is a period if a second copy of X, shifted right p positions and placed over the original, matches in the overlapping positions.

The main aim of this talk is to investigate necessary and sufficient conditions for a set of integers to be the set of all periods of some string. In Section 2 we introduce a representation of the set of all periods of some string as another binary string. Sections 3 and 4 contain the statements of necessary and sufficient conditions, along with a motivating discussion. Section 5 presents our main result, namely a theorem establishing that the above mentioned conditions actually characterize the sets which are periods of strings. In Section 6 we discuss the enumeration of the sets of periods of strings of a given length, and in Section 7, we count the number of strings having a given set of periods. The material presented here is based on [0], which represents joint work with A.M. Odlyzko. The reader is referred there for the proof of all theorems not explicitly given a reference.

Investigations into the periodicities of strings have most recently been motivated by research into string searching algorithms. Such algorithms work by attempting to match a pattern string at various places in a text string. The more sophisticated of these algorithms extract information from an unsuccessful match and use it to rule out other matches which have no chance of succeeding. These decisions invariably require knowledge of the periods of the pattern, or of its prefixes or suffixes. See [1,4,5,8].

In a more purely mathematical context, properties of the periods of strings have been investigated previously by Schützenberger and others in [2,9,11]. These papers, as well as the already referenced work [8] prove and make use of the so-called GCD rule for periods. Results less general than those we exhibit in Section 7 were previously obtained by Harborth [7], who studied the related problem of the enumeration of strings of a certain length with a given minimal period. An extensive collection of applications of our representation for the set of periods is given in [5].

NATO ASI Series, Vol. F12
Combinatorial Algorithms on Words
Edited by A. Apostolico and Z. Galil

2. BASIC DEFINITIONS AND NOTATIONS

If X and Y are two strings, we will define the *correlation* of X over Y to be a binary vector of the same length as X, composed as follows. The i-th bit (from the left) of the correlation is determined by placing Y under X so that Y's leftmost character is under the i-th character of X (from the left). Then, if all pairs of characters that are directly over each other match, the i-th bit of the correlation is 1, else it is 0. For example, if $\Sigma = \{H, T\}$, $X = H\ H\ T\ T\ H\ H$, and $Y = T\ H\ H\ T\ H\ H$, then the correlation is $0\ 0\ 0\ 1\ 0\ 0$, as depicted below:

```
X        H H T T H H
Y        T H H T H H.............0
           T H H T H H ............0
             T H H T H H ............0
               T H H T H H ............1
                 T H H T H H............0
                   T H H T H H............0
```

Figure 1. A correlation

We denote the correlation of X over Y by XY, a notation first introduced in [3]. Note that in general $XY \neq YX$. Let v denote the correlation of X over itself, i.e., $v = XX$. Let $n = |X|$ be the length of X, and number the bits of n from 0 to $n-1$. Then note that $np = 1$ if and only if p is a period of X. Thus the (auto) correlation of X provides a convenient encoding of the set of periods of X.

The smallest non-zero period of a string X will be called its *basic period*. By convention, the basic period of a string X of correlation $XX = 100 \cdots 0$ (i.e., no non-zero periods) is $|X|$. Two periods p and q of X will be called *independent* if neither is a multiple of the other. If X has period p, then we will often use the fact that X can be written as $X = UU \cdots UU'$, where $|U| = p$ and U' is a prefix of U. Given any binary vector v, indexed by $[0, n)$, we will denote by $\pi(v)$ the smallest positive i such that $v_i = 1$. If no such i exists, we set $\pi(v) = n$. Thus $\pi(XX)$ denotes the basic period of X. Note that if $|X| = |Y|$, but $X \neq Y$, then the correlation XY has the form $0 \cdots 0z$, where z is the auto correlation of the first match Z of Y into X, i.e., the longest Z such that $X = UZ$ and $Y = ZV$.

Conversely, any bit vector of the form $0 \cdots 0z$, where z is the auto correlation of some string Z, can obviously arise as the correlation of two easily constructed strings X and Y. For this reason we will confine ourselves from now on to the properties and characterization of auto correlations. For brevity we will use term correlation synonymously with auto correlation.

3. THE PROGRAM RULES

In this section we define two abstract properties of binary vectors that reflect necessary conditions satisfied by the sets of periods of strings. This fact will not be proved until Section 5, but for convenience of language we will continue to use the terminology of the previous section. We show that these two properties imply a previously known and very useful result on periods, the so called GCD rule.

The forward propagation rule essentially asserts the transitivity of matching (or equality): if X has periods $p, q, p < q$, then it also has $1+(q-p)=p+2(p-p)$ as a period.

Definition 3.1: (Forward Propagation Rule). A bit vector $v=(v_0, v_1, \ldots, v_{n-1})$ of length n satisfies the forward propagation rule if, whenever we have $v_p = v_q = 1$, with $p > q$, we also have $v_t = 1$, for all t of the form $p+i(q-p)$, $i=0,1,2,\ldots$, and t on the range $[p,n)$.

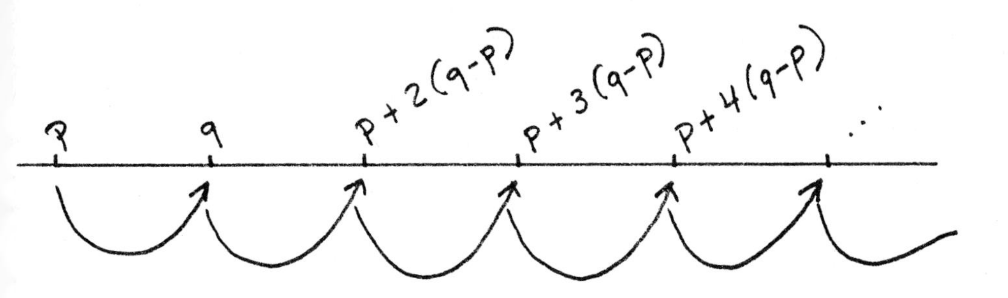

Figure 2. The forward propagation rule

The backward propagation rule asserts that if we follow the arithmetic progression defined by periods p and q to the *left*, and find that $p-(q-p)$ is *not* a period, then in proceeding to the left (unless we fall off the beginning) we must encounter at least as many 0's as we encountered 1's (really, full periods) in going to the right from q on.

Definition 3.2. (Backward Propagation Rule). A bit vector $v=(v_0, v_1, \ldots, v_{n-1})$ of length n satisfies the backward propagation rule if the following condition holds. For every p and q such that $p < 1 \leq 2p$ with $v_p = v_1 = 1$, but $v_{2p-q} = 0$, let $s = \lfloor (n-p)/(q-p) \rfloor$. Then for all t in the range $[0, 2p-q]$ of the form $p-i(q-p)$, $i=1,2,\ldots,s$, we have $v_t = 0$.

Lemma 3.1. If v satisfies the forward and backward propagation rules, then so does any prefix or suffix of v.

The following theorem shows that the classical GCD (greatest common divisor) result follows from the propagation rules. For additional discussion on this remarkable result on the periods of strings see [2,8].

Theorem 3.1. (The GCD Rule). Let $v=(v_0, v_1, \ldots, v_{n-1})$ be a non-empty bit vector of length n satisfying the forward and backward propagation rules, and having $v_0 = 1$.

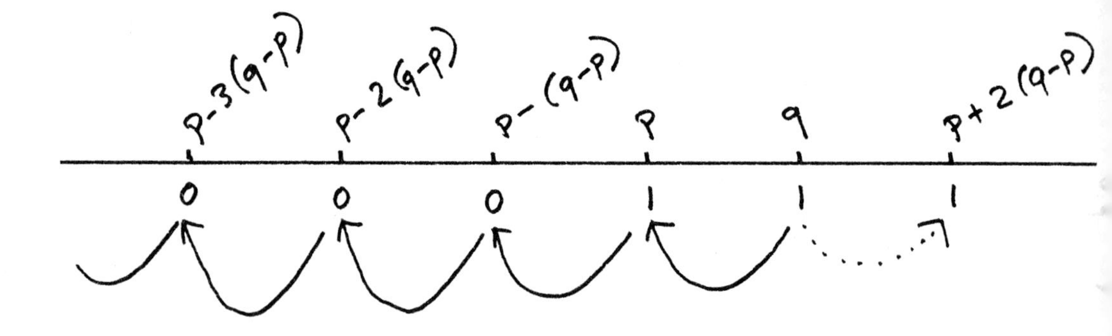

Figure 3. The backward propagation rule

Consider a pair of indices p and q in $[1,n)$ and let $t = \mathrm{GCD}(p,q)$. If $v_p = v_q = 1$ and $p+1 \leq n+t$, then we also have $v_t = 1$.

The illustration below depicts the case $p=7$, $q=12$, $n=18$. Then $\mathrm{GCD}(7,12)=1$, and $1+18 \leq 7+12$, so 1 must also be a period. In the figure we see how this follows from the successive applications first of the forward propagation rule, and then of the backward propagation rule.

4. THE RECURSIVE DEFINITION

We now introduce a recursive predicate on binary vectors which, as the next section shows, also turns out to be equivalent to the condition that the binary vector is a correlation.

Definition 4.1. (The Recursive Predicate Ξ Let v be a bit vector of length n. Define $p = \pi(v)$. The vector v satisfies predicate Ξ iff v is empty (equivalently, $n=0$), or v can be written as $v = (v_0, v_1, \ldots, v_{n-1})$ and satisfies the following constraints:

(1) $v_0 = 1$, and

(2) one of the following two mutually disjoint conditions holds

Case (a) $[n \geq 2p]$. Let $r = n - p(\lfloor \frac{n}{p} \rfloor - 1)$. Then $v_i = 0$ for $i \in [1, n-r)$ except at multiples of p (where $v_i = 1$). Further, if $w = (w_0, \ldots, w_{r-1}) = (v_{n-r}, \ldots, v_{n-1})$ then

- (i) $w_p = 1$ or $r = p$
- (ii) $\pi(w)$ is not a proper divisor of p, and
- (iii) w satisfies predicate Ξ.

Case (b) $[n < 2p]$. Let $r = n - p$. Then $v_i = 0$ for $i \in [1,p)$, and if $w = (v_0, \ldots, w_{r-1}) = (v_p \cdots v_{n-1})$, then w satisfies predicate Ξ.

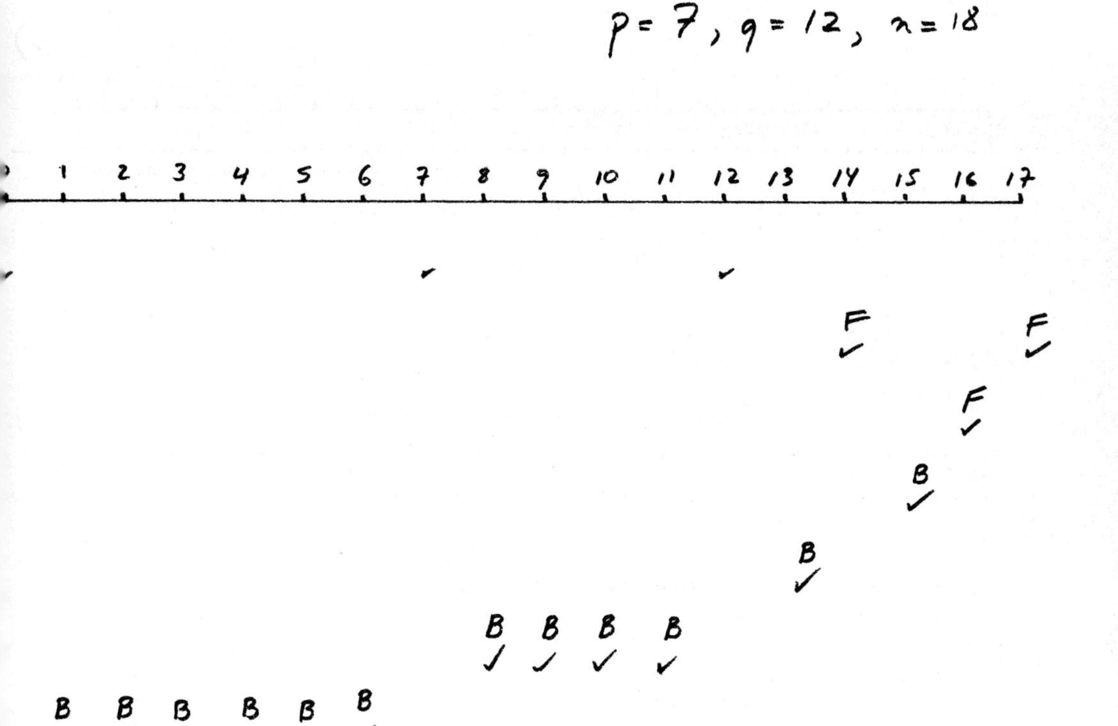

Figure 4. The GCD Lemma

In case (a) we are choosing r so that $p \le r < 2p$ and $n - r$ is a multiple of p. By the GCD rule, condition (ii) of case (a) is equivalent to: if $\pi(w) < p$, then $\pi(w) > (r - p) + \mathrm{GCD}(p, \pi(w))$. The figure below illustrates the two cases. We now give a recursive procedure *ksi*, given below in pseudo ALGOL, that implements the above predicate. Predicate Ξ is true of vector v iff $ksi[0,n]{:}1$ (e.g., the first component of the returned record) is true. (We must remember to set the boundary condition $v[n]=1$). The second component of the returned record is the length of the basic period of v. Note that the number of bit position examinations done by *ksi* is bounded by $2n$, so *ksi* runs in linear time. This bound follows since each bit position is examined at most once

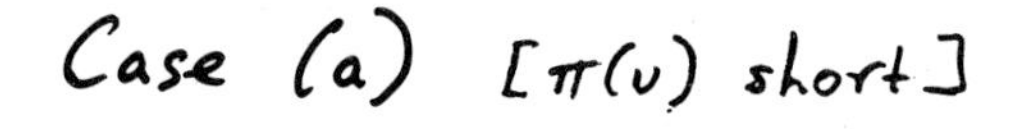

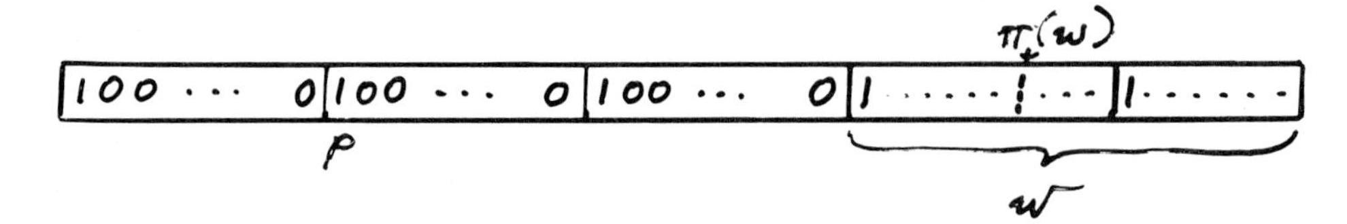

Case (b) [π(v) long]

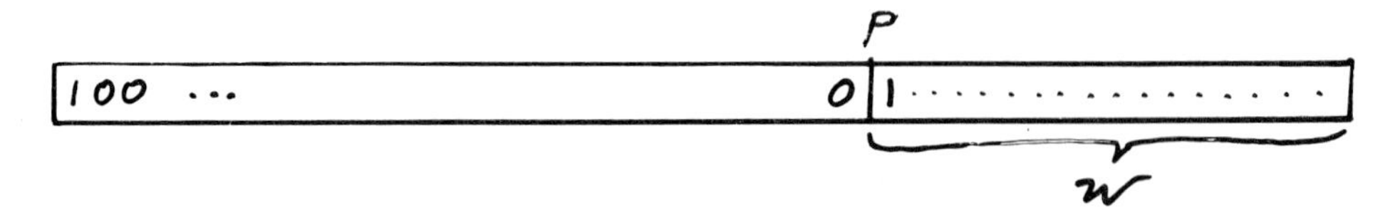

Figure 5. The two cases of Ξ

before the recursive call is made. *Ksi* is then invoked recursively on a virgin substring consisting of bits that have not yet been examined. There is one additional bit examined in the final test, but the total number of such examinations is certainly bounded by the number of recursive calls, which is at most n.

5. PROOF OF EQUIVALENCE

In this section, we give a theorem that proves that the sets of conditions stated in Sections 3 and 4 each characterize the bit vectors that arise as the correlation of some string. Note that condition (1) below refers to *binary* strings. This implies the non-obvious fact that an alphabet of size two gives rise to all sets of periods that can arise with strings over an alphabet of arbitrary size.

Theorem 5.1. (Characterization of Periods). Let $v = (v_0, v_1, \ldots, v_{n-1})$ be a non-empty bit-vector. Then the following four statements are equivalent:

(1) v is a correlation of a binary string

(1') v is a correlation of some string

(2) $v_0 = 1$ and v satisfies the forward and backward propagation rules

(3) v satisfies predicate Ξ.

```
PROCEDURE Ksi[INTEGER l, r] RETURNS RECORD[BOOLEAN; INTEGER];
  BEGIN INTEGER i, p, q, s, newp; BOOLEAN flag;
  IF r = l THEN RETURN[TRUE, 0];
  IF v[l] = 0 THEN RETURN[FALSE, UNDEFINED];
  p ← 1;
  WHILE l + p < r AND v[l + p] = 0 DO p ← p + 1;
  COMMENT now p is the basic period of v[l..r];
  s ← (r − l) DIV p;
  IF s ⩾ 2 THEN
    BEGIN COMMENT case (a);
    q ← p * (s − 1);
    flag ← TRUE;
    FOR i IN [1..q) DO
      BEGIN
      IF (i MOD p) = 0 THE flag ← flag AND (v[l + i] = 1)
        ELSE flag ← flag AND (v[l + i] = 0);
      END;
    IF NOT flag THEN RETURN[FALSE, UNDEFINED]
    [flag, newp] ← Ksi[l + q, r];
    RETURN[flag AND (v[l + q + p] = 1) AND
        ((p = newp) OR (newp > r − l − q − p + GCD(newp, p))), p];
    COMMENT here we made the additional bit examination;
    END COMMENT case (a);
  ELSE BEGIN COMMENT casr (b);
    q ← p;
    [flag, newp] ← Ksi[l + q, r];
  RETURN[flag, p];
  END COMMENT case (b);
END.
```

Figure 6. The procedure *ksi*

6. COUNTING THE CORRELATIONS

In this section we use the recursive predicate Ξ to obtain bounds on the number of distint correlations of length n.

Theorem 6.1. (the number of correlations). The number $k(n)$ of distinct correlations of length n satisfies

$$\left(\frac{1}{2\ln 2} + \theta(1)\right)\ln^2 n \leq \ln k(n) \leq \left(\frac{1}{2\ln(3/2)} + \theta(1)\right)\ln^2 n$$

as $n \to \infty$. We conjecture that $\ln k(a)$ is in fact asymptotic to a constant times $\ln^2 n$, but we have been unable to prove this. A table of some values of $k(n)$ follows.

n	1	2	3	4	5	6	7	8	9	10	11	12	13	14	15	16	17	18	19	20
$\kappa(n)$	1	2	3	4	6	8	10	13	17	21	27	30	37	47	57	62	75	87	102	116

n	25	30	35	40	45	50	55
$\kappa(n)$	220	392	664	1005	1552	2240	3226

Figure 7. Counting the correlations

7. POPULANCE

In this section we count the number of strings of length n over an alphabet of size q, which have a given (auto) correlation. We do this by obtaining a recurrence on n for $Ln(C)$, the population of strings with correlation $K = 100 \cdots OOC$, where K is of length n and consists of a 1 followed by all O's until the final suffix C, which is itself a correlation and assumed fixed. As usual c will denote the length of C. For simplicity of notation we will often write Ln instead of $Ln(C)$, C being implicitly understood. Before we can state our result we need one additional definition. Let ψ denote a sequence (depending on C) defined for all integers by

$$\left.\begin{array}{rl} \psi_k = 0 & \text{for } k > C \\ C[c-k] & \text{for } 1 \le k \le c \\ q^{-k} & \text{for } k \le 0 \end{array}\right\}$$

Thus, for $1 \le k \le c$, $\psi_k = 1$ if and only if $c - k$ is a period of C.

Our first theorem states the recurrence on Ln:

Theorem 7.1. (basic population recurrence). The number Ln of q-ary strings of length n

which have correlation $10 \cdots 0C$ satisfies the recurrence

$$Ln + \sum_{\substack{k,l \\ n+l=2k}} L_k \psi_l = 2\psi_{2c-n} Lc ,$$

where we set $Ln = 0$ for $n < c$.

An immediate consequence of this theorem is that Lc / Ln.

To continue our analysis, we introduce the generating functions

$$L(z) = \sum_{n=0}^{\infty} Ln\, z^{-n} , \text{ and}$$

$$\Psi(z) = \sum_{n=-\infty}^{\infty} \psi_n z^n .$$

In this notation the result of Theorem 7.1. can simply be stated as

$$L(z) + \Psi(z) L(z^2) = 2Lc\, \Psi(z) z^{-2c}$$

It will be convenient to introduce the normalized generating function

$$\bar{L}(z) = L(z)/L(z)/Lc$$

Thus we can rewrite (2) as

$$\bar{L}(z) + \Psi(z)\bar{L}(z^2) = 2\Psi(z) z^{-2c}$$

The following theorem describes the asymptotics of Ln as $n \to \infty$ with C fixed.

Theorem 7.2 (the asymptotics of populations). The number Ln of q-any strings of length n and correlation $10 \cdots 0C$ has the asymptotic value

$$\frac{Ln}{Lc} = \left(\frac{2}{q^{2c}} - \bar{L}(q^2) \right) q^n + \theta \left((q+\epsilon)^{n/2} \right),$$

where $\bar{L}(z)$ satisfies the functional equation

$$\bar{L}(z) + \Psi(z)\bar{L}(z^2) = 2\Psi(z) z^{-2c} .$$

The above expansion is valid as $n \to \infty$, for any positive ϵ, and the implied θ constant is independent of the underlying correlation C (but depends on ϵ).

Remark: The above expansion can be continued as far as we please. By the same technique, we can recursively compute the asymptotics of Lc in terms of a still smaller correlation c' ($c = 10 \cdots 0C$), and so on. With the aid of a symbolic manipulation system, we could carry out this process to obtain good numerical estimates for Ln. Let us define

$$\theta(z) = z^{-2c}\Psi(z) - z^{-c}, \text{ and}$$

$$\beta = 2q^{-2c} - \bar{L}(q^2) = q^{-2c} - \sum_{k=1}^{\infty} (-1)^{k-1}\theta(q^{2^k}) \prod_{r=1}^{k-1} \Psi(q^2).$$

The coefficient β is the asymptotic limit of $Ln/(Lcq^n)$. The above formula gives us a very efficient way to compute β, via an alternating series whose terms decrease on absolute value. Some interesting values are given below.

	C	β	$L_c\beta$
$q=2$	Λ	0.26771654	0.26771654
	1	0.150203882	0.300407764
	10	0.055000309	0.110000618
	11	0.04445766	0.08891532
$q=3$	Λ	0.55697974	0.55697974
	1	0.094234491	0.282703474
	10	0.0121190452	0.072714272
	11	0.0109175415	0.0327526242
$q=24$	Λ	0.95659723	0.95659723
	1	0.001732971	0.041591309

Figure 8. The asymptotics of some population

Here Λ denotes the empty string. Thus approximately 27 of all binary strings have the trivial correlation [10]. Slightly more, 30 percent, have the correlation $10 \cdots 01$, which can be shown to be the most popular correlation in base 2. Our results extend the bounds devised by Harborth in [7]. Note that for $q=3$ or larger, the trivial correlation is the most popular. For large q, the above results show that this fraction is essentially $(q-2)/(q-1)$. For $q=24$, more than 95 percent of all strings have that correlation.

A result follows that allows us to compare β's corresponding to C's of the same length.

Theorem 7.3 (asymptotic comparison) Let A, B denote correlation of length c. If $A_q \geq B_q$, then $\beta_A \le \beta_B$. Here T_q denotes the correlation T, viewed as a number to the base q.

This result is only true asymptotically. For example, if $q=2$, then

$$L(10^{11}100011)/L(100011) = 62, \text{ but}$$
$$L(10^{11}100100)/L(100100) = 63.$$

(By 0^{11} we mean the string of 11 0's). The table below gives the population, over a binary alphabet, of all legal correlations of length 20. There are 116 different correlations, each followed by the number of *binary strings* giving rise to it.

Correlation	Population	Correlation	Population
10000000000000000000	281076	10000000000100001111	8
10000000000000000001	315322	10000000000100010001	40
10000000000000000010	115226	10000000000100010011	8
10000000000000000011	93146	10000000000100100100	16
10000000000000000100	63568	10000000000100100101	8
10000000000000000101	29874	10000000000101010101	8
10000000000000000111	24234	10000000000111111111	6
10000000000000001000	24318	10000000001000000000	284
10000000000000001001	23940	10000000001000000001	318
10000000000000001010	7612	10000000001000000010	110
10000000000000001111	6094	10000000001000000011	88
10000000000000010000	12276	10000000001000000100	64
10000000000000010001	10190	10000000001000000101	30
10000000000000010010	4024	10000000001000000111	24
10000000000000010011	2004	10000000001000001000	24
10000000000000010101	1918	10000000001000001001	18
10000000000000011111	1528	10000000001000001010	6
10000000000000100000	5080	10000000001000001111	6
10000000000000100001	5588	10000000001000100010	6
10000000000000100010	1518	10000000001000100011	6
10000000000000100011	1512	10000000001001001001	6
10000000000000100100	1000	10000000010000000010	232
10000000000000100101	500	10000000010000000011	182
10000000000000101010	476	10000000010000000111	46
10000000000000111111	382	10000000010000001010	16
10000000000001000000	2560	10000000010000001111	12
10000000000001000001	2432	10000000010000010010	4
10000000000001000010	1024	10000000010000010011	2
10000000000001000011	768	10000000010000011111	2
10000000000001000100	512	10000000010000100011	6
10000000000001000101	128	10000000010001000111	2
10000000000001000111	126	10000000100000001000	90
10000000000001001001	378	10000000100000001001	90
10000000000001010101	120	10000000100000001010	30
10000000000001111111	96	10000000100000001111	24
10000000000010000000	1184	10000000100000011111	6
10000000000010000001	1312	10000001000000100000	40
10000000000010000010	390	10000001000000100001	44
10000000000010000011	330	10000001000000100010	12
10000000000010000100	256	10000001000000100011	12
10000000000010000101	128	10000001000000100100	8
10000000000010000111	90	10000001000000100101	4
10000000000010001000	90	10000001000000101010	4
10000000000010001001	90	10000001000000111111	2
10000000000010010010	60	10000010000010000010	26
10000000000010010011	30	10000010000010000011	22
10000000000010101010	30	10000010000010000111	6
10000000000011111111	24	10000100001000010000	12
10000000000100000000	592	10000100001000010001	10
10000000000100000001	616	10000100001000010010	4
10000000000100000010	240	10000100001000010011	2
10000000000100000011	184	10000100001000010101	2
10000000000100000100	112	10001000100010001000	6
10000000000100000101	48	10001000100010001001	6
10000000000100000111	48	10010010010010010010	4
10000000000100001000	48	10010010010010010011	2
10000000000100001001	48	10101010101010101010	2
10000000000100001010	16	11111111111111111111	2

Figure 9. The population of correlations of length 20 over a binary alphabet

8. REFERENCES

[0] L.J. GUIBAS and A.M. ODLYZKO, Periods in Strings, *J. Comp. Theory*, Series A, Vol. 30, No. 1, 1981.

[1] S.B. BOYER and J.S. MOORE, A fast string searching algorithm, *Comm. Assoc. Comput. Mach.* **20** *(1977), 762-771.*

[2] N.J. FINE and H.S. WILF, Uniqueness theorems for periodic functions, *Proc. Amer. Math. Soc.* **16** (1965), 109-114.

[3] M. GARDNER, On the paradoxical situations that arise from nontransitive relations. *Sci. Amer.* (October 1974), 120-125.

[4] L.J. GUIBAS and A.M. ODLYZKO, A new proof of the linearity of the Boyer-Moore string searching algorithm, in "Proceedings, 18th Annual Symposium on Foundations of Computer Science Proceedings," IEEE Computer Society, New York, 1977. Revised version *SIAM J. Comput.*, in press.

[5] L.J. GUIBAS and A.M. ODLYZKO, String overlaps, pattern matching and nontransitive games. *J. Combinatorial Theory Ser. A.*, in press.

[6] Z. GALIL and J. SEIFERAS, Saving space in fast string matching, in "Proceedings. 18th Annual Symposium on Foundations of Computer Proceedings." IEEE Computer Society, New York, 1977.

[7] H. HARBORTH, Endliche 0-1-Folgen mit gleichen Teilblocken, *Crelle's J. 271* (1974), 139-154.

[8] D.E. KNUTH, J.H. MORRIS and V.R. PRATT, Fast pattern matching in strings, *SIAM J. Comput.* **6** (1977), 323-350.

[9] R.C. LYNDON and M.P. SCHUTZENBERGER, The equation $a^M = b^N C^P$ in a free group. *Michigan Math. J.* **9** (1962), 289-298.

[10] T.P. NIELSEN, On the expected duration of a search for a fixed pattern in random data. *IEEE Trans. Inform Theory* (1973), 702-704.

[11] M.P. SCHUTZENBERGER, A property of finitely generated submonoids of free monoids. "Actes de Colloque sur le semigroups a Szeged, 1976," North-Holland, Amsterdam, in press.

LINEAR TIME RECOGNITION OF SQUAREFREE STRINGS

Michael G. Main*
Department of Computer Science
University of Colorado
Boulder, CO 80309 USA

Richard J. Lorentz
Computer Science Group
Harvey Mudd College
Claremont, CA 91711 USA

ABSTRACT

A *square* is an immediately repeated nonempty string, *e.g.*, *aa*, *abab*, *abcabc*. This paper presents a new $O(n \log n)$ algorithm to determine whether a string of length n has a substring which is a square. The algorithm is not as general as some previous algorithms for finding all squares [1,7,8,13], but it does have a simplicity which the others lack. Also, for a fixed alphabet of size k, the algorithm can be improved by a factor of $\log_k(n)$, yielding an $O(n)$ algorithm for determining whether a string contains a square.

1. INTRODUCTION

A *square* is an immediately repeated nonempty string, *e.g.*, *aa*, *abab*, *wallawalla*. A string is called *squarefree* if it has no square as a substring. Axel Thue was the first to wonder how long squarefree strings could be [17,18]. Of course, the answer depends on how many different characters are allowed to appear in the string. For example, if *a* and *b* are the only characters allowed, then *aba* is the longest squarefree string. Thue's remarkable result was that by using only three different characters, squarefree strings of any length can be constructed. Since Thue's work, others have rediscovered his result, applying it in diverse fields [2,6,10,11,14,15,19].

Interest in squarefree strings among formal language theorists was revitalized about six years ago with the appearance of Berstel's paper on squarefree homomorphisms [4,5], as well as a more general paper by Bean, *et. al* [3]. Shortly after these papers, there appeared an $O(n \log n)$ algorithm to determine whether a string of length n was squarefree [12]. Following this, there were at least three independently designed (and quite different) algorithms for finding all squares in time $O(n \log n)$[1,7,8,13]. These algorithms are interesting, both from a practical standpoint and as an example of pattern-matching algorithms. (On the practical side, the earlier paper of Bean, *et. al* needed to prove the squarefreeness of a set of strings with lengths around 210.)

This paper presents a new $O(n \log n)$ algorithm to determine whether a

*Research supported by the University of Colorado Council on Research and Creative Work.

NATO ASI Series, Vol. F12
Combinatorial Algorithms on Words
Edited by A. Apostolico and Z. Galil

string is squarefree. The algorithm has two prominent features:

(1) It is by far the simplest of the four algorithms that we know of. The pseudo-code presentation of the algorithm in section 2 is complete and concise. Apart from the input string itself, there is only a constant amount of additional memory required, due to the fact that the algorithm only determines squarefreeness, and doesn't find all squares.

(2) For a fixed alphabet of size k, the algorithm is easily sped up by a factor of $\log_k(n)$. Thus, for a fixed alphabet, we have an $O(n)$ algorithm for determining whether a string is squarefree. This speed-up is effected in section 3.

Throughout the paper we will use s for the string of length n, in which we are trying to find a square. A substring of s from position i through position j is denoted $s[i...j]$. We denote the length of a substring x by $|x|$.

2. THE NEW $O(n \log n)$ ALGORITHM

The algorithm is best described by the pseudo-code presentation in figure 1. The key to the algorithm is this: if a string $s = s[1...n]$ is partitioned into blocks of size l, then any square xx in s, with $|x| \geq 2l-1$, must completely contain one of the blocks within the first occurrence of x. Also, if $|x| < 4l-1$, then the block appearing in the first half of x must reappear in s, no further than $4l-2$ characters away from the start of the first block, and no less than $2l-1$ characters away.

The steps of the pseudo-code are explained here:

(1) The integer l is the size of the blocks which s is partitioned into.

(2) The loop invariant is: s contains no square xx with $|x| < 2l-1$. Inside the loop we will check for squares with $2l-1 \leq |x| < 4l-1$. Also, one iteration of the loop takes time $O(n)$, as will become evident.

(3) The **for** loop (3-10) examines each block of length l. The current block begins at position $[i]$. Recall that if s contains a square xx with $2l-1 \leq |x| < 4l-1$, then one of these blocks will occur completely within the first occurrence of x. Moreover, if the contained block begins at position $[i]$, then the block must reappear with a starting position in the range $[i+2l-1]...[i+4l-2]$.

(4) This step finds the places where the current block reappears, in the appropriate range. These reappearances indicate possible squares. Note that the block can reappear at most twice in this range (of size $2l-1$), otherwise two occurrences of the block would overlap, implying a square yy with $|y| \leq l$. We know that such a square is impossible by the loop invariant (see (2) above).

Function to determine if a string $s[1...n]$ contains a square. The function returns **true** iff the string contains a square.

```
 1. l ← 1;
 2. while (2l - 1) ≤ n/2 do begin
 3.    for i ← 1 to n - 3l + 2 by l do begin
 4.       Search for the substring s[i...i+l-1] with starting position between
                [i+2l-1] and [i+4l-2]. Set m to the number of such oc-
                currences (at most 2) and put the starting positions of
                these occurrences in x[1] and x[2].
 5.       for j ← 1 to m do begin
 6.          left ← the length of the longest common
                suffix of s[1...i-1] and s[1...x[j]-1].
 7.          right ← the length of the longest common
                prefix of s[i...n] and s[x[j]...n].
 8.          if (left + right) > (x[j] - i) then return(true)
 9.       end
10.    end
11.    l ← 2l
12. end
13. return(false)
```

Figure 1.

It is also important to note that the search for reappearances need not be "clever", *i.e.*, a sequential search of the target area suffices. Specifically, we can start at $s[i+2l-1]$ and match characters with the current block. If we find a mismatch before the end of the block, then we can restart the matching *after* the mismatch. This is possible because if the block occurs starting at or before the mismatch, then there are adjacent prefixes of the block, which contradicts the loop invariant (see (2) above). Similarly, if we find a reappearance of the block, then we can record this occurrence, and restart the matching *after* the reappearance.

Using this technique, step (4) requires $O(l)$ time.

(5-9) For each case where the current block reappeared, we check adjoining characters to see if a square is found. There are at most $3l-2$ characters to check for each reappearance, and at most two reappearances, so this takes time $O(l)$.

(10) Processing each block took time $O(l)$. As there are $\frac{n}{l}$ blocks, the entire loop (3-10) takes time $O(n)$.

(11-12) Since l doubles each time through the **while** loop, there are at most $\log_2(n)$ iterations of (2-12). So, the entire algorithm takes time $O(n \log n)$.

(13) If the **while** loop finally ends, then the loop invariant tells us that there is no square xx with $|x| \leq \frac{n}{2}$. Hence, s is squarefree and we may return **false**.

3. THE $\log_k(n)$ SPEED-UP

The algorithm of the previous section determined whether a string $s[1..n]$ was squarefree in time $O(n \log_2 n)$. If we know nothing of the size of the alphabet, then this is optimal for an algorithm that depends solely on comparisons of characters [12]. However, if we have a finite alphabet of k characters, then we can improve our $O(n \log n)$ algorithm by a factor of $\log_k(n)$. The new complexity becomes $O(\frac{n \log_2 n}{\log_k n}) = O(n \log_2 k)$, which is $O(n)$ for constant alphabet size k.

The main idea is this: rather than work with individual characters in the string, we will work with clusters of characters, of size proportional to $\log_k(n)$. This will allow us to speed up the work in step (4) and the loop (5-9) by $\log_k(n)$. To do this requires some preliminary work which we show takes $O(n)$ time.

The first preliminary step is to build a table in lexical order of all strings of size $c \equiv \lceil \log_k(n^a) \rceil$, where a is an arbitrary value in the range $0 < a < 1$. There are $k^c = O(n^a)$ such strings of length c, which we will call *clusters*. We then label each string in the table as squarefree or not. Using a naive $O(c^2)$ algorithm on each of these strings requires $O(log^2(n^a))$ time on each string. As there are $O(n^a)$ strings, the total time to build the table is $O(n^a \log^2(n^a)) < O(n)$, since $a < 1$.

Next we introduce two new components to each position in the string $s[1...n]$. The first will be a pointer into the above table. The pointer will point to the cluster corresponding to the c characters beginning at the given position. (The pointer is undefined for the last $c-1$ characters of s.) This can be done in $O(n)$ time because given the pointer, $s[i].ptr$, for position i in the string, we can calculate the next pointer in constant time, as follows: Let $s[i].ch \in \{0,...,k-1\}$ be the lexical order of the character at position $s[i]$. Then:

$$s[i+1].ptr = (k \times s[i].ptr + s[i+c].ch) \bmod k^c.$$

Having established these pointers, n table lookups (through each of the n

pointers) can establish whether the string has any squares xx with $|x| \le \frac{c}{2}$.

The second component to be added to each position in the string will be denoted $s[i].next$. Define $s[i].next$ to be the smallest $j > i$ such that $s[i].ptr = s[j].ptr$ (or ∞ if no such j exists). Intuitively, $s[i].next$ is the next occurrence of the cluster at position i. Given the *ptr* component, *next* can easily be calculated in $O(n)$.

There is one more preliminary step. Having established that there are no squares xx with $|x| \le \frac{c}{2}$, we set $l \leftarrow \frac{c}{4}$ in our original algorithm from section 2. This makes the loop invariant true at (2), and allows us to perform the body of the while loop twice ($O(n)$ time) to establish that s has no square xx with $|x| < 2c-1$. We now set $l \leftarrow c$ in step (1) and perform the same algorithm, with the following modifications to steps (4), (6) and (7):

(4) In this step we must find all occurrences of the current block with starting position in the range $[i+2l-1] \ldots [i+4l-2]$. This can be accomplished by the following two steps, each requiring $O(\frac{l}{c})$ time.

(4a) Find the first position f in $s[i+2l-1 \ldots i+4l-2]$ with $s[i].ptr = s[f].ptr$. (If there is no such position, then $f = \infty$.) This is done here:

$f \leftarrow i$
while $f < (i+2l-1)$ **do** $f \leftarrow s[f].next$

Each iteration of this **while** loop advances f by at least c positions, otherwise there would be a square xx with $|x| < c$. Thus, (4a) takes $O(\frac{l}{c})$ time.

(4b) Match the characters of $s[i \ldots n]$ with $s[f \ldots n]$. Using the *ptr* components, we can carry this out c characters at a time. If we find a mismatch before l characters, then we can use the value $s[f].next$ to advance f beyond the mismatch (as in (4a)). If we find a match of length l (remember l is a multiple of c), then we can record this match and advance f beyond the match. Since we are matching in clusters of size c, the matching in this step can take only $O(\frac{l}{c})$ time. Advancing f also takes time $O(\frac{l}{c})$, as in (4a). Hence, the total time for (4b) is $O(\frac{l}{c})$.

(6-7) Using the *ptr* components, it is easy to calculate *left* or *right* in time $O(\frac{p}{c}) + O(c)$, where p is the length of the match. This is accomplished by matching characters c at a time until a mismatch occurs, then matching characters one at a time after that. We can restrict p so that

$p \le x[j]-i = O(l)$, so the first part becomes $O(\frac{l}{c})$, as desired.

In general, the second part, of time $O(c)$, might not be $O(\frac{l}{c})$, but we can still show that the *total* done in this $O(c)$ part, throughout the entire algorithm, will be $O(n)$. This is done as follows: Each time a block is checked, *left* and *right* may be calculated twice each. Thus, each block contributes at most $O(4c) = O(c)$ work to the total computation. How many blocks do we check in all? The first time through the **while** loop we check $\frac{n}{c}$ blocks (remember l started at c). Each subsequent iteration of the **while** loop has half as many blocks as the previous iteration. So the total number of blocks is:

$$\frac{n}{c} + \frac{n}{2c} + \frac{n}{4c} + \cdots + \frac{n}{n} = O(\frac{n}{c})$$

So, the total time spent in calculating *left* and *right* is $O(\frac{n}{c} \times c) = O(n)$, as required.

We have shown how to effect a speed-up $c = \log_k(n^a) = a \log_k(n)$ in all the deepest loops ((4), (6) and (7)) of the algorithm. So, we have sped up the original algorithm by $O(\log_k n)$, yielding a new complexity of $O(n)$.

4. A QUESTION

A *permuted square* or (*permutation*) is a non-empty string of the form xy, where x has the same characters as y, but possibly in a different order. A string is called *permutation-free* if it has no permutation as a substring. In 1968, Evdokimov [9] showed that there are infinite permutation-free sequences on a 25 letter alphabet. Two years later, Pleasants [16] reduced the alphabet size to five. (The case of a four letter alphabet remains open.)

The question we pose is this: how quickly can an algorithm determine if a string is permutation-free? Little is known about the complexity of this problem beyond the obvious $O(n^2)$ algorithm and the even more obvious $O(n)$ lower bound. For a fixed alphabet of size k, the algorithm can be improved to $O\left(\left(\frac{n}{\log_k n}\right)^2\right)$, using a technique similar to that of section 3.

References

(1) A. Apostolico and F.P. Preparata, Optimal off-line detection of repetitions in a string, *TCS* 22 (1983) 297-315.

(2) S.E. Arshon, Dokazatel'stov suscestvovanija n-znacnyh beskonecnyh asimmetricnyh posledovatel'nostei, *Mat. Sb.* 44 (1937), 769-779.

(3) D.R. Bean, A. Ehrenfeucht and G.F. McNulty, Avoidable patterns in strings of symbols, *Pacific J. Math.*, 85 (1979), 261-294.

(4) J. Berstel, Sur les mots sans carre definis par un morphisme, in *Automata, Languages and Programming* (A. Maurer, Ed.), LNCS 71, Springer-Verlag, Berlin, 1979, 16-25.

(5) J. Berstel, Mots sans carre et morphismes iters, *Discrete Math.* 29 (1979), 235-244.

(6) C.H. Braunholtz, Solution to problem 5030, *Am. Math. Mo.* 70 (1963), 675-676.

(7) M. Crochemore, Determination optimale des repetitions dans un mot, Technical Report 80-53, Laboratoire Informatique Theorique et Programmation, University of Paris (1981).

(8) M. Crochemore, An optimal algorithm for computing repetitions in a word, *IPL* 12 (1981), 244-250.

(9) A.A. Evdokimov, Strongly asymmetric sequences generated by a finite number of symbols, *Soviet Math. Dokl.*, 9 (1968), 536-539.

(10) D. Hawkins and W.E. Mientka, On sequences which contain no repetitions, *Math. Student* 24 (1956), 185-187. MR 19 (1958) #241.

(11) G.A. Hedlund, Remarks on the work of Axel Thue on sequences, *Nord. Mat. Tidskr.* 16 (1967), 148-150. MR 37 (1959), #4454.

(12) M.G. Main and R.J. Lorentz, An $O(n \log n)$ algorithm for recognizing repetition, Washington State University Technical Report CS-79-056, Pullman, WA 99164 (1979).

(13) M.G. Main and R.J. Lorentz, An $O(n \log n)$ algorithm for finding all repetitions in a string, *J. of Algorithms*, to appear (1984).

(14) M. Morse, A solution of the problem of infinite play in chess, *Bull. Amer. Math. Soc.*, 44 (1938), 632.

(15) M. Morse and G.A. Hedlund, Unending chess, symbolic dynamics and a problem in semigroups, *Duke Math. J.* 11 (1944), 1-7. MR 5 (1944) #202.

(16) P.A.B. Pleasants, Nonrepetitive sequences, *Proc. Cambridge Phil. Soc.* 68 (1970), 267-274.

(17) A. Thue, Uber unendliche Zeichenreihen, *Norske Videnskabers Selskabs Skrifter Mat.-Nat. Kl. (Kristiania)* (1906), Nr. 7, 1-22.

(18) A. Thue, Uber die gegenseitige Lage gleicher Teile gewisser Zeichenreihen, *Norske Videnskabers Selskabs Skrifter Mat.-Nat. Kl. (Kristiania)* (1912), Nr. 1, 1-67.

(19) T. Zech, Wiederholungsfreie Folgen, *Z. Angew. Math. Mech.* 38 (1958), 206-209.

DISCOVERING REPETITIONS IN STRINGS

Michael O. Rabin†

Division of Applied Sciences, Harvard University
Institute of Mathematics, Hebrew University

INTRODUCTION

There are many variants of problems involving string matchings. The basic problem is to determine whether a pattern string x appears as a (contiguous) substring of a text y, i.e. whether for some strings u, v, we have $y = uxv$. The classical string matching algorithm due to Knuth, Morris, Pratt [5] solves this problem in time linear in the length $|y|$ of y. The KMP algorithm requires $|x|$ auxiliary storage locations. Galil and Seiferas [3, 9] reduced the auxiliary storage requirements first to $O(\log|x|)$ and then to a constant number.

Another variety of string matching problems involves questions such as: whether y is a square, i.e. $y = xx$ for some x; whether y is a palindrome, i.e. $y = xx^R$ for some x (here x^R is the string obtained by reversing the order of symbols in x); whether y has a repetition i.e. $y = wxuxz$ subject to the restrictions $|w| = r|x|$, $|u| = s|x|$, $|z| = t|x|$ for fixed numbers r, s, t, etc.

All of these problems are solvable in real-time on appropriate machine models such as multi-tape Turing machine (see [3, 9, 10]). The algorithms are combinatorial in nature, examining prefixes, suffixes and overlaps of subwords of y. As a rule these algorithms tend to be complicated to describe and difficult to prove correct.

In [4, 8] R. Karp and the present author solved the basic string matching problem using randomization. A hashing ϕ function from strings to a finite domain D is randomly chosen from a set H of such functions. The functions $\phi \in H$ have the following informally stated properties:

1. For $x = \sigma u$, $\sigma, \tau \in \Sigma$, $u \in \Sigma^*$, the values $\phi(u)$ and $\phi(x\tau)$ are computable in constant time from $\phi(x)$.

†This work was supported by NSF Grant MCS-8121431.

NATO ASI Series, Vol. F12
Combinatorial Algorithms on Words
Edited by A. Apostolico and Z. Galil

2. Let y be a string of length m. The probability that for a randomly chosen $\phi \in H$ there will be two *unequal* substrings x_1 and x_2 of y for which $\phi(x_1) = \phi(x_2)$ is suitably small; for example smaller than m^{-2} or 2^{-m}.

Within the framework of a random-access machine model with machine word size of about $\log_2 m$ bits (similar assumptions are used for KMP [5] and some of the other classical results in this field), we obtained very simple and fast algorithms for all the aforementioned string matching problems. An important advantage of the fingerprinting method is that the algorithms based on it are rather insensitive to the particular detailed form of the string matching task.

In the present paper we employ the fingerprinting method to solve yet another string matching problem. Given a string y we want to find the earliest repetition, i.e. the shortest w and x such that $y = wxxz$. We shall call this the repetition problem.

This question was treated by Apostolico and Preparata [1] who gave an $O(|y| \log |y|)$ algorithm for this problem and some of its extensions. Lorentz and Main [6] give a linear time algorithm for recognition of square-free strings. Both papers use the suffix tree method [1 , 7 ,11]. The Lorentz-Main further reduction of complexity to linear running time, employs the $O(\log |y|)$ size of the computer word to encode information and string operations.

We give a very simple algorithm for discovering repetitions by use of fingerprints. If $y \in \Sigma^*$, $|\Sigma| = s$, $|y| = m$ then the expected running time is $O(m \log_2 m \log_s m)$. But by further use of the power of encoding several operations into one machine-word operation, the running time can be reduced to $O(m \log_2 m)$, and perhaps even further. An additional feature of this algorithm is its parallelizability. An $n \leqslant m$ processor machine would produce approximately an n^{-1} reduction in running time. This would produce practical improvements even for small values of n.

1. FINGERPRINTS AND THEIR PROPERTIES

Let the alphabet Σ have s symbols and let $s \leqslant q$ be the smallest prime number greater or equal to s. Practical cases include $s = q = 2$ or $s = 256$, $q = 257$. Denote by Z_q the field of residues $\{0,1,..,q-1\}$. Without loss of generality we may assume that $\Sigma = \{0,1,...,s-1\}$, so that

$\Sigma \subseteq Z_q$.

We associate with a string $y = y_1 y_1 \cdots y_m \in \Sigma^*$ the polynomial

$$f(t) = y_1 t^{m-1} + y_2 t^{m-2} + \cdots + y_m \in Z_q[t] \quad .$$

If x_1 and x_2 are any strings and g_1, g_2 their associated polynomials, then $x_1 = x_2$ if and only if $g_1 = g_2$. But the latter comparison requires as many operations as comparing x_1 with x_2, so that no work is saved. A vast improvement is effected by the following device.

Let k be small prime which will depend on $m = |y|$. For all practical purposes, $k \in \{5,7,11,13\}$ will suffice.

Let

$$p(t) = t^k + b_1 t^{k-1} + \cdots + b_k \in Z_q[t] \tag{1}$$

be an irreducible polynomial with coefficients in Z_q, i.e. $0 \leqslant b_i \leqslant q-1$.

Denote by $\bar{f} = f \bmod p$ the residue of $f(t)$ modulu $p(t)$. Thus

$$\bar{f}(t) = a_1 t^{k-1} + \cdots + a_k \quad .$$

In computations, $\bar{f}$ will be represented by the k-tuple $(a_1, \ldots, a_k)$. This $\bar{f}$ will be viewed as a "fingerprint" of f. The saving in computational labor will result from employing fingerprints.

Assume for some $\ell < m$ that $\bar{f}_\ell$, where $y(\ell) = y_1 \cdots y_\ell$, is given. To compute $\bar{f}_{\ell+1}$, the fingerprint for $y(\ell) y_{\ell+1}$, note that

$$f_{\ell+1}(t) = f_\ell(t) \cdot t + y_{\ell+1} \quad .$$

Thus if $\bar{f}_\ell = (a_1, \ldots, a_k)$ then

$$\bar{f}_{\ell+1}(t) = ((a_1 t^{k-1} + \cdots + a_k) \cdot t) \bmod p + y_{\ell+1} \quad .$$

From (1) it follows that

$$t^k \equiv c_1 t^{k-1} + \cdots + c_k \mod p \ ,$$

where $c_i = -b_i \bmod q$, $1 \leqslant i \leqslant k$. Hence

$$\bar{f}_{\ell+1} = (a_2, \ldots, a_k, y_{\ell+1}) + (a_1 c_1, \ldots, a_1 c_k)$$

where, again, $a_1 c_i$ is computed in Z_q, and the vectors are added component-wise. Hence it follows that computing $\bar{f}_{\ell+1}$ from $\bar{f}_\ell$ requires $O(k)$ operations in Z_q. Since we view k as fixed, we have a fixed number of operations.

Thus we now compute $\bar{f}_1,\ldots,\bar{f}_m = \bar{f}$ in $O(m)$ operations. In the same way we compute $t^i \bmod p$, $1 \leqslant i \leqslant m$ in $O(m)$ operations.

For $1 \leqslant i < j \leqslant n$, denote by $\bar{f}(i,j)$ the polynomial of the substring $y(i,j) = y_i y_{i+1} \ldots y_j$, i.e.

$$f(i,j)(t) = y_i t^{j-i} + y_{i+1} t^{j-i-1} + \cdots + y_j \quad . \tag{2}$$

It is readily seen that

$$\bar{f}(i,j) = \bar{f}_j - (\bar{f}_{i-1} \cdot t^{j-i+1} \bmod p) \bmod p \quad .$$

Thus for any given pair (i,j), the fingerprint of $y(i,j)$ can be computed, from the precomputed values given above, by a constant number of operations.

In solving the substring repetition problem, we shall replace tests for equality of $y(i,j)$ with $y(i_1,j_1)$, which require $|y(i,j)|$ operations, by tests for equality of the fingerprint $\bar{f}(i,j)$ with $\bar{f}(i_1,j_1)$, which require just a fixed number of operations.

This process may give rise to *false matches*. For the given $p(t) \in Z_p[t]$ and $y \in \Sigma^*$, it may occur that for some $i < j$, $i_1 < j_1$,

$$y(i,j) \neq y(i_1,j_1) \quad \text{and} \quad \bar{f}(i,j) = \bar{f}(i_1,j_1) \quad . \tag{3}$$

Such a false match may lead to an erroneous determination of a repetition in the string y. We overcome this by choosing $p(t)$ randomly and showing that the probability for a false match will be small.

<u>Lemma 1</u>. *There is an algorithm for choosing, with equal probabilities, an irreducible polynomial* $p(t) \in Z_q[t]$ *of degree* k. *The number of operations required for effecting such a choice is* k^3.

<u>Lemma 2</u>. *The number of irreducible polynomials of degree* k, *where* k *is prime, is* $(q^k - q)/k$, *i.e. nearly* q^k/k.

The proofs for these Lemmata can be found in [8].

In Section 2 we shall give the algorithm for discovering the first repetition in y. It will turn out that it will involve $\log_2 m$ phases. In the i-th phase there will be at most m substrings each of length 2^i. Potentially, every pair of substrings at the i-th stage may be tested for equality. This means that for at most $m^2 \log_2 m$ pairs (of pairs) (i,j), (i_1,j_1), we shall test whether $f(i,j)(t) = f(i_1,j_1)(t)$ (see (2)) by testing whether $\bar{f}(i,j) = \bar{f}(i_1,j_1)$.

We shall now give an estimate for the probability that any one of these tests will result in a false match.

<u>Theorem 3</u>. *Let* $y \in \Sigma^*$, $|y| = m$, *and assume that we use fingerprints modulu a randomly chosen irreducible polynomial of degree* k. *Let* $\Pr$ *(false match) denote the probability for* $p(t)$ *that the repetition algorithm applied to this particular* y *will produce one or more false matches. Then*

$$\Pr(\text{false match}) \leqslant \frac{m^3}{q^k} \quad .$$

<u>Proof</u>. Define

$$H(t) = \prod \left(f(i,j)(t) - f(i_1,j_1)(t)\right) \quad , \tag{4}$$

where the product is taken over all pairs (of pairs) (i,j), (i_1,j_1), for which tests whether $y(i,j) = y(i_1,j_1)$ may be done in running the repetition algorithm on y, and for which actually $y(i,j) \neq y(i_1,j_1)$.

If a fingerprinting polynomial $p(t)$ produces a false match when used in the algorithm on y, then for some (i,j), (i_1,j_1) appearing in the product (4), we have $\bar{f}(i,j) = \bar{f}(i_1,j_1)$. This means that $p(t) \mid (f(i,j) - f(i_1,j_1))$ and hence $p(t) \mid H(t)$.

Let $p_1(t),\ldots,p_r(t)$ be the list of all different monic irreducible polynomials of degree k dividing $H(t)$. The irreducibility of the $p_i(t)$ implies

$$p_1(t) \cdot \ldots \cdot p_r(t) \mid H(t) \quad .$$

Hence

$$\sum \deg p_i = rk \leqslant \deg H \leqslant \frac{m^2}{2} \cdot \sum_{1 \leqslant i \leqslant \log m} 2^i \leqslant m^3 \quad .$$

The rightmost inequality following from the analysis of the substring comparisons arising in the algorithm.

Hence $r \leqslant m^3/k$. Now, there are r choices of polynomials leading to a false match, and the total number of polynomials from which the choice is made is, by Lemma 2, q^k/k. Hence

$$\Pr(\text{false match}) = r/(q^k/k) \leqslant m^3/q^k \quad . \qquad \square$$

Note that in the spirit of randomizing algorithms, our result applies to *every* string y, $|y| = m$, individually, and not just to the average string y.

By way of illustration, assume that $|\Sigma| = 256$, $q = 257$, $m = 100{,}000$. Taking $k = 7$, we get m^3/q^k to be about $16 \cdot 10^{-3}$. We shall see that this will suffice to produce a fast, and errorless, algorithm.

2. THE STRING REPETITION ALGORITHM

Algorithm FSR: Finds shortest repetition.

Input: A string $y \in \Sigma^*$, $|y| = m$.

Output: A pair of indices $1 \leqslant i < j \leqslant m$ such that $y(i,j-1) = y(j,2j-i-1)$ and $|j-i|$ is minimal, or else "No repetition".

Stage 1. In the first stage of FSR we randomly select an irreducible polynomial $p(t) \in Z_q[t]$ of degree k, where k is such that $m^3/q^k < 1/2$. Next we compute for y all the fingerprints $\bar{f}_i$, $1 \leqslant i \leqslant m$, (see Section 1) as well as all the residues $t^i \bmod p$, $1 \leqslant i \leqslant m$.

As explained in Section 1, these precomputations allow us to compute every desired fingerprint $\bar{f}(i,j)$ and perform tests $\bar{f}(i,j) = \bar{f}(i_1,j_1)$ in a constant (strictly speaking k) number of steps.

Stage 2. Stage two of FSR has $\lceil \log_2 m \rceil$ phases. At the start of the j-th phase we assume that there is no repetition $y = wxxz$ with $|x| \leqslant 2^j$. Denote $2^j - 1$ by L.

The set of indices $\{1,2,\ldots,m-L\}$ is already partitioned into sets $S_1,\ldots,S_h$, $h \leq m$ with the following properties. Let $S_\ell = \{i_1,\ldots,i_n\}$ where $i_1 < i_2 < \ldots < i_n$, then

1. $y(i_a,i_a+L) = y(i_b,i_b+L)$, $1 \leq a, b \leq n$.
2. If $y(i,i+L) = y(i_a,i_a+L)$ for some $i_a \in S_\ell$, then $i \in S_\ell$.
3. We have $i_a + L < i_{a+1}$ for $1 \leq a < n$, i.e. no two of the substrings $y(i_a,i_a+L)$ overlap.

Each of the sets S_ℓ is treated separately. During Phase j, either a repetition $y = wxxz$ with $2^j < |x| \leq 2^{j+1}$ will be found, or else each S_ℓ will be further partitioned so that 1-3 will hold, with $L = 2^{j+1} - 1$, for the new partition of $\{1,\ldots,m-L\}$.

Step 1. For each $i_a \in S_\ell$ such that $i_{a+1} \leq i_a + 2\cdot L + 1$, test whether

$$y(i_a,i_{a+1} - 1) = y(i_{a+1},2i_{a+1} - i_a - 1) \quad .$$

The test is done using the fingerprints for these strings.

If in one or more of these tests there is equality of fingerprints, then test for equality of the shortest actual substrings for which we had equality of fingerprints.

If this further test confirms equality, return the indexes identifying the repetition.

If a false match was discovered then return to Stage 1 of FSR.

Step 2. If no repetition or false match were discovered in Step 1, then update $L := 2^{j+1} - 1$.

By use of the randomized hashing functions of Carter and Wegman [2], find all the repetitions of fingerprints among $\bar{f}(i_a,i_a + L)$, $i_a \in S_\ell$, (for the new L). Place all the i_a with the same fingerprint into the same subset of S_ℓ, thus partitioning S_ℓ. If $j + 1 \leq \lceil \log_2 m \rceil$, return to Step 1.

Theorem 4. *For* $y \in \Sigma^*$, $|y| = m$, *if* $k \geq 3\cdot\log_q m + 1$, *then algorithm FSR will correctly find the shortest repetition in* y, *or determine that there are no repetitions in* y, *in expected time* $O(m \log_2 m \log_q m)$, *or in expected time* $O(m \log_2 m)$ *if* k *operations of* Z_q *can be considered as one machine operation.*

Proof. The preparatory computations in Stage 1 require $k^3 + O(m \log_q m)$ operations.

In Stage 2, we assume first that the fingerprinting polynomial does not produce false matches in any one of the equality tests for which it is employed.

We claim that the number of steps in Phase j is $O(m \log_q m)$. This is clear since for each $i_a \in \{1,\ldots,m-L\}$ at most three substrings $y(i_a, j_1)$, $y(i_a, j_2)$, and $y(i_a, i_a+L)$ (for $L = 2^{j+1} - 1$) are fingerprinted. Each of the first two fingerprints is compared to just one other fingerprint. The last fingerprint is placed in a hash table, which in the Carter-Wegman scheme requires a constant expected number of steps.

Since there are $\lceil \log_2 m \rceil$ phases, it follows that the expected number of steps is $O(m \log_2 m \log_q m)$ under the assumption that no false matches occurred in any of the equality tests. We shall presently show that false matches can only produce a candidate-repetition $y = wx_1x_2z$ with $x_1 \neq x_2$, but never cause FSR to miss the shortest repetition. This erroneous candidate will be detected in the actual equality test $x_1 = x_2$. When this occurs, FSR returns to Stage 1. By Theorem 3 and the choice of k, the probability for this to happen is smaller than q^{-1}. The event of having a false match in using one polynomial for fingerprinting substrings is independent of the event of having a false match using another randomly chosen polynomial. Thus the total expected time for FSR is $O(m \log_2 m \log_q m)$.

To prove correctness of the algorithm, assume again that using the $p(t)$ chosen in Stage 1 produces no false matches.

The reader can work out for himself the details of initialization of Stage 2 and check that at the start of Phase 0, the conditions 1-3 hold.

Assume that at the start of Phase j conditions 1-3 hold. By an inductive assumption, since we do execute Phase j, there are no repetitions x in y with $|x| \leq 2^j$. Thus the shortest possible repetition $y = wxxz$ in y satisfies $2^j < |x|$. We shall show that if $|x| \leq 2^{j+1}$ this repetition will be discovered within Step 1 of Phase j. A repetition of this form means that for some $k_1 < k_2$, $x = y(k_1, k_2-1) = y(k_2, 2k_2-k_1-1)$ and $2^j < |x| = k_2 - k_1 \leq 2^{j+1}$. But then also $y(k_1, k_1+2^j-1) = y(k_2, k_2+2^j-1)$. It follows from conditions 1-2 that for some ℓ, $k_1 \in S_\ell$ and $k_2 \in S_\ell$. Thus $k_1 = i_a$ and $k_2 = i_b$, with $a < b$. From $|i_b - i_a| \leq 2^{j+1}$ and Condition 3 holding for Phase j (with $L = 2^j$), it follows that $b = a+1$. This means that the shortest repetition is $y(i_a, i_b - 1) = y(i_b, 2i_b - i_a - 1)$. Now, this repetition will be detected in Step 1 of Phase j.

Assume next that there is no repetion x with $|x| \leq 2^{j+1}$. Then FSR passes to Step 2 of Phase j. Surely every S_ℓ will be correctly partitioned so that Conditions 1-2 again hold.

Assume that at the end of Phase j, Condition 3 does not hold. This can only happen if at the beginning of Phase j there is a set S_ℓ in the partition and an $i_a \in S_\ell$ such that

$$i_{a+1} \leq i_a + 2^{j+1} - 1 \quad \text{and} \quad y(i_a, i_a + 2^{j+1} - 1) = y(i_{a+1}, i_{a+1} + 2^{j+1} - 1) .$$

But then also $y(i_a, i_{a+1} - 1) = y(i_{a+1}, 2i_{a+1} - i_a - 1)$, contradicting the assumption that there is no repetition x with $|x| \leq 2^{j+1}$.

Thus the inductive assumption concerning the effect of FSR in Phase j has been established.

3. CONCLUSIONS

We see that for the string repetition problem, the fingerprinting method again produces a very simple algorithm. All that is required is a combination of the use of fingerprints for testing equality of substrings with a process that essentially doubles in each phase the lengths of the strings to be tested.

The algorithm can be easily parallelized in a useful way. If we have n processors, both Stage 1 and each phase of Stage 2 can be sped up almost n-fold. This is not completely obvious for Stage 1, but the details can be worked out.

Finally, the fingerprinting method and the suffix tree method are powerful, each on its own merits. It would be interesting to see whether these two approaches can be combined.

BIBLIOGRAPHY

1. Apostolico, A. and Preparata, F.P. Optimal off-time detection of repetitions in a string. *Theoretical Computer Science* 12 (1983), 297-315.

2. Carter, J.L. and Wegman, M.N. Universal classes of hash functions. 9th Annual STOC, ACM, 1977, pp. 106-112.

3. Galil, Z. and Seiferas, J. Saving space in fast string matching. *SIAM J. on Computing* 9 (1980), 417-438.

4. Karp, R.M. and Rabin, M.O. Efficient randomized pattern matching algorithms. Harvard University, Center for Research in Computing Technology, 1981. TR-31-81.

5. Knuth, D.E., Morris, J.H., and Pratt, V.R. Fast pattern matching in strings. *SIAM J. on Computing* 6, (1977), 323-350.

6. Lorentz, R. and Main, M. Linear time recognition of square free strings. This volume.

7. McCreight, E.M. A space economical suffix tree construction algorithm. *JACM* 33 (1976), 262-276.

8. Rabin, M.O. Fingerprinting by random polynomials. Harvard University, Center for Research in Computing Technology, 1981. TR-15-81.

9. Seiferas, J. and Galil, Z. Real-time recognition of substring repetition and reversal. *Math. Sys. Theory*, vol. 11 (1977), pp. 111-146.

10. Slisenko, A.O. Recognition of palindroms by multihead Turing machines. Proceedings of Steklov Inst. of Math., No. 129 (1973), pp. 30-202.

11. Weiner, P. Linear pattern matching algorithms. Proc. 14th Annual Symp. on Switching and Automata Theory, 1973, pp. 1-11.

SOME DECISION RESULTS ON NONREPETITIVE WORDS

ANTONIO RESTIVO *and* SERGIO SALEMI
Istituto di Matematica dell'Universitá di Palermo
Via Archirafi 34, Palermo (Italy)

ABSTRACT

The paper addresses some generalizations of the Thue Problem such as: given a word u, does there exist an infinite nonrepetitive overlap free (or square free) word having u as a prefix? A solution to this as well as to related problems is given for the case of overlap free words on a binary alphabet.

1. INTRODUCTION

At the beginning of this century, Axel Thue [8,9], considered problems of "finding as long words as possible with as little repetitiveness as possible". He proved in particular the existence of an infinite word on the binary alphabet having no two overlapping occurrences of the same factor, and the existence in a three letter alphabet of an infinite square free word. His proofs provided also an explicit construction of such words. This phenomenon has been later rediscovered and used in various situations, e.g. in group theory, symbolic dynamics, games, and in formal language theory. References can be found in [4]. We shall call here *Thue problem* all questions concerning the study and the construction of arbitrarily large nonrepetitive words.

In this paper we consider some generalizations of the Thue problem by introducing supplementary constraints. The main questions we deal with are the following: i) Given a word u, does there exist an infinite nonrepetitive (overlap free or square free) word having u as a prefix? ii) Given two words u and v, does there exist a nonrepetitive word having u as a prefix and v as a suffix? These questions arise naturally if one interprets a word as describing a process: in this context the supplementary constraints introduced play the role of "initial" or "terminal" conditions on the process Deciding the existence and constructing such words both constitute hard algorithmic problems.

A solution to these and related problems is given here in the case of overlap free words over a binary alphabet. The results are based on a structural theorem which gives, for any overlap free word, a unique factorization in a given standard form. It turns out that the case of square free words is more difficult to handle: in this case we have only partial results and most of the problems considered remain open.

2. DEFINITIONS AND PROBLEMS

A word w is *overlap free* if it does not contain as factors two overlapping occurrences of a nonempty word u. It is well known (see, for example, [4]) that a word is overlap free if and only if it does not contain a factor of the form $xvxvx$, with x a letter and v a word. We denote by S the set of overlap free words in the binary alphabet $\{a,b\}$. Thue proved that S is infinite by giving an explicit construction of an

NATO ASI Series, Vol. F12
Combinatorial Algorithms on Words
Edited by A. Apostolico and Z. Galil

infinite overlap free word, which is now known as the *Thue word*. This latter is obtained by iterating the morphism h defined as follows: $h(a)=ab, h(b)=ba$ (cfr. [4]).

A word w is *square free* if it does not contain a square, i.e. a factor of the form uu, where u is a nonempty word. We denote by T the set of square free words on the alphabet $\{a,b,c\}$. As proved by Thue, also the set T is infinite.

A useful notion in our investigations is that of *center of a language*, introduced by Nivat in [5]: the center of a language is the set of words having infinite right completions in the language. This notion has been extended in [3] to accommodate also left and two sided completions. Many of the questions we ask have their natural formulation in terms of some of the centers of the languages S and T.

We give now the formal definitions for the centers of T. The *left center* $C_l(T)$, the *right center* $C_r(T)$ and the *(bilateral) center* $C(T)$ of the language T are defined respectively as follows:

$$C_l(T)=\{w \in T \mid wA^* \cap T \text{ is infinite}\}$$

$$C_r(T)=\{w \in T \mid A^*w \cap T \text{ is infinite}\}$$

$$C(T)=\{w \in T \mid \text{for any } n \geq 1, \text{there are } u_n, v_n, |u_n|, |v_n| \geq n, \text{such that } u_n w v_n \in T\}.$$

These definitions extend naturally to the centers of S. A notion closely relate to the above is that of *depth*. A word $u \in T$ has depth $d(u)=k \geq 0$ if there exists a word v of length $|v|=k$ such that $uv \in T$ and, for all words w of length greater than k, $uw \notin T$. Clearly, the elements of $C_l(T)$ are exactly the words of infinite depth.

Example 1. Consider the square free word on the alphabet $\{a,b,c\}$: $w=cacbacabcacbac$. Clearly, wb and wc have a square, whereas wa is square free. The word wax has a square for any symbol x. Thus w has depth 1.

Our last definition is that of *transition*. Let $u,v \in T$. We say that there is a transition from u to v if there exists a word w such that $uwv \in T$.

Example 2. Let $u=acaba$ and $v=abaca$ be two square free words. There is a transition from u to v: indeed $ucbcv$ is again square free.

The problems which we consider are the following:

Problem 1: Given a word $u \in T$, decide whether $u \in C_l(T)$ (resp. $u \in C_r(T)$, $u \in C(T)$).

Problem 2: Given a word $u \in T$, construct - if it exists (cfr. Problem 1) - an infinite square free word having u as prefix.

Closely related to Problem 1 is the following:

Problem 3: Given an arbitrary integer $k \geq 0$, does there exist a word $u \in T$ such that $k \leq d(u) < \infty$? (i.e., does there exist square free words of arbitrarily large, finite depth?)

Problem 4: Given two words u, $v \in T$, decide whether there is a transition from u to v.

Problem 5: Given $u, v \in T$, find explicitly a word w - if it exists (cfr. Problem 4) - such that $uwv \in T$.

3. OVERLAP FREE WORDS

In this section we answer the questions posed in the case of overlap free words. A more detailed exposition of the results reported here can be found in [6].

For any integer $n \geq 0$, let $A_n = \{a_n, b_n\}$ be a pair of words of A^* defined inductively as follows:

$$a_0 = a \; ; \qquad b_0 = b$$
$$a_{n+1} = a_n b_n b_n a_n \; ; \; b_{n+1} = b_n a_n a_n b_n \; .$$

For any word $w \in A^*_n$, the *length* of w *relative to* A_n, denoted $|w|_n$, is defined by: $|w|_n = m$ iff $w \in A_n^m$. One has $|w| = 4^n |w|_n$.

Notice that a_n and b_n can also be expressed in terms of the morphism h defining the infinite Thue word. Indeed, one has:

$$a_n = h^{2n}(a), \quad b_n = h^{2n}(b).$$

Thus a_n and b_n are factors of length 4^n of the Thue word. For any positive integer k and for any word $w \in S$, consider the set:

$$P(w,k) = \{u \in A^k \mid wux \in S \text{ for some } x \in A\}.$$

The proof of the following lemma is obtained by induction on n.

Lemma 1. For any integer $n \geq 0$, one has:

$$P(a_n a_n, 4^n) = \{b_n\}$$
$$P(b_n b_n, 4^n) = \{a_n\}$$
$$P(a_n b_n, 4^n) = P(b_n a_n, 4^n) = \{a_n, b_n\}.$$

For any word $w \in A^*$, we denote by $\overline{w}$ the *reversal* of w, i.e., if $w = a_1 ... a_n$ then $\overline{w} = a_n ... a_1$. If L is a subset of A^*, denote by $\overline{L}$ the set of reversals of words of L. For any integer $n \geq 0$, let G_n (resp. D_n) be the set of *left* (resp. *right*) *borders* of order n defined as follows:

$G_n = \{1, \quad a_n, b_n, a_n b_n, b_n a_n, \quad a_n a_n, b_n b_n, a_n a_n b_n, \quad a_n b_n a_n, b_n a_n b_n, \quad b_n b_n a_n, a_n a_n b_n a_n, a_n b_n a_n b_n, b_n a_n b_n a_n, b_n b_n a_n b_n\}$;

$D_n = \overline{G}_n$;

where 1 denotes the empty word of A^*. The following lemma states that any overlap free word of A^*_n of sufficient length can be factorized uniquely into words of A^*_{n+1} up to a left border and a right border of order n.

Lemma 2. Let $w \in A^*_n$ be a overlap free word such that $|w|_n \geq 12$. Then w can be factorized uniquely as follows

$$w = g_n u d_n ,$$

with

$$g_n \in G_n , \; d_n \in D_n \text{ and } u \in A^2_{n+1} A_{n+1} .$$

As a consequence of previous lemmas we obtain the following theorem concerning the structure of overlap free words.

Theorem 1. Let $w \in A^*$ be a overlap free word. There exists an integer $k \geq 0$ (the *rank* $r(w)$ of w) such that w can be uniquely factorized as follows:

$$w = g_0 g_1 \ldots g_{k-1} u d_{k-1} \ldots d_1 d_0$$

with

$$g_i \in G_i , d_i \in D_i \text{ , and } u \in [A_k]_2^{11} = A_k^2 \cup A_k^3 \cup \ldots \cup A_k^{11} .$$

Informally, this theorem states that the "central part" u of a (finite) overlap free word behaves like a factor of the infinite Thue word, whereas the "borders" have a different structure. We remark that the factorization is effective and that the algorithm is obtained by iterating the factorization defined in Lemma 2.

Example 3. The unique factorization of the following overlap free word w is obtained by two successive applications of lemma 2:

$$\begin{aligned} w &= aabaabbabaababbabaabbaababbabaababbaabbabaabbaababbabaabbaa \\ &= a_0 a_0 b_0 a_0 a_1 b_1 a_1 b_1 b_1 a_1 b_1 a_1 a_1 b_1 b_1 a_1 b_1 b_0 a_0 a_0 \\ &= a_0 a_0 b_0 a_0 a_1 b_1 a_2 b_2 b_1 a_1 b_1 b_0 a_0 a_0 . \end{aligned}$$

The rank of w is $r(w) = 2$.

As a consequence of Theorem 1, we obtain also a polynomial bound on the number of overlap free words of length n. We remark that, contrary to this, the number of square free words on a three- symbols alphabet (as well as that of cube free words on a two-symbols alphabet) grows exponentially with n. Let $\Pi_S(n)$ denote the density function of the language S:

$$\Pi_S(n) = \text{Card} -(S \cap A_n).$$

Theorem 2. There exists a constant C such that, for any positive integer n, $\Pi_S(n) \leq C\, n^t$, where $t = \log_2 15$.

The proof is obtained by computing all possible factorizations for words of length n. By using Theorem 1 we are now able to answer the question posed in the preceding section.

Theorem 3. For any positive integer m, there exists an overlap free word w such that $m \le d(w) < \infty$.

A word verifying this property is a word of the form $w = b_0 a_k a_k b_k a_k a_k$, with k such that $4^k - 1 \ge m$. By Lemma 1, $P(w, 4^k) \subseteq P(a_k a_k, 4^k) = \{b_k\}$. However, $wb_k = b_0 a_k a_k b_k a_k a_k b_k$ is not an overlap free word. Theorem 3 is proved by showing that all proper prefixes of wb_k are overlap free.

The word w exhibited above also shows that, in a very long word, the first symbol can play a crucial role in determining its depth. Indeed, w has finite depth, whereas w with the first symbol crossed out, that is, $a_k a_k b_k a_k a_b$, has infinite depth.

The following technical lemma gives a handle in deciding whether a word $w \in S$ has infinite depth.

Lemma 3. Let w be an overlap free word of rank $r(w) = k$. If wa_{k+2} (resp. wb_{k+2}) is overlap free, then wa_{k+3} (resp. wb_{k+3}) is overlap free.

Theorem 4. Given an overlap free word w, it is effectively decidable whether w has an infinite depth (i.e. whether $w \in C_l(S)$). Moreover, in the affirmative case, an infinite overlap free word having w as prefix can be effectively constructed.

Proof. First, factorize w as per Theorem 1:

$$w = c_1 u c_2,\ c_1 = g_0 \dots g_{k-1},\ c_2 = d_{k-1} \dots d_0,\ r(w) = k.$$

Next, search for factorizations $u = u_1 u_2$ of u such that $u_2 c_2$ is a prefix of a_{k+2} or b_{k+2}. If such factorization does not exist, then $w \notin C_l(S)$. If such factorization exists, then, by the previous lemma, $c_1 u_1 a_n \in S$ for all $n \ge k+1$ or $c_1 u_1 b_n \in S$ for all $n \ge k+2$.

Example 4. Consider the word w in example 3; we have the factorization

$$w = a_0 a_0 b_0 a_0 a_1 b_1 a_2 b_2 b_1 a_1 b_1 b_0 a_0 a_0.$$

By setting:

$$a_0 a_0 b_0 a_0 a_1 b_1 = c_1$$
$$a_2 b_2 = u$$
$$b_1 a_1 b_1 b_0 a_0 a_0 = c_2$$

it is easy to see that, for all factorizations $u = u_1 u_2$, $u_2 c_2$ is neither a prefix of a_4 nor of b_4. Hence $w \notin C_l(S)$.

The following theorem, a straightforward consequence of Theorem 1, gives a procedure to decide whether a given word $w \in S$ belongs to $C(S)$.

Theorem 5. Let $w \in S$ be a word of rank $r(w) = k$. Then $w \in C(S)$ iff w is a factor of a_{k+2} or a factor of b_{k+2}.

Problems 4 and 5 have been solved by A. Carpi [2]. The proof of Theorem 4 implies in particular that if $u \in C_l(S)$ then there exists a word $u' \in A^*$ and a positive integer h, such that $u'a_n \in uA^* \cap S$ or $u'b_n \in uA^* \cap S$ for all $n \geq k$. Symmetrically, if $v \in C_r(S)$, then there exists a word $v' \in A^*$ and a positive integer k such that $a_n v' \in A^*v \cap S$ or $b_n v' \in A^*v \cap S$ for all $n \geq k$. Carpi proved that, if $u \in C_l(S)$, $v \in C_r(S)$ and u', v', h and k are as above, then, for all $n \geq \max(h, k, \log_4 u'v')$, at least one of the four words $u'a_n v'$, $u'b_n v'$, $u'a_n b_n v'$, $u'b_n a_n v'$ belongs to the set $uA^*v \cap S$. Hence:

Theorem 6 (Carpi). If $u \in C_l(S)$ and $v \in C_r(S)$, then there exist infinitely many words $w \in A^*$ such that $uwv \in S$.

As a consequence, we obtain the following

Corollary. Given two words u, $v \in S$, it is effectively decidable whether there is a transition from u to v. Moreover, in the affirmative case, a word w verifying $uwv \in S$ can be constructed effectively.

4. SQUARE FREE WORDS

In this section we consider the case of square free words. As mentioned, we have in this case only partial results and most of the questions posed in Section 2 shall be left unanswered. In brief, the only result we know of is contained in a paper by Shelton and Soni [7] and answers our Problem 1.

Theorem 7 (Shelton, Soni). There exists a constant k such that for any word $u \in T$ of length $|u| = n$, if there exists a word v of length $k\, n^{3/2}$ such that $uv \in T$, then $u \in C_l(T)$.

This result gives a procedure to decide whether a given word $u \in T$ belongs to $C_l(T)$. However, the paper by Shelton and Soni, which is also hard to read, does not contain an explicit construction of an infinite square free word having u as a prefix, in case such a word exists. Also, it leaves the question open as to whether there exist square free words of arbitrary large, finite depth (Problem 3).

In connection with the decision problem concerning the (bilateral) center $C(T)$, one might wonder whether the intersection $C_l(T) \cap C_r(T)$ coincides with $C(T)$. The answer is negative, as shown by the following:

Example 5. Consider the square free word:

$$w = acabacbcabacabcbacabacbcaba$$

By using Theorem 7, one shows that $w \in C_l(T)$ and $w \in C_r(T)$. However, one can verify that $w \notin C(T)$.

This example shows that the decision problems concerning the (bilateral) center cannot be reduced to decision problems for left and right centers.

As for Problems 4 and 5, we are unable to state whether or not the analogous of Theorem 6 for square free words holds. Thus these problems are still open.

5. REFERENCES

[1] **F.J. BRANDENBURG**, Uniformly growing k-th power free homomorphism, *Theoretical Computer Science*, 23 (1983) 69-82.

[2] **A. CARPI**, A result on weakly square free words related to a problem of Restivo, *Bulletin of the EATCS*, 22 (1984) 116-119.

[3] **A. deLUCA, A. RESTIVO, S. SALEMI**, On the centers of a language, *Theoretical Computer Science*, 24 (1983) 21-34.

[4] **M. LOTHAIRE**, *Combinatorics on words*, Addison-Wesley (1983).

[5] **M. NIVAT**, Sur les ensembles de mots infinis engendres par une grammaire algebrique, *RAIRO J. Theoretical Informatics* 12 (1980) 259-278.

[6] **A. RESTIVO, S. SALEMI**, On weakly square free words, *Bulletin of the EATCS*, 21 (1983) 49-56.

[7] **R. SHELTON, R. SONI**, Aperiodic words on three symbols, III, *J. reine angew Math.*, 330 (1982) 44-52.

[8] **A. THUE**, Uber unendliche Zeichernreihen, *Norske Vid. Selsk. Skr. I, Mat. Nat. Kl.* Christiana 7 (1906) 1-22.

[9] **A. THUE**, Uber die gegenseitie Lage gleicher Teile gewisser Zeichernreihen, *Norske Vid. Selsk. Skr. I. Mat. Nat. Kl.* Christiana 1 (1912) 1-67.

6 - MISCELLANEOUS

On the Complexity of Some Word Problems That Arise in Testing the Security of Protocols

Shimon Even*

ABSTRACT

Consider the class of protocols, for two participants, in which the initiator applies a sequence of operators to a message M and sends it to the other participant; in each step, one of the participants applies a sequence of operators to the message received last, and sends it back. This "ping-pong" action continues several times, using sequences of operators as specified by the protocol. The set of operators may include public-key encryptions and decryptions.

A theorem of Dolev and Yao is presented with a new proof.

An $O(n^3)$ algorithm, of Dolev, Even and Karp, for testing the security of two user ping-pong protocols, is described. It is shown that their results are easily extendible from the security of secret messages to more general questions of security, such as security against forged signatures.

The results of Even and Goldreich concerning the security of multi-party protocols are briefly described.

Finally, comments are made concerning recently published attacks on RSA. It is argued that it makes no sense to criticize a cryptographic system without reference to the protocols that use it, and that in fact, the attacks suggested assume an insecure protocol. Also, a theorem is stated to the effect that if a protocol (or set of protocols) is secure on the abstract level, then the algebraic properties used in these attacks will never convert a secure system into an insecure one.

I. Introduction

The use of public key encryption (Diffie and Hellman, 1976; Rivest et al., 1978) for secure network communication has received considerable attention. Such systems are effective against a "passive" eavesdropper, that is, one who merely taps the communication line and tries to decipher the intercepted message. However, as pointed out by Needham and Schroeder (1978), an improperly designed protocol can be vulnerable to "active" sabotage.

The saboteur may be a legitimate user in the network. He can intercept and alter messages, impersonate other users or initiate instances of the protocol between himself and other users, in order to use their responses. It is possible that through such complex manipulations he can read messages, which are supposed to be protected, without cracking the cryptographic systems in use.

* On leave from Technion, Haifa, Israel.

NATO ASI Series, Vol. F12
Combinatorial Algorithms on Words
Edited by A. Apostolico and Z. Galil

In view of this danger it is desirable to have a formal model for discussing security issues in a precise manner, and to investigate the existence of efficient algorithms for checking the security of protocols.

Dolev and Yao (1983) investigated the security of what we call here "ping-pong protocols." These protocols involve two participants, the sender S and the receiver R. Let M be a message generated by S. First, S applies a sequence of operators to M and sends it to R. Next, R applies a sequence of operators to the message received, and sends the result back to S. In each step, the participant applies a sequence of operators to the last message received, and sends it back. The number of times this is done, as well as the sequences of operators used, is defined by the protocol.

Dolev and Yao considered the security of two such families of protocols, assuming only a few limitations on the behavior of the saboteur.

In the first family, it is assumed that the only operators used are the public-key encryption and decryption of users. They present a theorem that allows one to determine the security of such a protocol by inspection. A new and simpler proof is presented here.

Their second family was extended by Dolev, Even and Karp (1982) to allow more general operators for which the relations between them can all be expressed in terms of cancellation rules. The algorithm of Dolev, Even and Karp for testing the security of protocols of two users in $O(n^3)$ time is described. Also, their results are extended to answer more general question of security, such as checking whether a saboteur can forge signatures. Results of Even and Goldreich (1983) on the security problem of multi-user protocols are briefly stated.

Attacks on the RSA (Rivest, Shamir and Adleman, 1978) and the Rabin signature system (1978), such as described by Denning (1984) are criticized. It is pointed out that the underlying protocol assumed in these attacks is insecure and therefore it is no surprise that a failure occurs. It is stated that if the protocols in use are secure (assuming ideal cryptosystems) then the use of the algebraic properties, of systems such as RSA, cannot render the system insecure.

II. The Security Problem of Two-User Protocols

Let us briefly recall the essence of public-key systems (see Diffie and Hellman, 1976 or Rivest et al., 1978 for more details). Every user X has an encryption function E_X and a decryption function D_X. Both are mappings from $\{0,1\}^*$ into $\{0,1\}^*$. There is a public directory containing all (X, E_X) pairs, while the decryption function D_X is known only to X. The main requirements on E_X, D_X are:

(1) $E_X D_X = D_X E_X = \lambda$, where λ is the identity function, and
(2) knowledge of $E_X(M)$ does not reveal anything about the value M.

Before we attempt any formal definitions of protocols or security let us consider several simple examples of protocols, and discuss informally their security.

EXAMPLE 1. Consider the following protocol:

(1) $(X, E_Y(M), Y)$,
(2) $(Y, E_X(M), X)$,

which simply means this: X wants to send M to Y and get an echo in order to verify that M has reached Y. He computes $E_Y(M)$, using Y's public encryption key and sends via the network $(X, E_Y(M), Y)$, which stands for "X sends to Y the message $E_Y(M)$." Clearly, no one but Y can apply D_Y to $E_Y(M)$, in order to recover M. After doing so, Y computes $E_X(M)$ and sends $(Y, E_X(M), X)$. When X gets it he can compare the echo, $D_X E_X(M)$, with the original M in order to verify that M has indeed reached Y.

This innocent-looking protocol is insecure. A saboteur Z may intercept $(X, E_Y(M), Y)$ and replace it by $(Z, E_Y(M), Y)$. Y will get M, and respond, according to the protocol by sending $(Y, E_Z(M), Z)$. Z can now read M by applying his secret key, D_Z. He can then even produce the echo $(Y, E_X(M), X)$ and send it over to the satisfied and unsuspecting X. Clearly, this works only if M does not include information about the original sender's identity. Indeed, this observation leads to the technique of name-appending suggested by Needham and Schroeder.

EXAMPLE 2.

(1) $(X, E_Y(MX), Y)$,
(2) $(Y, E_X(M), X)$.

The word MX is formed by appending to M the name X. Now, after Y applies D_Y to $E_Y(MX)$ to get MX, he checks whether the suffix of the string matches the declared name of the sender, i.e., X. If it does not, he knows that someone has meddled with the message and simply terminates his participation in this instance of the protocol. Otherwise, he computes $E_X(M)$ and sends $(Y, E_X(M), X)$.

This protocol is indeed secure. A formal way to prove it will be shown in Section 3.

One may be led to believe that name-appending is the cure for all evils, but consider this seemingly "even safer" protocol:

EXAMPLE 3.

(1) $(X, E_Y(E_Y(M)X), Y)$,
(2) $(Y, E_X(M), X)$.

Here the name X is appended to $E_Y(M)$ instead of to M itself. This protocol is insecure!

One can use the algorithm of Section 3 to verify that this protocol is insecure. There are two natural ways to crack this protocol. One method is as follows: Z sends $(Z, E_Y(E_Y(E_Y(M)X)$

$Z), Y)$, receives $(Y, E_Z(E_Y(M)X), Z)$, sends $(Z, E_Y(E_Y(M)Z), Y)$ and receives $(Y, E_Z(M), Z)$. Another method: Z sends $(Z, E_X(E_X(M)Z), X)$ and receives $(X, E_Z(M), Z)$.

In addition to the operator E_X and D_X, we have used in the last two examples two more operators, which we shall denote i_X and d_X. If X is a string (name of user X) and M is a string (message) then $i_X(M) = MX$. Let S be a string, $d_X(S)$ is defined as follows. If X is a suffix of S, i.e., $S = MX$, then $d_X(S) = M$; else, $d_X(S)$ is undefined, which means that the participation in this instance of the protocol is terminated. Clearly,

$$d_X i_X = \lambda$$

but $i_X d_X(S)$ is not even defined (unless $S = MX$ for some M).

Let us also define an operator d, which is simply the removal of the appended user name. This is easy to do if, for example, all names use exactly the same number of bits. Therefore, it is natural to assume that a saboteur can perform d. Thus, for every user name X,

$$d i_X = \lambda$$

but again $i_X d(S) = S$ only if $S = MX$.

In general we shall assume that there is a set of operators Σ that can be used by the participants in the network. Some operators may have a user name subscript (such as E_X, D_X, i_X, and d_X in our examples). The subset of operators that user X can perform will be denoted by Σ_X and will be called X's *vocabulary*. The vocabularies of all users are similar in the sense that if one replaces the index X by Y, and Y by X, in Σ_X, the result is Σ_Y.

Also, there will be a given set of *cancellation rules* of the form $\sigma\tau = \lambda$, where σ and τ are elements of Σ. If both σ and τ are indexed then the indices are the same. The cancellation rules are similar for all users. Thus, if one or both operators are indexed then the same cancellation rule holds for every user name index.

In our examples,

$$\Sigma = \{ d \} \cup \{ E_X, D_X, i_X, d_X \mid X \text{ is a user name} \}.$$

$$\Sigma_X - \{ d, D_X \} \cup \{ E_Y, i_Y, d_Y \mid Y \text{ is a user name} \}.$$

and the cancellation rules are

$$E_X D_X = \lambda,$$

$$D_X E_X = \lambda,$$

$$d_X i_X = \lambda,$$

and

$$d i_X = \lambda.$$

Note that if $a, b, c \in \Sigma$, $ab = \lambda$ and $bc = \lambda$ then $a = c$. This follows from the fact that members of Σ are operators: Let $w \in \{0,1\}^*$. $abc(w) = a(bc(w)) = a(w)$, since $bc = \lambda$, but on the other hand, $abc(w) = ab(c(w)) = c(w)$. Thus $a = c$.

Given a string $\alpha \in \Sigma^*$, one may repeatedly apply cancellation rules until no cancellation rule is applicable any more. By the previous paragraph, the reduction process has the Church-Rosser property (Rosen, 1973), and thus the end result is unique. Let us denote this *reduced form* of α by $\overline{\alpha}$.

An underlying assumption in our analysis is that the set Σ is free from any relations other than those implied by the cancellation rules. That is, two strings of operators α and β are equivalent if and only if both have the same reduced form.

DEFINITION. A ping-pong protocol $P(S, R)$ is a sequence $\Gamma = (\alpha_1, \alpha_2, \ldots, \alpha_l)$ of operator-words, such that if i is odd then $\alpha_i \in \Sigma_S^*$ and if it is even then $\alpha_i \in \Sigma_R^*$.

The structure of the protocol is similar for every ordered pair of (different) users. Thus, if in $P(V, W)$ we replace the index V by X, and W by Y, we get $P(X, Y)$. We assume that for every two users X and Y, $P(X, Y)$ may be initiated, i.e., there are no restrictions, imposed by the network or the users, on communication via P.

In Example 1, $\alpha_1[S, R] = E_R$, $\alpha_2[S, R] = E_S D_R$ and $l = 2$. In Example 2, $\alpha_1[S, R] = E_R i_S$, $\alpha_2[S, R] = E_S d_S D_R$ and again $l = 2$. In both examples, $\alpha_1(M)$ is sent by S to R and $\alpha_2 \alpha_1(M)$ is sent by R, back to S.

In general the interpretation is as follows: S invents a message-word $M \in \{0,1\}^*$. He applies α_1 to it and sends it to R; i.e., the first step is $(S, \alpha_1(M), R)$. Next $(R, \alpha_2\alpha_1(M), S)$, etc. If l is odd the last step is $(S, \alpha_l \alpha_{l-1} \ldots \alpha_1(M), R)$, and if L is even, then it is $(R, \alpha_l \alpha_{l-1} \ldots \alpha_1(M), S)$.

In this paper, we are not concerned with the purpose of using P. Instead, we are interested only in the question of whether a saboteur (or a group of them) can extract M.

Thus, we assume that some user S has invented a message M, chosen a user R and initiated $P(S, R)$ on M. We assume that neither S nor R is a saboteur. We have to define what are the actions that the saboteur(s) can take.

We shall assume that for every $1 \le i \le l$, for every two different users X and Y and for every $W \in \{0,1\}^*$ the saboteur can effect $\alpha_i[X, Y]$ on W. We shall explain shortly why we make this assumption, but if one believes that this is too conservative one may restrict the saboteur actions, and as long as these restrictions are symmetric (not username dependent), an $O(n^3)$ algorithm for checking security still exists. For example, one could assume that a saboteur cannot effect $\alpha_1[S, R]$ if he is not S.

Let us denote the saboteur by Z. If $X = Z$, then Z has no difficulty to get $\alpha_1(W)$, since $\alpha_1 \in \Sigma_Z^*$. If $X \neq Z$ (and X is not one of the collaborating saboteurs) Z may be able to persuade X to initiate $P(X,Y)$ on W. By tapping the message $(X, \alpha_1(W), Y)$, Z will get $\alpha_1(W)$.

In order to effect $\alpha_i[X,Y]$ on W, for $1 < i \leq l$, Z can wait for $P(X,Y)$ to occur (or somehow persuade X to initiate it), wait for the $(i-1)$th message, $(X, \alpha_{i-1}\alpha_{i-2}\ldots\alpha_1(M), Y)$—assuming i is even—intercept it and replace it with (X, W, Y). Now, Y responds with $(Y, \alpha_i(W), X)$, as expected of him (assuming $\alpha_i(W)$ is defined) and sends it through the network, where Z can tap it. It follows that the language of operator-words that a single saboteur Z can effect (on any $W \in \{0,1\}^*$) is

$$\Delta = \left[\Sigma_Z \cup \{\, \alpha_i[X,Y] \,\middle|\, 1 \leq i \leq l, \ X \text{ and } Y \text{ are different users} \}\right]^*.$$

DEFINITION. Let $\alpha_1[S,R]$ be the first operator-word of $P(S,R)$ and $Z \notin \{S,R\}$. P is *insecure* if there exists an operator-word $\gamma \in \Delta$ such that $\overline{\gamma\alpha_1} = \lambda$.

Observe that it is not necessary to consider $\alpha_i\alpha_{i-1}\ldots\alpha_1(M)$, for $1 < i \leq l$, which is also heard over the network. For if a $\gamma \in \Delta$ exists that satisfies $\overline{\gamma\alpha_i\alpha_{i-1}\ldots\alpha_1} = \lambda$ then there is a $\gamma' \in \Delta$ (in fact $\gamma' = \gamma\alpha_i\alpha_{i-1}\ldots\alpha_2$ will do) for which $\overline{\gamma'\alpha_1} = \lambda$.

In the definition of security given above, we called P insecure if for some ordered pair of users (not including saboteurs), a $\gamma \in \Delta$ exists for which $\overline{\gamma\alpha_1} = \lambda$. In fact, such a γ exists for one pair (S,R) if and only if it exists for every set of users. This follows immediately from the fact that change of names of users does not change the pattern of cancellations. Thus, in what follows we shall restrict our attention to a fixed pair of users, (S,R), free of saboteurs, and only consider the question of whether for $\alpha_1[S,R]$ a $\gamma \in \Delta$ exists that satisfies $\overline{\gamma\alpha_1} = \lambda$.

One may wonder why we have defined Δ to include Σ_Z, but have not allowed a set of saboteurs $\{Z_1, Z_2, \ldots, Z_m\}$ and put $\bigcup_{i=1}^m \Sigma_Z$ in Δ instead. Let us show that this is not necessary, since whatever a set of saboteurs can do, a single saboteur can do also.

Assume $\gamma = \delta_1\beta_1\delta_2\beta_2\ldots\beta_k\delta_{k+1}$, where for every $1 \leq j \leq k+1$,

$$\delta_j \in \left[\bigcup_{p=1}^{m} \Sigma_{Z_p}\right]^*$$

and for every $1 \leq j \leq k$, β_j is some $\alpha_i[X,Y]$, where $X \neq Y$ and

$$\overline{\gamma \cdot \alpha_1[S,R]} = \lambda.$$

We may assume that if $\beta_j = \alpha_i[X,Y]$ then the performer of α_i (X for odd i, Y for even i) is not a saboteur, for otherwise $\alpha_i \in [\bigcup \Sigma_{Z_p}]^*$ and there is no need to single out β_j, which can be absorbed in δ_j.

If we now replace saboteur Z_p by Z, for all p, in γ, the cancellation will still occur, and all α_i's used (for β_j's) will still be legitimate, since its two users will be different. Thus, if a γ exists for a set of saboteurs, it exists for one.

Our next goal is to restrict Δ even further, in order to simplify the security decision problem. Let us show that if a $\gamma \in \Delta$ exists, for which $\overline{\gamma \cdot \alpha_1[S,R]} = \lambda$, then the same statement holds for

$$\Delta' = \{ \Sigma_Z \cup \{ \alpha_i[X,Y] \mid 1 \leq i \leq l, \quad X \neq Y \quad \text{and} \quad \{X,Y\} \subset R,S,Z \} \}^*.$$

If we replace each user $U \notin \{R,S\}$ who appears in γ by Z, the cancellation pattern is maintained while each α_i either remains legitimate (with two different users X, Y, $\{X,Y\} \subset \{R,S,Z\}$) or becomes an operator-word in Σ_Z^*. This proves that we can replace Δ by Δ' in the security decision problem.

III. The Theorem of Dolev and Yao for the Security of Protocols of the First Family

The *first family* of (two-user) protocols is defined by

$$\Sigma_X = \{ D_X \} \cup \{ E_Y \mid Y \text{ is a user name} \};$$

namely, the only operators are the public-key encryptions and decryptions.

Let us assume that every $\alpha_i \in \Gamma$ is in reduced form. An operator word α, in reduced form, is said to be *balanced* if the appearance of D_X in α implies the appearance of E_X in α; the number of times each appears is irrelevant as far as balancing is concerned.

THEOREM 3.1. *A protocol $P(S,R)$, of the first family, is secure if and only if the following two conditions hold:*
(1) α_1 contains at least one appearance of E_S or E_R.
(2) For every $1 \leq i \leq l$, α_i is balanced.

This theorem, of Dolev and Yao (1983), allows the determination of the security of a protocol, of the first family, by inspection.

Consider the protocol of Example 1. As is easily observed, $\alpha_1[S,R]$ $(= E_R)$ satisfies condition (1), but $\alpha_2[S,R]$ $(= E_S D_R)$ is not balanced since E_R does not appear in it. And indeed, the protocol is insecure.

It is easy to see that both conditions are necessary: If condition (1) does not hold then the message M is only operated on by decryptions that any user can remove. If condition (2) does not hold for some $\alpha_i[S,R]$, $1 \leq i \leq l$, and i is even, then a saboteur Z can initiate $P(Z,X)$. By prepadding w with the proper operators, affecting $\alpha_i[Z,X]$ on the result, and properly postpadding, he can gain access to $D_X(w)$, for every word w. If i is odd, he has to persuade X to initiate $P(X,Z)$ and again, by prepadding w, affecting $\alpha_i[X,Z]$ on the result and properly postpadding, he can gain access to $D_X(w)$. Clearly, if a saboteur can get $D_X(w)$ for every $w \in \{0,1\}^*$ and every user X then the protocol is insecure.

We shall present a new and simpler proof of the sufficiency of the conditions, by stating a more "general" theorem and proving it.

THEOREM 3.2. *Let L be the set of reduced operator words in*

$$\{E_S, E_R, E_Z, D_S\}^* \cup \{E_S, E_R, E_Z, D_R\}^*$$

that are balanced. There is no word $\beta \in (L \cup \{E_S, E_R, E_Z, D_Z\})^*$ that contains an appearance of E_S or E_R and $\overline{\beta} = \lambda$.

Proof. By contradiction. Let β be a shortest word in $(L \cup \{E_S, E_R, E_Z, D_Z\})^*$ that contains an appearance of E_S or E_R and for which $\overline{\beta} = \lambda$.

Consider a sequence of cancellations on β that leads to λ, and let us pause at the first cancellation of an E_S or E_R.

$$\beta = \beta_1 \; D_X \; \beta_2 \; E_X \; \beta_3$$

$$\uparrow \qquad \uparrow$$

first cancellation

W.L.O.G. we may assume that D_X is on the left of E_X, where X is either S or R and no cancellation has taken place other than those in β_2. Clearly, β_2 must cancel out first, and D_X is part of a word w_1 taken from L. Let ξ_1 be the left hand part of w_1, up to D_X. Clearly, ξ_1 is still balanced and in fact contains an appearance of E_X. In case the E_X, which is cancelled first, is also a part of a word w_2, taken from L, let ξ_2 be the right hand part of w_2 from E_X on. ξ_2 may be out of balance, having been deprived of E_X, but in this case $\overline{\xi_1 \xi_2}$ is balanced and belongs to L. In case ξ_2 is not out of balance, or if E_X is taken from $\{E_S, E_R, E_Z, D_Z\}^*$, there is a shorter word β' (deleting $D_X \beta_2 E_X$ from β) that is in $(L \cup \{E_S, E_R, E_Z, D_Z\})^*$, contains an appearance of E_S or E_R and $\overline{\beta'} = \lambda$. In the former case the same statement follows after the deletion of $D_X \beta_2 E_X$ and the cancellations in $\xi_1 \xi_2$. In either case we reach a contradiction of the minimality of β. Q.E.D.

This completes the proof of Theorem 3.1.

It is easy to see that Theorem 3.1 generalizes to protocols of k users, where $k \geq 2$. The proof technique remains essentially the same.

IV. The Algorithm of Dolev, Even and Karp

We proceed with the description of the algorithm for testing the security of a two-user ping-pong protocol, $P(S, R)$.

Construct a nondeterministic finite state automaton A, as follows:

(1) State 0 is the (unique) initial state and state 1 is the (unique) accepting state. The (input) alphabet is $\Sigma = \Sigma_Z \cup \Sigma_S \cup \Sigma_R$.

(2) There is a directed path from state 0 to state 1 whose (input) labels correspond to $\alpha_1[S, R]$.

(3) For every input letter (operator) $\sigma \in \Sigma_Z$, there is a self-loop from 0 to 0, labelled σ.

(4) For every $\alpha_i[X,Y]$, $1 \leq i \leq l$ and $\{X,Y\} \subset \{R,S,Z\}$ there is a loop from 0 to 0 whose edges are labelled, in sequence, by the letters of α_i.

Consider the protocol of Example 2. We have seen that $\alpha_1[X,Y] = E_Y i_X$ and $\alpha_2[X,Y] = E_X d_X D_Y$. Thus, the automaton A is as shown in Fig. 1. The self-loop labeled $\sigma \in \Sigma_Z$ represents 11 self-loops, i.e., one for each member of Σ_Z, where

$$\Sigma_Z = \{E_R, E_S, E_Z, D_Z, i_R, i_S, i_Z, d_R, d_S, d_Z, d\}.$$

The security question, therefore, translates into the following: Is there no accepting path, in A, whose corresponding input word w satisfies $\overline{w} = \lambda$? The protocol is secure if and only if no such collapsing word is accepted by A.

In our example, the loops whose intermediate states are 11 to 20 are all superfluous, since they correspond to words in Σ_Z^*. Thus, A can be simplified, as shown in Fig. 2.

Let us assume that the (simplified) automaton A has been constructed and that its set of states is $S = \{0, 1, \ldots, s\}$. We say that a directed path p in A *collapses*, if its corresponding word w collapses, i.e., $\overline{w} = \lambda$. Define the *collapsing relation* $C \subseteq S \times S$ as follows: $(i,j) \in C$ if there is a directed path from i to j in A that collapses. The security question is therefore reduced to the question of whether $(0,1) \in C$. The protocol P is secure if and only if $(0,1) \notin C$.

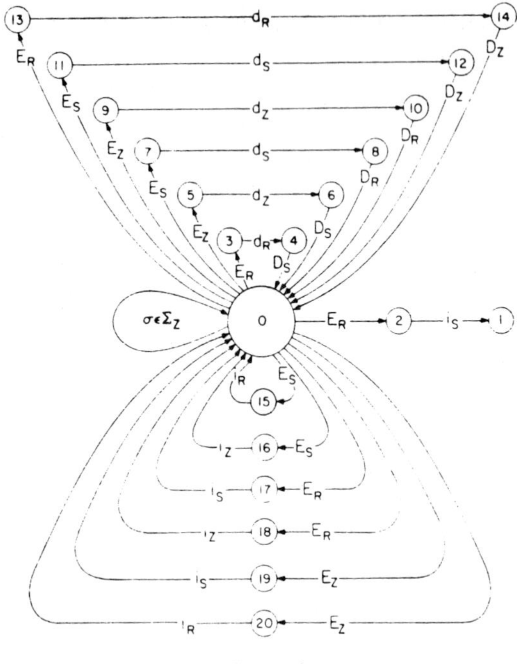

FIGURE 1

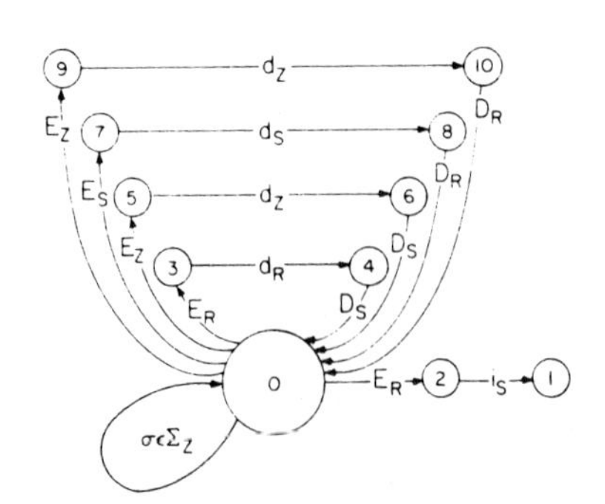

FIGURE 2

In what follows, $i \rightarrow^{\sigma} j$ stands for an edge from state i to state j, labelled σ. Q is a queue of pairs of states. Our algorithm for constructing C is as follows:
(0) $C \leftarrow \{ (i,i) \mid 0 \leq i \leq s \}$, $Q \leftarrow C$. [*Comment*: Each new pair of C enters Q once.]

while $Q \neq \emptyset$, *do*

(1) Delete the first pair, (i,j), from Q.

(2) *If* $(j,k) \in C$ and $(i,k) \notin C$ *then* put (i,k) in C and in Q.

(3) *If* $(k,i) \in C$ and $(k,j) \notin C$ *then* put (k,j) in C and in Q.

(4) *If* $k \rightarrow^{\sigma} i$ and $j \rightarrow^{\tau} l$ and $\sigma\tau = \lambda$ [is one of the cancellation rules] and $(k,l) \notin C$ *then* put (k,l) in C and in Q. *od*

The algorithm terminates, since there can be at most $(s+1)^2$ pairs in C and each can cause the loop to occur once: the number of operations in each pass of steps (1)–(4) is easily seen to be finite. We shall shortly examine the time complexity questions more closely.

THEOREM 4.1. *The algorithm generates the collapsing relation C of automaton A.*

Proof. For every $(i,j) \in C$ let $l(i,j)$ be the length of a shortest collapsing path from i to j. It is easy to see that each (i,j) that is put in C by the algorithm belongs there. We prove that if $(i,j) \in C$ then the algorithm will put it into C, by induction on $l(i,j)$. If $l(i,j) = 0$ (i.e., $i = j$) then (i,j) is put in C in step (0).

Assume now that all $(i,j) \in C$ for which $l(i,j) < L$ have been put in C; let us prove that if $l(i,j) = L$ then it is put in C also. Let w be the word that corresponds to some shortest collapsing path for (i,j). Let σ be the first letter of w. Eventually, in the process of collapsing w, σ is cancelled with some τ via a cancellation rule $\sigma\tau = \lambda$. Thus, $w = \sigma w_1 \tau w_2$.

If $w_2 = \lambda$, then for $\sigma\tau = \lambda$ to happen, w_1 must vanish first. Thus, $\overline{w_1} = \lambda$. Now if w_1 corresponds to a (p,q) path, since $l(p,q) = L - 2$, by the inductive hypothesis (p,q) has been put in C and in Q. When it leaves Q the pair (i,j) is discovered via step (4), if it has not been generated earlier.

If $w_2 \neq \lambda$ then both $\overline{\sigma w_1 \tau} = \lambda$ and $\overline{w_2} = \lambda$. Let k be the state on the path between $\sigma w_1 \tau$ and w_2. Clearly both (i,k) and (k,j) are in C, and since $l(i,k) < L$ and $l(k,j) < L$, by the inductive hypothesis both have been put in C and Q. When the last of them leaves Q, either through step (2) or step (3), (i,j) will be generated and put in C, if it has not been put in C earlier. Q.E.D.

If we apply the algorithm to the automaton of Example 2 (Fig. 2), the final matrix, describing C, is as follows:

	0	1	2	3	4	5	6	7	8	9	10
0	1					1				1	
1		1									
2			1								
3	1			1		1				1	
4	1	1			1	1		1		1	
5	1					1				1	
6	1	1				1	1	1		1	
7	1	1				1		1		1	
8	1		1	1		1			1	1	
9	1					1				1	
10	1		1	1		1				1	1

Since the $(0, 1)$ cell is empty (i.e., $(0, 1) \notin C$, the protocol of Example 2 is secure.

In the complexity analysis that follows, we assume the RAM model, and that the basic word-length is sufficient to accommodate all the operators. Thus, the test of whether $\sigma\tau = \lambda$ takes constant time.

THEOREM 4.2. *The time-complexity of the algorithm for constructing the collapsing relation of automaton A (of $s + 1$ states) is $O(s^3 + s|\Sigma_Z|)$.*

Proof. Note that for all states $v \neq 0$, the in-degree, $d_{\text{in}}(v)$, is exactly 1, and the out-degree, $d_{\text{out}}(v)$, is at most 1. For state 0, both $d_{\text{in}}(0)$ and $d_{\text{out}}(0)$ are bounded by $s + |\Sigma_Z|$. Now consider the loop (1)–(4) of the algorithm.

If $i \neq 0$, $j \neq 0$, then step (2) takes at most s steps, since all we have to do is compare the ith row of the matrix describing the current, C, with the jth row. A similar observation holds for step (3), using columns. For step (4) there is only one σ and one τ to check. Thus, in this case the loop takes time $O(s)$, and since the number of such pairs in C is bounded by s^2, the total time spent on such pairs is $O(s^3)$.

If $i = 0$ but $j \neq 0$, then steps (2) and (3) are still $O(s)$ time, while step (4) is $O(|\Sigma_Z| + s)$ since we have to check each incoming edge $k \to^{\sigma} 0$ against the $j \to^{\tau} l$ edge (assuming $j \neq 1$) to see if $\sigma\tau = \lambda$, and the number of incoming edges ($k \to^{\sigma} 0$) is bounded by $|\Sigma_Z| + s$. Since the number of such pairs is s, the total time spent on such pairs is $O(s(|\Sigma_Z| + s))$. The case of $j = 0$ but $i \neq 0$ is similar.

Finally, if both $i = 0$ and $j = 0$, steps (2) and (3) are redundant, while step (4) takes $O(s(|\Sigma_Z| + s))$ time, since each incoming edge $i \to^{\sigma} 0$ $(i \neq 0)$ has to be checked against each of the $d_{\text{out}}(0)$ $(\leq |\Sigma_Z| + s)$ outgoing edges, and each $0 \to^{\tau} j$ $(j \neq 0)$ has to be checked against each of the $d_{\text{in}}(0)$ $(\leq |\Sigma_Z| + s)$ incoming edges, but there is no need to check a self-loop against a self-loop. Thus, the time spent in this case is also $O(s(|\Sigma_Z| + s))$. Q.E.D.

Let us denote by n the *length of the protocol* P, which is measured as

$$n = \sum_{i=1}^{l} |\alpha_i|,$$

where $|\alpha_i|$ is the length of the operator-word α_i. Thus, n is the total number of operators used in P. Since each word $\alpha_i[X, Y]$ generates exactly 6 loops in the automaton A (one for each choice of an ordered pair of users (X, Y) out of the set $\{ S, R, Z \}$), the number of states s of A is $O(n)$, while the number of self-loops is $|\Sigma_Z|$. If the operators (and cancellation rules) are fixed and are not part of the input of the security problem, then $|\Sigma_Z|$ and the table of cancellation rules is of constant size.

Thus, Theorem 2 immediately implies the following corollary:

COROLLARY 4.1. *For fixed vocabulary and cancellation rules, there exists a security checking algorithm of ping-pong protocols (of two users). Its time-complexity is* $O(n^3)$, *where* n *is the length of the protocol.*

In fact, one may allow the definition of the generic vocabulary and cancellation rules to be part of the input, and still maintain the $O(n^3)$ bound on the time-complexity. One only needs to incorporate the preparation of the cancellation rules in form of a table into the algorithm (in time $O(n^2)$). Thus,

COROLLARY 4.2. *For ping-pong protocols of two users there exists a security checking algorithm whose input is the generic cancellation rules and the protocol. Its time complexity is* $O(n^3)$, *where* n *is the length of the input.*

The problem solved in this section is that of checking whether the intersection of a regular language and a certain context-free language is non-empty. Classically, if one is given a context-free language L, by a grammar G in *CNF*, and a regular language R, by a nondeterministic automaton A, one constructs a new grammar G' that defines $L \cap R$, and then one can check in linear-time whether G' defines the empty language. If the description of G is of length m and A is of n states then G' comes out of size $O(n^3 m)$. Thus, this leads to an $O(n^3)$-time, $O(n^3)$-space algorithm to solve the security problem, while our solution is $O(n^3)$-time, $O(n^2)$-space. In fact, our algorithm can be generalized to answer the question of whether $L \cap R$ is empty, in $O(n^3 m)$-time, $O(n^2 m)$-space.

V. Extension of the Testing Algorithm to Other Security Questions

So far, we followed the definition of Dolev and Yao of security, i.e., the security of the initial message M. However, security may be defined in many other ways. For example, we may consider the question of whether a saboteur Z can obtain $D_A(M)$ for a given message M and user A. This is equivalent to asking whether a saboteur can forge (simple) signatures.

It is easy to see that $D_A(M)$ can be obtained by the saboteur if and only if he can decode messages encoded by E_A. Thus, the question reduces to whether there exists a $\gamma \in \Delta$ such that $\overline{\gamma E_A} = \lambda$, where

$$\Delta = \left[\Sigma_Z \cup \{\, \alpha_i[A,Z],\ \alpha_i[Z,A] \,\middle|\, 1 \le i \le l\}\right]^*.$$

The algorithm of Dolev, Even and Karp still applies, with similar complexity.

Similarly, for every word of operators $\beta \in (\Sigma_S \cup \Sigma_R)^*$, the question of whether a saboteur Z can obtain $\beta(M)$ for a given message M can be reduced to whether there exists a $\gamma \in \Delta'$ such that $\overline{\gamma\beta^{-1}} = \lambda$, where β^{-1} is the right hand side inverse word of operators of β and Δ' is as in Section II. In case some of the operators in β do not have right hand side inverses, we can simply introduce them with the proper cancellation rules. Another approach to solve this problem by checking the emptiness of an intersection of a context-free language and a regular language was recently suggested by Rubinstein (1984).

VI. On Multi-User Protocols

In a recent paper, Even and Goldreich (1983) have examined the security of multi-user protocols, and their main results are briefly described here.

Let $P(U_1, U_2, \ldots, U_k)$, $k > 2$, be a k-party protocol. If one assumes that the saboteur can affect every α_i for k users, not necessarily distinct, then one saboteur is as powerful as many, and an $O(n^3)$ security-checking algorithm similar to the one shown in Section IV follows. However, it is natural to assume that this is not the case, since the user who is supposed to perform α_i, observing that not all k users are distinct, will become suspicious and will not cooperate.

Even and Goldreich show that $3(k-2)+1$ is a lower bound on the number of saboteurs that must be considered and no more than $3(k-2)+2$ saboteurs need be considered. Thus, for a fixed k, the security of k-party ping-pong protocols can be tested in $O(n^3)$ time and $O(n^2)$ space. Also, they show that if k is part of the input then the security problem becomes NP-hard.

Finally, even for 2-party protocols, they show that if the operators are assumed to have certain commutative properties, such as one gets if one assumes that operators are allowed to operate on half words, then the security problem becomes undecidable.

VII. On Certain "Attacks" on RSA and Alike Systems

My aim in this section is to clear away a certain popular misconception concerning certain "attacks" on RSA and similar systems, such as Rabin's (1979) and Williams' (1980). A clear case of this misconception can be seen in Denning (1984).

In the RSA implementation of a public-key cryptosystem, there are numeric operations such as $*$ (modular multiplication) and I (inverse operator) that satisfy a number of algebraic properties:

$$
\begin{aligned}
&M_1 * (M_2 * M_3) = (M_1 * M_2) * M_3, \\
&M_1 * M_2 = M_2 * M_1, \\
&I(M) * M = 1, \\
&II = \lambda, \\
&E(M_1 * M_2) = E(M_1) * E(M_2), \\
&D(M_1 * M_2) = D(M_1) * D(M_2), \\
&IE = EI, \\
&ID = DI.
\end{aligned}
$$

Denning shows (following Davida and Moore) that these algebraic properties can be used to "attack" an RSA system and alike. In order to obtain a victim's signature on C:

1. Pick any message $\widetilde{M}$ and compute $\widetilde{C} = E(\widetilde{M})$; clearly $\widetilde{M} = D(\widetilde{C})$.
2. Get the victim to sign the message $\widetilde{C} * C$. The signature is $D(\widetilde{C} * C) = \widetilde{M} * D(C)$.
3. Compute $I(\widetilde{M})$.
4. Compute $I(\widetilde{M}) * \widetilde{M} * D(C) = 1 * D(C) = D(C)$.

The key question is why should the victim sign (as in Step 2)? If the protocol contains D as an operator-word, it is trivially insecure! Indeed, Denning proceeds to suggest the use of hashing before signing. In essence, she converts the background protocol into a secure one (and this can be done in many ways); but there is a basic misconception here: If the protocol is insecure then it fails for ideal public-key systems as well, and no weakness of RSA has been exposed.

In fact the following theorem (to be proved in a forthcoming paper of Even, Goldreich and Shamir) shows that the whole discussion is meaningless:

THEOREM 7.1. *Let P be a ping-pong protocol that is secure (in the sense of the previous sections). Assume the public-key implementation satisfies the algebraic properties. The protocol implementation cannot be shown to be insecure on the basis of the cancellation rules and the algebraic properties alone.*

The point is that RSA may indeed be insecure; for example, if factoring can be done in polynomial time. However, the algebraic properties alone do not make a secure protocol insecure.

Acknowledgement

The author is indebted to Stuart Kurtz for several useful comments.

References

DIFFIE, W. AND HELLMAN, M. E. (1976). New directions in cryptography. *IEEE Trans. Inform. Theory* **IT-22** (6), 644–654.

RIVEST, R. L., SHAMIR, A., AND ADLEMAN, L. (1978). A method for obtaining digital signatures and public-key cryptosystems. *Comm. ACM* **21** (2), 120–126.

NEEDHAM, R. M. AND SCHROEDER, M. D. (1978). Using encryption for authentication in large networks of computers. *Comm. ACM* **21** (12), 993-999.

DOLEV, D. AND YAO, A. C. (1983). On the security of public key protocols. *IEEE Trans. Inform. Theory* **IT-30** (2), 198–208.

DOLEV, D., EVEN, S. AND KARP, R. M. (1982). On the security of ping-pong protocols. *Inform. and Contr.* **55**, 57-68.

EVEN, S. AND GOLDREICH, O. (1983). On the security of multi-party ping-pong protocols. 24th Annual Symp. of Found. of Comp. Sci., Nov. 1983. *IEEE Computer Society*, 34-39.

RABIN, M. O. (1979). Digital signatures and public key functions as intractable as factorization. MIT/LCS/TR-212.

DENNING, D. E. (1984). Digital signatures with RSA and other public-key cryptosystems. *Comm. of the ACM* **27**, April 1984, 388–392.

ROSEN, R. K. (1973). Tree-manipulating systems and church-rosser theorems. *J. Assoc. Comput. Mach.* 20 (1), 160–187.

RUBINSTEIN, R. S. (1984). Signatures and ping-pong protocols. Dept. of Comp. Sci., Iowa State Univ.

WILLIAMS, H. C. (1980). A modification of the RSA public-key encryption procedure. *IEEE Trans. Inform. Theory* **IT-26**, 726–729.

Code Properties and Derivatives of DOL Systems

Tom Head
Dept. of Math. Sciences
Univ. of Alaska, Fairbanks, 99701 U.S.A.

John Wilkinson
ELXSI
2334 Lundy Place, San Jose, California, 95131 U.S.A.

Abstract

The types of codes delt with in detail are prefix codes, suffix codes, and two special classes of biprefix codes called infix codes and outfix codes. Conditions are given under which polynomially bounded DOL languages form codes of each of these types.

A concept of homomorphism is defined for DOL systems. It is demonstrated that when E is a homomorphic image of a DOL system D and L(E) is infinite then: If L(E) is a code of any of the types listed above, L(D) is also a code of the same type. A concept of derivative is defined for DOL systems that is related to a special type of homomorphism based on the erasure of finite symbols. Code properties of linearly bounded DOL languages are studied in detail. The results are then extended to apply to polynomially bounded DOL languages through the use of the newly introduced derivative concept.

It is shown that for every polynomially bounded DOL language L, L\{1} is a commutative equivalent of a prefix code. Every DOL language is shown to be the union of (1) a finite set, (2) a finite number of DOL languages each of which has a singleton as alphabet, and (3) a commutative equivalent of a prefix code.

Acknowledgement: This research was supported in part by Grants MCS-8003348 and MCS-8303922 of the National Science Foundation of the United States of America.

NATO ASI Series, Vol. F12
Combinatorial Algorithms on Words
Edited by A. Apostolico and Z. Galil

1. Background and introduction.

A DOL *scheme* is a pair S=(A,h) where A is a finite non-empty set and h is a homomorphism from A^* into itself. A DOL *system* is a triple D=(A,h,w) where (A,h) is a DOL scheme and w is a non-empty string in A^*. Each such DOL system D generates a DOL *language* L(D) = $\{h^i(w):i\geq 0\}$. Let p(x) be a polynomial with integer coefficients. Then the DOL system D is said to be *bounded by* p(x) if, for each $i\geq 0$, length of $h^i(w) \leq p(i)$. A DOL system is *polynomially bounded* if the system is bounded by some polynomial p(x).

It was proved in [4] that each infinite polynomially bounded DOL language is either a prefix code or a suffix code. This result is the starting point of the present work. Here a detailed investigation is made of the connection between code properties and structural properties of polynomially bounded DOL languages. This is done by separating the problem into two parts: (1) the study of the linearly bounded DOL languages (see Sec.5), and (2) the study of what we have called the derivatives of DOL systems (see Sec.4). In Sec.6 these parts are re-integrated to give a new treatment of the relationship between the code properties and the structural properties of polynomially bounded DOL languages.

Our concern with code properties has been expanded to include not only prefix codes and suffix codes but also biprefix codes and what we have called infix codes and outfix codes. The main results in Sec.5 have the pattern: Let L be the language generated by a linearly bounded DOL system D. Then either L has extremely strong code properties or L has extremely restricted structural properties. Propositions 5-8 and Corollaries 5-7 have this form.

The central point of Sec.4 is that code properties possessed by the language generated by the derivative D′ of a DOL system D must hold for the language generated by D (see Cor.1). When Secs. 4&5 are combined in Sec.6 the main results take the form: Let L be the language generated by a polynomially bounded DOL system D. Then either L has extremely strong code properties or the language F generated by the final infinite derivative of D has extremely restricted structural properties. Theorem 3 and Corollary 9 have this form. The result from [4] that was stated above is obtained again in the present context as Corollary 8.

M.Schutzenberger and D.Perrin have called attention to the concept of commutative equivalents of prefix codes (see [1] and [9]). In Corollary 10 of Sec.7 we combine Corollary 8 with the results on finite DOL languages given in [4] and [5] to assert that for polynomially bounded DOL languages L, $L\setminus\{1\}$ is always a commutative equivalent of a prefix code. This leads to a result that links arbitrary DOL languages with commutative equivalents of prefix codes (see Thm.4). We close this article with a discussion of several unanswered decidability questions concerning code properties of DOL languages.

As background references on codes and on DOL languages [7],[9],[11], and [12] are recommended. No proofs are given in the present article. A longer version of this article that includes all proofs will be submitted shortly. See [6].

I. Definitions, Concepts, and General Principles

2. Codes and homomorphisms of free monoids.

Definitions. A subset C of a free semigroup $A^+=AA^*$ on a set A is:

(1) a <u>code</u> if whenever an equation $u_1 \dots u_m = v_1 \dots v_n$ holds where $u_1, \dots, u_m, v_1, \dots, v_n$ lie in C then $u_1=v_1$;

(2) a <u>prefix code</u> if whenever strings u and uv lie in C it follows that $v=1$;

(3) a <u>suffix code</u> if whenever strings u and vu lie in C it follows that $v=1$;

(4) a <u>biprefix code</u> if it is both a prefix code and a suffix code;

(5) an <u>infix code</u> if whenever strings v and uvw lie in C it follows that $u=w=1$; and

(6) an <u>outfix code</u> if whenever strings uw and uvw lie in C it follows that $v=1$.

Notice that a prefix code (resp., suffix code) is necessarily a code. Note also that an infix code (resp., outfix code) is necessarily a biprefix code. The empty set is a code but the empty string does not occur in any code.

Many of the assertions made in the present article are true for each of the six classes of codes defined above. For this reason we use the following convention: _Each sentence below that contains the words "special code" has the force of six sentences: Obtain each of these six sentences by replacing all occurrences of "special code" by "code", by "prefix code", ... ,by "outfix code"_.

Let A and B be finite sets and let $f:A^*\text{-->}B^*$ be a homomorphism. Let S be a subset of A^*.

Proposition 1. If the restriction of f to S is an injection and f(S) is a special code then S is a special code.

Definition. A subset T of A^* is a _commutative equivalent of a subset_ S _of_ A^* if there is a bijection $b:S\text{-->}T$ for which, for each s in S, b(s) is a permutation of s.

Let A and B be finite sets and let $f:A^*\text{-->}B^*$ be a homomorphism. Let C be the free commutative monoid on the set B. The elements of C may be regarded as Parikh vectors. Let $p:B^*$ -->C be the natural surjection, i.e. the Parikh map.

Proposition 2. If (1) the restriction of the composite homomorphism pf to S is an injection, (2) $f(A) = B \; \{1\}$, and (3) f(S) is a commutative equivalent of a special code, then S is a commutative equivalent of a special code.

3. Codes and homomorphisms of DOL systems.

Definition. _A homomorphism from a_ DOL _system_ D=(A,h,u) _to a_ DOL _system_ E=(B,k,v) is a monoid homomorphism $f:A^*\text{-->}B^*$ for which fh=kf and f(u)=v.

This definition of homomorphism for DOL systems is suggested by the general theory of algebras: As an instance of a general algebra, D consists of a set A^*, the binary operation of concatenation, the unary operation of applying h, and the nullary operation of choosing a base point u. The definition given above for a (homo)morphism f makes the usual demand on f: f must commute with the (three) operations of D.

Notice that for any such homomorphism f and any non-negative integer i, the i-th string of the sequence of D is mapped onto the i-th string of the sequence of E since $fh^i(u) = k^i f(u) = k^i(v)$. Consequently we have the following fundamental property of homomorphisms: f(L(D))=L(E).

Theorem 1. Let f be a homomorphism from a DOL system D=(A,h,u) to a DOL system E=(B,k,v) and assume that L(E) is infinite. Then L(D) must be a special code if L(E) is a special code. Assume further that f(A) = B {1}. Then L(D) must be a commutative equivalent of a special code if L(E) is a commutative equivalent of a special code.

The program of the present work is indicated by Thm.1: We demonstrate code properties of infinite languages generated by DOL systems by demonstrating them for infinite languages generated by proper homomorphic images of the systems. For this purpose a useful class of homomorphisms can be obtained through erasing appropriate subalphabets.

Definition. Let B be a subset of the alphabet A of a DOL system D=(A,h,w) and let S be the set of symbols in A that occur in at least one string in L(D). We say that B is <u>inaccessible</u> if, for each s in S for which h(s) contains an occurrence of a symbol in B, we have s in B.

Let D=(A,h,w) be a DOL system. Let B be a subset of A which is inaccessible and for which at least one symbol in B occurs in w. Define the homomorphism $e[B]:A^* \text{-->} B^*$ by e[B](a)=a if a is in B and e[B](a)=1 otherwise. We refer to such a homomorphism e[B] as <u>erasure down to</u> B. It is easily verified that E=(B,k,e[B](w)) is a DOL system, where k is the restriction of e[B]h to B. Then e[B] is a homomorphism from D to E.

4. Derivatives, polynomial growth, and exponential growth.

Let D=(A,h,w) be a DOL system. A symbol a in A is either _finite_ or _infinite_ relative to D according as the language generated by (A,h,a) is finite or infinite. Evidently the set A′ of all infinite symbols in A is inaccessible. Assume that A′ is not empty and that some symbol in A′ occurs in w. Then we may form a homomorphic image of D by erasing down to A′ as described in Sec.3: Let $e[A']:A^*\text{-->}A'^*$ be the homomorphism that erases finite symbols but leaves infinite symbols unchanged.

Definition. Let D=(A,h,w) be a DOL system. Let A′ be the subalphabet of infinite symbols and let e[A′] be the homomorphism that erases down to A′. When at least one symbol in A′ occurs in w we define _the derivative of_ D to be the homomorphic image D′=(A′,h′,w′) where h′ is the restriction of e[A′]h to A′ and where w′=e[A′](w). When no symbol in A′ occurs in w we do not define D′, but say instead that D′ _vanishes_.

Iteration of differentiation is meaningful. Successive derivatives of a DOL system D=(A,h,w) are denoted D′, D″,... with the n-th derivative also denoted $D^{(n)}$ and the alphabet of $D^{(n)}$ denoted $A^{(n)}$. When the derivative D′ of a system D vanishes we also say that $D^{(n)}$ _vanishes for all_ n>1. We consider that a vanishing derivative has empty alphabet. The following result is an immediate consequence of Thm.1 and the definition of the derivatives of a DOL system.

Corollary 1. Let D be a DOL system for which the language $L(D^{(n)})$ generated by the n-th derivative $D^{(n)}$ of D is infinite. If $L(D^{(n)})$ is either a special code or a commutative equivalent of a special code then so is L(D).

Ehrenfeucht and Rozenberg made a major study of polynomially bounded DOL languages in [2]. In this work an alternate concept of a DOL system with rank was developed and was shown to be equivalent to the concept of a polynomially bounded system. This work was reported upon briefly in [3]. The following result is merely an adaptation of results given in [2] to our present derivative terminology.

Theorem 2. A DOL system D=(A,h,w) is bounded by a polynomial of degree n if and only if the (n+1)st derivative of D vanishes.

The following known fact is an immediate consequence of Thm.2 and the definition of the derivative.

Corollary 2. If D=(A,h,w) is bounded by a polynomial of degree n but not of any polynomial of degree n-1 then A must contain at least n+1 elements.

For a DOL system D=(A,h,w) with a non-vanishing derivative $D' = (A',h',w')$, the alphabet A' of D' is a proper subset of A unless $D'=D$. The finiteness of A results in a dichotomy concerning the behavior of derivatives: Let k be the number of elements in A. Then either $A^{(i)}$ is empty and $D^{(i)}$ vanishes for some $i \leq k$, or $A^{(i+1)}=A^{(i)}$ and $D^{(i+1)}=D^{(i)}$ for some $i<k$. From Thm.2 it follows that _a system D that is not polynomially bounded possesses a non-vanishing higher derivative E that satisfies the differential equation $E'=E$ of exponential growth_. Thus the classical dichotomy of polynomial vs. exponential growth of DOL systems is expressed attractively in the terminology of differentiation.

II. DOL Languages and Codes of Special Types

5. Code properties of linearly bounded DOL languages.

A DOL system is said to be _linearly bounded_ if it is bounded by a linear polynomial. From Thm.2 we have the following characterizations of linear boundedness.

Proposition 3. For a DOL system D that generates an infinite language the following conditions are equivalent:

(1) D is linearly bounded;

(2) D' generates a finite language;

(3) D'' vanishes;

(4) There is a non-negative integer B for which no string in L(D) contains more than B infinite symbols.

Let (A,h) be a DOL scheme. A symbol a in A is mono-infinite if each string in the language generated by (A,h,a) contains only one infinite symbol. A symbol a in A is left-finite (resp., right-finite) if it is mono-infinite and there is a positive integer B for which, for each non-negative integer i, the infinite symbol in $h^i(a)$ occurs within the first B (resp., last B) symbol occurrences.

Proposition 4. Let D=(A,h,w) be a linearly bounded DOL system that generates an infinite language L. Then L is a prefix code (resp., suffix code) unless the axiom of D contains only a single occurrence of an infinite symbol and that symbol is left-finite (resp., right-finite).

Corollary 3. If the language generated by a linearly bounded DOL system is infinite then the language must be either a prefix code or a suffix code.

Proposition 5. Let D=(A,h,w) be a linearly bounded DOL system that generates an infinite language L. Then one of the following mutually exclusive conditions holds:

(1) L is a prefix code.

(2) There are strings $u_1,\dots,u_n,p_1,\dots,p_n$ for which

$$L = u_1 p_1^* \; \dots \; u_n p_n^* \qquad [I]$$

and at least one of the p_i is not 1.

Proposition 6. Let D=(A,h,w) be a linearly bounded DOL system that generates an infinite language L. Then one of the following mutually exclusive conditions holds:

(1) L is a suffix code.

(2) There are strings $q_1,\dots,q_n,u_1,\dots,u_n$ for which

$$L = q_1^* u_1 \; \dots \; q_n^* u_n \qquad [II]$$

and at least one of the q_i is not 1.

Corollary 4. There is no infinite DOL language that has both of the forms given by [I] and [II].

Proposition 7. Let D=(A,h,w) be a linearly bounded DOL system that generates an infinite language L. Then one of the following mutually exclusive conditions holds:

(1) L is an infix code.

(2) There are strings $u_1,\dots,u_n,m_1,\dots,m_n,v_1,\dots,v_n$ for which

$$L = \{u_1{}^k m_1 v_1{}^k : k \geq 0\} \; \dots \; \{u_n{}^k m_n v_n{}^k : k \geq 0\} \qquad [III]$$

and at least one of the u_i or v_j is not 1.

Proposition 8. Let D=(A,h,w) be a linearly bounded DOL system that generates an infinite language L. Then one of the following mutually exclusive conditions holds:

(1) L is an outfix code.

(2) There are strings $p_1,\dots,p_n,u_1,\dots,u_n,v_1,\dots,v_n,q_1,\dots,q_n$ for which:

$$L = \{p_1 u_1{}^k v_1{}^k q_1 : k \geq 0\} \; \dots \; \{p_n u_n{}^k v_n{}^k q_n : k \geq 0\} \qquad [IV]$$

and at least one of the u_i or v_j is not 1.

Proposition 9. Let D=(A,h,w) be a linearly bounded DOL system that generates an infinite language L. If L is a biprefix code then it must be either an infix code or an outfix code.

Corollary 5. A non-rational linearly bounded DOL language is either an infix code or an outfix code.

Corollary 6. A non-context free linearly bounded DOL language is both an infix code and an outfix code.

The pumping property of context free languages assures that any context free DOL language is linearly bounded [4].

Consequently Cor.3 and Cor.5 yield the following result, the first assertion of which was also noted in [4]:

Corollary 7. An infinite context free DOL language is either a prefix code or a suffix code. A non-rational context free DOL language is either an infix code or an outfix code.

6. Code properties of polynomially bounded DOL languages.

In this section we apply Cor.1 of Sec.4 to amplify the results of Sec.5. Let D=(A,h,w) be a polynomially bounded DOL system that generates an infinite language. Then there is a positive integer n for which D is bounded by a polynomial of degree n but by no polynomial of degree n-1. According to Thm.2 the (n+1)st derivative of D will be the first one to vanish. Then $D^{(n)}$ generates a finite language and $D^{(n-1)}$ is linearly bounded and generates an infinite language. We call this system $D^{(n-1)}$ _the final infinite derivative of_ D.

The following result was previously given in [4]. Here it is an immediate joint consequence of Cors.1 and 3.

Corollary 8. Every infinite polynomially bounded DOL language is either a prefix code or a suffix code.

Combining Cor.1 with Props. 5,6,7, and 8 we have our central result:

Theorem 3. Let D be a polynomially bounded system that generates an infinite language L. Let F be the language generated by the final infinite derivative of D. Then:

(1) L is a prefix code unless F has the form [I].
(2) L is a suffix code unless F has the form [II].
(3) L is an infix code unless F has the form [III].
(4) L is an outfix code unless F has the form [IV].

Corollary 9. Let D, L, and F be as in Thm.3. If F is not rational then L is either an infix code or an outfix code. If F is not context free then L is both an infix code and an outfix code.

7. DOL languages and commutative equivalents of prefix codes.

On combining results of [4] and [5] with Cor.8 we obtain the following result which introduces the topic of this section.

Corollary 10. For each polynomially bounded DOL language L, L\{1} is commutatively equivalent to a prefix code.

A DOL system D=(A,h,w) _has constant alphabet_ if, for every u,v in L(D), the same set of symbols of A occur in u as in v.

Proposition 10. Let D=(A,h,w) be a reduced DOL system which has constant alphabet and assume that A contains at least two symbols. Then L(D) is a commutative equivalent of a prefix code.

Theorem 4. Every DOL language is the union of

(1) a finite set, possibly empty,

(2) a finite number, possibly zero, of DOL languages each of which has a singleton as alphabet, and

(3) a commutative equivalent of a, possibly empty, prefix code.

The language L generated by the DOL system ({a,b},h,b), where h(a)=ab and h(b)=a, is not a commutative equivalent of a prefix code. Neither L\{a} or L\{b} is a commutative equivalent of a prefix code but L\{a,b} is.

III. Problems On The Existence Of Algorithms

Let D=(A,h,w) be a DOL system and let L be the language it generates.

Recall that M.Linna has given in [8] a procedure for deciding whether L is a prefix code, hence also for deciding whether L is a suffix code or a biprefix code.

Problem 1. Are there procedures for deciding whether L is an infix code or an outfix code ?

Problem 2. Is there a procedure for deciding whether L is a commutative equivalent of a prefix code ?

Perhaps Thm.4 provides the first step toward a solution of problem 2.

We have shown that when D is polynomially bounded $L\setminus\{1\}$ must be a code [4]. Except for the case in which L contains 1, this was done by showing that L must be either a prefix code or a suffix code (Cor.8). The situation is more complicated when D grows exponentially since examples show that L may be a code even though it is neither a prefix code nor a suffix code. In fact the familiar system $TM=(\{a,b\},h,a)$, where $h(a)=ab$ and $h(b)=ba$, generates a code which is neither a prefix code nor a suffix code. (The adherence of L(TM) - the limit language of L(TM) is the classical cube-free sequence studied by A.Thue and M.Morse. See[12].)

Problem 3. Is there a procedure for deciding whether L is a code ?

This problem may be at least as difficult as the classical unsolved problem of DOL theory: Is there a procedure for deciding whether L is locally catenative ? (See [10] for a discussion of this latter problem.) For L to be locally catenative is for L to fail to be multiplicatively independent, which is merely a specific manner of failing to be a code. This suggests:

Problem 4. Is the problem of deciding whether L is a code reducible to the problem of deciding whether L is locally catenative ? Is the latter problem reducible to the former ?

References

[1] J. Brzozowski, Open problems about regular languages, in R.V. Book, Ed., Formal Language Theory (Academic Press, New York, 1980).

[2] A. Ehrenfeucht and G. Rozenberg, On the structure of polynomially bounded DOL systems, Report #CU-CS-023-73, Dept. of C.S., Univ. of Colorado, Boulder, Colorado, July 1973.

[3] A. Ehrenfeucht and G. Rozenberg, DOL systems with rank, in G.Rozenberg and A.Salomaa, Eds., L Systems (Springer-Verlag, New York, 1974).

[4] T. Head and G. Thierrin, Polynomially bounded DOL systems yield codes, in L.J. Cummings, Ed.,Combinatorics on Words: Progress and Perspectives (Academic Press, New York, 1983)

[5] T. Head and J. Wilkinson, Finite DOL languages and codes, Theoretical Computer Science, 21 (1982) 357-361.

[6] T. Head and J. Wilkinson, Code properties and homomorphisms of DOL systems, Theoretical Computer Science (to appear).

[7] G. Lallement, Semigroups and Combinatorial Applications, (Wiley, New York, 1979).

[8] M. Linna, The decidability of the DOL prefix problem, Intern. J. of Computer Math., 6 (1977) 127-142.

[9] D. Perrin, Ed., Theorie des Codes, (Ecole Nationale Superieure de Techniques Avancees, 1980).

[10] G. Rozenberg, A survey of results and open problems in the mathematical theory of L systems, in: R.V.Book, Ed., Formal Language Theory (Academic Press, New York, 1980).

[11] G. Rozenberg and A. Salomaa, The Mathematical Theory of L Systems, (Academic Press, New York, 1980).

[12] A. Salomaa, Jewels of Formal Language Theory, (Computer Science Press, Rockville, Maryland, 1981).

WORDS OVER A PARTIALLY COMMUTATIVE ALPHABET

Dominique Perrin
L.I.T.P., Université Paris 7

INTRODUCTION

Many interesting combinatorial problems on words deal with rearrangements of words. One of the goals of such rearrangements is to provide bijective mappings between sets of words satisfying certain properties and therefore give some enumeration results on words. The interested reader may consult the Chapter by D. Foata in Lothaire's book [11] where some examples of rearrangements are developped. The algorithms involved in such rearrangements are, by the way, close to the more popular ones since many sorting problems can usefully be formulated in terms of rearrangements. The study of rearrangements has lead D. Foata to consider words over an alphabet in which some of the letters are allowed to commute. And this, in turn, could have raised the interest for studying "in abstracto" problems concerning these words and the structure of the commutation monoids which is their habitat. Nonetheless, it happened on the contrary that words on partially commutative alphabets became of interest to computer scientists studying problems of concurrency control. Roughly speaking, the alphabet considered in this framework is made of functions and the commutation between these functions corresponds to the commutation of mappings under composition. A typical problem is then to decide wether, up to the commutation rule, a given word is equivalent to one in a special form (see [13], chapter 10 for an exposition of this problem). My own interest in such questions was motivated by the work of M.P. Flé and G. Roucairol [9] who proved a surprising result on finite automata in commutation monoids motivated by problems of concurrency control. The aim of this paper is to present a survey of results obtained recently on commutation monoids including a generalization of the above mentionned. It does not intend to be a comprehensive exposition and many facets of the question have been left in the dark. The first section introduces some terminology and definitions. The second section contains the discussion of two normal form theorems in commutation monoids. The third section contains some results on the structure of commutation monoids. Finally, in the last section, I will discuss the problem of finite automata and commutation monoids.

1. FREE PARTIALLY COMMUTATIVE MONOIDS

We shall consider a finite alphabet A and a binary relation Θ on the set A. We suppose that Θ is symmetric. For two letters $a,b \in A$, we denote :

$$ab \equiv ba$$

whenever the pair (a,b) is in relation by Θ. We consider the congruence of A^* generated by this relation and we denote

$$u \equiv v \mod.\Theta$$

if the two words u,v are congruent. This means that there exists words $w_0, w_1, \ldots, w_k$ with $w_0 = u, w_k = v$ and for each i $(0 \leq i \leq k-1)$ $w_i = r_i abs_i$, $w_{i+1} = r_i bas_i$ with $(a,b) \in \Theta$.

NATO ASI Series, Vol. F12
Combinatorial Algorithms on Words
Edited by A. Apostolico and Z. Galil

We denote $M(A,\Theta)$ the quotient of A^* by the congruence $\equiv$. It is called the *free partially commutative monoid* generated by A with respect to the relation Θ. Such a monoid will also be called a *commutation monoid.*

A simple example of a commutation monoid is obtained by considering a direct product

$$M = A^*_1 \times A^*_2 \times \ldots \times A^*_n$$

of free monoids.

A useful characterization of equivalent words, proved in [6] is the following.

We denote by $|u|_a$ the number of occurrences of the letter a in the word u. For a subset B of A, we denote by $\pi_B(u)$ the projection of u on B^* which is obtained by erasing all letters which are not in B.

PROPOSITION 1.1. *One has*

$$u \equiv v \ \text{mod.}\ \Theta$$

iff

(i) *for each* $a \in A$, $|u|_a = |v|_a$

(ii) *for each pair* a,b *of letters such that* $(a,b) \notin \Theta$ *one has*

$$\pi_{a,b}(u) = \pi_{a,b}(v)$$

Proof : The conditions are obviously necessary. Conversely, we can use an induction on the common length of u,v. The property is clear if u,v are letters. For the induction step, let $u = au'$, $v = bv'$ with $a,b \in A$ and $u',v' \in A^*$. If $a = b$, we are done by the induction hypothesis. We suppose $a \neq b$. Since $|u|_a = |v|_a$, we have $|v|_a \geq 1$. Let $v' = ras$ with $|r|_a = 0$. Let c be any letter occurring in br. If we had $(a,c) \notin \Theta$, then $\pi_{a,c}(u)$ would begin by a and $\pi_{a,c}(v)$ would begin by c, a contradiction. Therefore $bra \equiv abr$. Now the pair u', brs satisfies the conditions (i) and (ii) and therefore the two words are equivalent by the induction hypothesis. Hence

$$u = au' \equiv abrs \equiv bras = v. \qquad \square$$

There is a representation of the elements of the monoid $M(A,\Theta)$ by labeled directed graphs which is obtained as follows.

We shall associate to each word $u \in A^*$ a graph $G(u)$ called its *dependency graph*. Let $u = a_1a_2\ldots a_n$ with $a_i \in A$. The vertices of G are the integers $1,2,\ldots,n$ and integer i is labeled by the letter a_i. There is an edge (i,j) iff

(i) $i < j$
(ii) $(a_i,a_j) \notin \Theta$ or $a_i = a_j$

(iii) for each k with $i < k < j$ we have $a_i \neq a_k$

The number of edges in $G(u)$ is at most kn where $k = Card(A)$. In fact, for each j, the number of edges (i,j) is at most k since $a_i \neq a_k$ when (i,j) and (k,j) are edges.

PROPOSITION 1.2. *One has* $G(u) = G(v)$ *iff* $u \equiv v$.

Proof : If $u \equiv v$ then certainly $G(u) = G(v)$ since this implication is obvious for $u = rabs$, $v = rbas$ with $(a,b) \in \Theta$. Conversely, we use Proposition 1.1. Condition (i) is trivially satisfied. Condition (ii) is also satisfied since there is an edge between any two consecutive positions of each word $\pi_{a,b}(u)$, $\pi_{a,b}(v)$. □

A simple algorithm to compute the dependency graph of u consists in scanning u from left to right and keeping for each letter an integer indicating its last occurrence in the left factor of u that has been scanned.

An example of a commutation monoid can be obtained as follows. Let I, B be two sets. The set M of mappings

$$m : I \to B^*$$

has a structure of a monoid with a product defined by

$$mn(i) = m(i)n(i)$$

It is isomorphic to the direct product of $Card(I)$ copies of B^* and therefore is a commutation monoid.

When $I = B$, this monoid is called in [3] the *flow monoid* and its elements are called *flows*.

2. NORMAL FORMS

We present in this section two normal form theorems. The first one is due to D. Foata and the second one to D. Knuth.

We begin with Foata Normal Form. Let $M(A, \Theta)$ be a free partially commutative monoid. We suppose that the alphabet A is totally ordered.

We say that a word $w \in A^+$ is in *Foata Normal Form* if

$$w = u_1 u_2 \ldots u_n \qquad (2.1)$$

where the words $u_1, u_2, \ldots, u_n$ satisfy the following conditions :

(i) each word u_i is a non empty product of distinct letters commuting with each other in increasing order

(ii) if a is a letter of u_{i+1} not appearing in u_i then there is a letter b of u_i such that

$$(a,b) \notin \Theta.$$

Roughly speaking, a word is in normal form when all its letters have been pushed at the left as much as possible to form blocks of distinct letters.

For instance, in the free commutative monoid over $A = \{a,b\}$, the word $w = (ab)(ab)(a)(a)$ is in normal form with blocks indicated by the parenthesis.

If w is in formal form, there is a unique way of factorizing w in (2.1). In fact u_1 is the largest left factor of w composed with distinct letters that commute with each other.

The following result appears in [3] (see also [10]).

THEOREM 2.1. *For each word* $u \in A^*$ *there is a unique word* $v \in A^*$ *in normal form equivalent to* u.

Proof : The proof consists in an algorithm computing the normal form which will be commented upon later on.

Let Γ be the set of words $w \in A^*$ which are in Foata Normal Form, that is satisfying conditions (i) and (ii) above. We define a mapping from $\Gamma \times A$ into Γ denoted $(u,a) \longmapsto u.a$ in the following way. Let $u = u_1u_2...u_n$ be the decomposition of u as in Eq.(2.1). Let i be the greatest integer $(0 \leqslant i \leqslant n)$ such that the word u_i contains the letter a or a letter b such that $(a,b) \notin \Theta$. We use the rule that $i = 0$ if no such integer exists. We then define

$$u.a = u_1u_2...u_iv_{i+1}u_{i+2}...u_n$$

where v_{i+1} is the result of inserting a in u_{i+1} to obtain a word with its letters in increasing order. When $i = n$, we just set $v_{i+1} = a$.

It is obvious that, according to the definition, we have $u.a \sim ua$. Also, it is easy to verify that $u.a \in \Gamma$. This proves the existence of a normal form for any word $w \in A^*$. Furthermore, let us extend the mapping $(u,a) \longmapsto u.a$ to a mapping $(u,v) \longmapsto u.v$ from $\Gamma \times A^* \to \Gamma$ by associativity. We have for each $u \in \Gamma$ and $(a,b) \in \Theta$ the equality

$$u.ab = u.ba$$

as readily verified from the definitions. Therefore, for any words $w,w' \in A^*$ such that $w \sim w'$ we have $1.w = 1.w'$. Moreover, if w is in normal form, we have $1.w = w'$. This proves the uniqueness of the normal form since if w,w' are two words in normal form that are equivalent, we have

$$w = 1.w = 1.w' = w'. \qquad \square$$

The proof given above leads to a simple algorithm to compute the normal form of a word. The normal form of a word $w[1], w[2], \ldots, w[n]$ is computed as a sequence of blocks $s[1], s[2], \ldots, s[k]$ each block being a subset of the alphabet. At the same time, one keeps track for each letter a of the alphabet of a pointer $p[a]$ indicating the index of the rightmost block in which either a or a letter b with $(a,b) \notin \Theta$ occurs in the sequence of blocks. It is then straightforward to obtain the normal form of a word by scanning its letters from left to right. The only point to be made precise is the updating of the pointers $p[b]$ after inserting a letter a on the right. This is simply obtained by the following :

$p[a] := p[a] + 1$;

for each b such that $(a,b) \notin \Theta$ **and** $p[b] \leq p[a]$ **do** $p[b] := p[a]$;

The resulting algorithm computes the normal form of a word of length n in time $O(kn)$ where k is the number of pairs (a,b) such that $(a,b) \notin \Theta$.

It is interesting to note that there is also the possibility of computing the normal form of a word by another algorithm which operates from right to left. The algorithm is somehow simpler since it does not require the use of pointers as before.

The idea is to use a stack for each letter of the alphabet. When processing a letter a in the right-to-left scanning of w, the letter is pushed on its stack and a marker is pushed on the stack of all the letters b such that $(a,b) \notin \Theta$. The normal form is obtained by recursively "peeling" the tops off the stacks as in the following example.

EXAMPLE 2.1 Let $A = \{a,b,c,d\}$ *and let the complement* $\bar{\Theta}$ *of* Θ *be given by the graph*

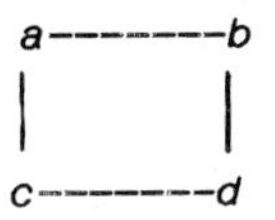

We consider the word $w = abdacbab$. The completed form of the four stacks is :

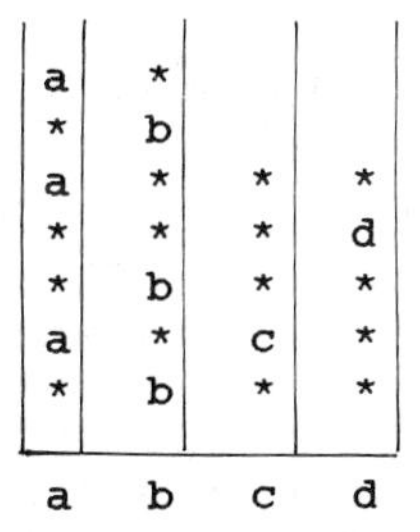

The normal form is obtained as :

$(a)\ (b)\ (ad)\ (bc)\ (a)\ (b)$ □

The algorithm can be implemented in the following way : we use for each letter a in A a stack $s[a]$ of boolean with the convention that "false" is the marker. The first part which consists in filling the stacks is the following :

```
for i:=n downto 1 do begin
        a:=w[i] ; push(true, s[a]) ;
        for each b in A such that (a,b) ∉ Θ do push
        (false, s[b])
 end ;
```

The second part which computes the normal form as a word u initialized to 1 uses a set variable B initialized to the whole alphabet A. It looks like

```
while B not empty do
    for a in B do
        if empty (s[a]) then B:=B-a else begin
            if top (s[a]) then begin
                    u:=ua ; pop (s[a]) ;
                    for b in B and (a,b)∉ Θ do pop (s[b])
            end
        end ;
```

The idea of this algorithm is related to Viennot's theory of "pileups" which gives an interesting geometrical model of partially commutative monoids [8]. The algorithm uses several slacks. From the point of view of automata theory, the question could be raised wether it could be performed by a finite automaton. The answer is easily seen to be negative. Consider indeed the mapping

$$\alpha : u \longmapsto v$$

assigning to each word its unique equivalent word in normal form. For $A = \{a,b\}$ with $ab \equiv ba$ we have

$$\alpha^{-1}((ab)^*) = \{u \in A^* \mid |u|_a = |u|_b\}$$

This shows that α is not a rational function since the right-hand side is not a rational language.

There is, however, something related to finite automata in all this : the set of words in normal form is a rational language. This gives the following corollary, whose idea was suggested to me by Philippe Flajolet.

PROPOSITION 2.2. *For each $n \geq 0$, let α_n be the number of equivalence classes mod. Θ represented by words of length n. The series $\Sigma\alpha_n z^n$ is rational.* □

EXAMPLE 2.2. We consider again the monoid of Example 2.1. The set of words in normal form is precisely the set of words having no factor equal to one of the eight words

$aad, \ acb, \ bbc, \ bda, \ ccb, \ cda, \ dcb, \ dda$

and that do not begin by cb or da. The computation of the generating series $\Sigma\alpha_n z^n$ is complicated and will be treated in Example 2.3. □

There is another normal form theorem for commutation monoids . It consists in a characterization of the smallest lexicographic element in each congruence class. The result is due to Anisimov and Knuth [1].

THEOREM 2.2. *A word* $w \in A^*$ *is the minimal lexicographic element of its class mod.* Θ *iff the following condition is satisfied : for each factorization* $w = rbsat$ *with* $r,s,t \in A^*$, $a,b \in A$ *and* $a < b$ *there exists a letter* c *of* bs *such that* $(a,c) \notin \Theta$.

Proof : First suppose that w is minimal. Let $w = rbsat$ with $r,s,t \in A^*$ and $a,b \in A$, $a < b$. If we had $bsa \equiv abs$ then also

$$w \equiv rabst$$

but $rabst < rbsat$ whence a contradiction. Hence there exists a letter c of s that does not commute with a. This proves that w satisfies the condition of the statement.

Conversely, suppose that there is a word $u \equiv w$ with $u < w$. Let r be the longest common left factor of u,w. Then $u = rau'$, $w = rbw'$ with $a < b$. Since $|u|_a = |w|_a$ there is a factorization of w' in $w' = sat$ with $|s|_a = 0$. Then all the letters of s commute with a since, otherwise, we would not have $u \equiv v$. Therefore w does not satisfy the condition of the statement. □

An algorithm to compute the lexicographic normal form could use the dependency graph of section 1. It consists in selecting recursively the smallest letter without predecessor and deleting the corresponding vertex.

The set of words in lexicographic normal form is again a rational language as illustrated in the following example.

EXAMPLE 2.3. With the same monoid as in Example 2.1, the set of words in lexicographic normal form is the set of words having no factor equal to cb or da. This gives the generating series of the numbers α_n of classes mod. Θ of words of length n

$$\alpha = \frac{1}{1 - 4z + 2z^2}$$

Indeed, let X denote the set of words in normal form. We have

$$X(1 + cb + da) = XA + 1$$

whence the above equality. □

3. STRUCTURE OF COMMUTATION MONOIDS

There is a family of partially commutative free monoids which is well-known and simple enough. It is the family of monoids which are a direct product of free monoids. The following result, due to M. Clerbout and M. Latteux [4], shows that this family of monoids gives, in a sense, the generic case.

THEOREM 3.1. *Any finitely generated commutation monoid can be embedded into a finitely generated monoid which is a direct product of free monoids.*

Proof : Let $M = M(A,\Theta)$ with A finite. Let $A' \subset A$ be the set of letters which commute with all other letters in A.

$$A' = \{a \in A \mid \forall b \in A,\ (a,b) \in \Theta\}$$

Let $A' = \{a_1, a_2, \ldots, a_k\}$ and $A'' = A - A'$. The monoid M is isomorphic with the direct product

$$M(A'',\Theta) \times a^*_1 \times a^*_2 \times \ldots \times a^*_k$$

The proof is therefore only needed for a monoid $M(A,\Theta)$ such that $A' = \emptyset$. Let $B = A \times A$ and let $\alpha : A^* \to B^*$ be the morphism defined for $a \in A$ by

$$\alpha(a) = (a,a_1)(a,a_2) \ \ldots \ (a,a_k)$$

where $\{a_1, a_2, \ldots, a_k\}$ is, in some order, the set of letters b such that $(a,b) \notin \Theta$. Since $A' = \emptyset$, the morphism α is injective. We define a relation τ on the alphabet B by

$$\tau = B \times B - \{((a,b),(b,a)) \mid a,b \in A\}$$

The monoid $M(B,\tau)$ is clearly isomorphic to the direct product of the free monoids with two generators $\{(a,b),(b,a)\}^*$ for $a,b \in A$.

To prove that the morphism α induces a isomorphism from M into $M(B,\tau)$ we need to prove that for $u,v \in A^*$

$$u \equiv v \ \ mod.\ \Theta \iff \alpha(u) \equiv \alpha(v) \ mod.\ \tau$$

The direct implication follows directly from the fact that $ab \equiv ba$ implies $\alpha(ab) \equiv \alpha(ba)$ for $a,b \in A$. For the converse, we use Proposition 1.1. Let $u,v \in A^*$ be such that $\alpha(u) \equiv \alpha(v)$ *mod.* τ. *Let* $a \in A$ be a letter. Since $A' = \emptyset$ by the discussion above, there is a letter $b \in A$ such that $(a,b) \notin \Theta$. *Then*

$$|u|_a = |\alpha(u)|_{(a,b)} = |\alpha(v)|_{(a,b)} = |v|_b$$

Therefore, condition (i) is satisfied. Let now $a,b \in A$ be such that $(a,b) \notin \Theta$. Then clearly $\pi_{a,b}(u) = \pi_{a,b}(v)$ since the projections of $\alpha(u)$, $\alpha(v)$ on $\{(a,b),\ (b,a)\}^*$ are equal. Hence, by Proposition 1.1, we have $u \equiv v$ mod. Θ. □

The "embedding theorem" has several consequences on the structure of commutation monoids. Among them, there is the following result which was proved by R. Cori and Y. Metivier [5].

THEOREM 3.1. *For any word* $u \in A^*$ *the set*

$$C(u) = \{u \in A^* \mid uv \equiv vu \text{ mod. } \Theta\}$$

is a rational language.

Proof : Let α be a morphism from A^* into a direct product $A^*_1 x \ldots x A^*_n$. Since, in a free monoid, the commutator of an element is a cyclic submonoid, the commutator of any element of $A^*_1 x \ldots x A^*_n$ is rational. The set $C(u)$ is the inverse image by α of a rational subset and is therefore rational. □

Another results of [5] also proved independantly by C. Duboc [7] is that the image of $C(u)$ in the monoid $M(A,\Theta)$ is a finitely generated submonoid. The proof is based on the fact that if the graph of the complement of Θ is connected, the set $C(u)$ is a cyclic submonoid.

4. RECOGNIZABILITY IN COMMUTATION MONOIDS

We will be interested in this section in the following problem : how can one describe the *recognizable* subsets of a commutation monoid ? Recall that, given a monoid M, a subset X of M is said to be recognizable if there exists a morphism ϕ from M into a finite monoid F which saturates X, i.e. such that $\phi^{-1}\phi(X) = X$. The family of recognizable subsets of M is denoted $Rec(M)$. When M is a free monoid, the recognizable sets coincide with the rational sets, i.e. the sets described by rational expressions. In the general case, a recognizable set is always rational but the converse is not true.

In a monoid which is a direct product of free monoids, the family of recognizable sets has a simple structure : it consists of finite unions of direct products of recognizable subsets of each component. This comes from the well-known result asserting that the recognizable subsets of a direct product of two monoids are finite unions of recognizable subsets of the components (see [2] e.g.).

The situation is more complicated in general commutation monoids. We shall concentrate here on the closure properties of recognizable sets under rational operations : union, product, star.

The closure of the family of recognizable sets under union holds, in any monoid. For products, we have the following result [6] :

THEOREM 4.1. *The product* XY *of two recognizable subsets* X, Y *of a commutation monoid is recognizable.*

Proof : We use the "embedding theorem" 3.1. According to this theorem, it is enough to prove the closure under product in the case of a direct product of free monoids. Indeed if M, N are two monoids with $M \subset N$ a set $X \subset M$ which is recognizable in N is also recognizable in M. Let M be the family of all monoids in which the family of recognizable sets is closed under product. If M and N belong to **M** then also MxN belongs to **M**. In fact, if X, Y are recognizable in MxN we have

$$X = \bigcup_{i=1}^{n} R_i \times S_i \quad , \quad Y = \bigcup_{j=1}^{m} T_j \times V_j$$

with $R_i, T_j \in Rec(M)$ and $S_i, V_j \in Rec(N)$. Then

$$XY = \bigcup_{i,j} R_i T_j \times S_i V_j$$

and therefore $XY \in Rec(MxN)$. Since any free monoid belongs to the family **M**, any direct product of free monoids also belongs to **M**. □

The closure under the star operation presents a difficulty. In fact, it is not true in general that the star X^* of a recognizable subset X of a commutation monoid is recognizable. It is enough to consider the case of $X = \{(1,1)\}$ in the monoid NxN since

$$X^* = \{(n,n) \mid n \geqslant 0\}$$

which is the diagonal of NxN and which is not recognizable.

The following theorem is the best result known at present on the closure under the star operation. We present it without proof.

THEOREM 4.2. *Let $M = M(A,\Theta)$ be a commutation monoid. Let X be a recognizable subset of M satisfying the following condition : for each word $x \in A^*$ representing an element of X, the restriction of the complement of Θ to the set of letters appearing in x is connected. Then the set X^* is recognizable in M.*

This result is due to Y. Métivier [12]. It has been obtained by successive generalizations of previous ones. First of all, the result was proved by M.P. Flé and G. Roucairol [9] under the stronger hypothesis that :

(i) X is finite
(ii) for each element x in X there is only one word in A^* representing X.

In a further work, done by R. Cori and myself [6], we were able to replace hypothesis (i) by the less restrictive hypothesis that X is recognizable. The result was also proved independantly by M. Latteux. Finally, Y. Métivier has succeeded to adapt our proof to obtain Theorem 4.2 in which condition (ii) is stated in a much weaker form.

EXAMPLE 4.1. Let $A = \{a,b,c,d\}$ and Θ be given by the graph of its complement by the figure below

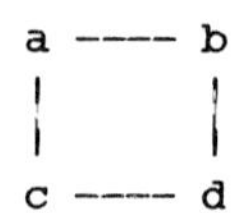

The commutation monoid $M(A,\Theta)$ is the same as in Example 2.1. Consider the sets

$$X = \{ab, cd\}$$
$$Y = \{y \in A^* \mid \exists x \in X^*,\ y \equiv x\}$$

By Theorem 4.2, the set Y is recognizable. It is in fact recognized by the finite automaton given below

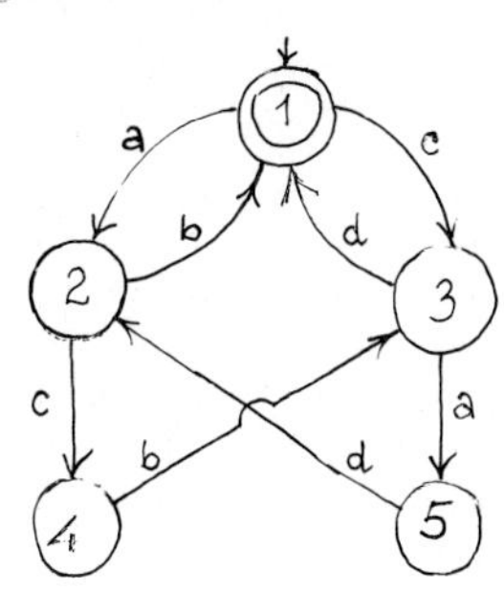

REFERENCES

[1] Anisimov, A.V., Knuth D., Inhomogeneous sorting, unpublished manuscript

[2] Berstel, J., *Transductions and Context Free Languages*, Teubner, 1979.

[3] Cartier, P., Foata, D., *Problèmes Combinatoires de Commutation et Réarrangements*, Lecture Notes in Math., 85, Springer Verlag, 1969.

[4] Clerbout, M., Latteux, M., Partial Commutations and faithful rational transductions, research report, Univ. Lille.

[5] Cori, R., Métivier, Y. Rational subsets of some partially commutative monoids, *Theoret. Comput. Sci.*, to appear.

[6] Cori, R., Perrin, D., Sur la reconnaissabilité dans les monoides partiellement commutatifs libres, *Rairo Informat. Theor.*, to appear

[7] Duboc, C., Some properties of commutations in free partially commutative monoids *Inform. Processing Letters*, to appear.

[8] Dulucq, S., Viennot, G., Bijective proof and generalizations of McMahon Master Theorem, unpublished manuscript.

[9] Flé, M.P., Roucairol, G., Maximal serializability of iterated transactions, *Theoret. Comput. Sci.* to appear (see also ACM SIGACT SIGOPS, 1982, 194–200).

[10] Lallement, G., *Semigroups and Combinatorial Applications*, Wiley, 1979.

[11] Lothaire, M., *Combinatorics on Words*, Addison Wesley, 1983.

[12] Métivier, Y., Une condition suffisante de reconnaissabilité dans les monoides partiellement commutatifs, *Rairo Informat. Theor.*, to appear.

[13] Ullman, J., *Principles of Database Systems*, Computer Science Press, 1980.

THE COMPLEXITY OF TWO-WAY PUSHDOWN AUTOMATA AND RECURSIVE PROGRAMS

Wojciech Rytter,

Institute of Informatics,Warsaw University
PKiN VIII p, 00-901 WarszawaP.O.Box 1210,Poland

1. Introduction

We survey the techniques used to speed up recursive and stack (pushdown) manipulating algorithms on words. In the case of the parallel speedup new algorithmic tools are presented: the operation "bush" acting on path systems and a new parallel pebble game. We show how these tools can be applied to some dynamic programming problems related to combinatorial algorithms on words and to some language recognition problems. The techniques are illustrated mostly on pushdown automata (2pda's, for short) which can be treated as limited algorithms acting on words. The history of the discovery of the fast string matching algorithm [15] shows that 2pda's can be useful in the design of efficient algorithms on words. In this paper we investigate one aspect of 2pda's (which in our view is the most important): algorithmic techniques for fast simulation of 2pda's and their generalisations to recursive programs.
Two important facts about 2pda's are: linear time simulation of 2dpda's [4], and cubic time simulation of 2npda's [1]. As it was written in [7]: "Cook's result is of a great significance and justifies further effort to study this class". Cook's theorem [4] is important because it shows that for a variety of word problems fast algorithms do exist. However the constant factor arising from the use of Cook's simulation directly is quite large. It is natural because of the generality of the method. We are interseted in generality of designed algorithms. We review some simulations of 2pda's, present new angles in classical algorithms [1,4] ,and develop new ones.

NATO ASI Series, Vol. F12
Combinatorial Algorithms on Words
Edited by A. Apostolico and Z. Galil

2. Sequential algorithms

We refer the reader to [1,30] for the definitions of 2pda's and their surface configurations (configurations, for short). Let K denote the set of configurations of a given 2pda A for a given input word of the length n. Subcomputations are pairs of configurations (x,y) such that there is a computation of A from x to y, initially and at the end of such computation the stack is one-element. If y is here a pop configuration then y is called a terminator of x. The pair (x,y) is below (x1,y1) iff A can go from x to x1 by a push move and from y1 to y by a pop move, top symbols in x, y are the same. If (x,y) is below some subcomputation then y is said to be a successor of x. In the case of recursive programs we have analogous concepts of surface configurations and terminators (see [22]). The pair (x,y) is there a subcomputation iff x, y are endpoints of a computation in the same instance of the recursive procedure.

The first general approach to the simulation of 2pda's is based on path systems and the idea of merging smaller subcomputations into larger ones. The path system G is given by a 4-tuple $G=(N,T,S,\otimes)$, where N is a finite set of nodes, T is the set of terminal nodes, S is a goal (source) node, and $\otimes$ is a binary operation whose arguments are elements of N and values are subsets of N. Denote by closure(V) the least set containing V and closed under the operation $\otimes$. The elements of closure(T) are called admissible nodes. A given path system is solvable iff the goal node is admissible. The system is said to be unambiguous iff $(y1,z1)\neq(y2,z2)$ implies that $y1 \otimes z1$ and $y2 \otimes z2$ are disjoint sets for every admissible y1,y2,z1,z2.

We introduce operations gener(x,Z) and genernew(x,Z) for $x \in N, Z \subseteq N$, $\text{gener}(x,Z) = \{z:\ z \in x \otimes y \text{ or } z \in y \otimes x \text{ for } y \in Z\}$. genernew(x,Z)=gener(x,Z)-Z. The operation insert(y,Z) inserts the element y into Z and the operation delete(Z) deletes an element from Z, the deleted element is the value of the last operation. ACTIVE and CLOSURE are subsets of N, both initially empty. The algorithm below checks if the system G is solvable. The final value of CLOSURE is closure(T).

Algorithm 1

```
CLOSURE:=T;ACTIVE:=T;
while ACTIVE ≠ ∅ do
 begin
 x:=delete(ACTIVE); NEW:=genernew(x,CLOSURE);
 for each y ∈ NEW do
    begin insert(y,CLOSURE);insert(y,ACTIVE) end;
 end;
if S ∈ CLOSURE then ACCEPT
endalgorithm.
```

The time complexity of this algorithm is dominated by the total cost of executed instructions genernew.

For a given 2npda A and an input word w we define the path system $G(A,w)=(N,T,S,\otimes)$, where N is the set of subcomputations, T is the set of subcomputations of the form (x,x), where x is a pop configuration, S is the pair (initial configuration, accepting configuration) and the operation $\otimes$ is defined as follows: $(x,y)\otimes(y,z)=\{(x,z)\}$ if $y\neq z$, and $(x,y)\otimes(x,y)=\mathrm{below}(x,y)$, where below(x,y) is the set of pairs which are below (x,y). In other cases the value of the operation $\otimes$ is $\emptyset$. The constructed system is solvable iff A accepts w.

Now the set CLOSURE can be implemented as $m\times m$ boolean table, where $m=|K|$ and (i,j)th entry contains true iff $(x_i,x_j)\in$ CLOSURE (assuming that configurations are numbered). The operation below is easily implemented in constant time, hence it is enough to implement efficiently the following operation:

$$\mathrm{new}(i,j,T) = \{(i,k): T(j,k) \text{ and not } T(i,k)\}\cup\{(k,j):T(k,i) \text{ and not } T(k,j)\}.$$

$O(m)$ time implementation is straightforward. In order to reduce the time complexity we divide rows and columns of the table into small sections and encode them by integers. After some preprocessing the operation new can be computed in $O(m/\log m + v)$ time, where v is the size of the result (number of generated elements, see [30]. We have $O(n^2)$ pairs of configuration and each pair is inserted into ACTIVE at most once. We have $m= O(n)$. This proves the following theorem [30].

Theorem 1

Every 2npda A can be simulated in $O(n^3/\log n)$ time.

Now we design a linear time simulation of 2dpaa's in the framework of path systems. For a given 2dpaa A and input w define the path system G1(A,w)=(N,T ,S , ⊗), where N is now the set of subcomputations (x,y) such that y is a terminator of x or a successor of x. T and S are the same as before.

(x,y) ⊗ (x,y) = below(x,y) if y is a pop configuration,

(x,y) ⊗ (y,z) = {(x,z)} if y≠z.

Fact

The system G1(A,w) is unambiguous and its size is O(n).

Now Algorithm1 can be simplified due to the unambiguity.

Algorithm2

```
ACTIVE:=T;
while ACTIVE≠ ∅ do
   begin
   a:=delete(ACTIVE) ;insert(a,CLOSURE);
   for each b ∈ gener(a,CLOSURE) do insert(b,ACTIVE)
   end;
if S ∈ CLOSURE then ACCEPT
end.
```

Observe that the operation gener is easier than genernew. Assume that the system is unambiguous. The algorithm halts and its cost is proportional to the size of the system if the cost of gener is O(v+1), where v is the number of generated elements. This follows from the fact that each node is inserted into ACTIVE at most once. In our case the set CLOSURE can be implemented by two tables:

TERM(x)={y: (x,y) ∈ CLOSURE and y is terminator of x}

PRED(x)={y: (y,x) ∈ CLOSURE and x is not a pop configuration}.

The operation gener can be now easily implemented to guarántee the required complexity[20]. We arrive at the algorithm which is a variation of Cook's simulation by exploring the key property of the underlying system - unambiguity. Moreover this shows a close relationship between linear time simulation of 2dpaa's and fast recognition of unambiguous cfl's.

A more algebraic approach based on computations of some closures gives fast simulations of some subclasses of 2npda's [23]

Theorem 2. Every 2npda(k) A can be simulated in time $O(BM(n^k) \cdot F(n))$, if F(n) is a bound on the height of the stack ,or on the number of reversals of input head, or it is a bound on the number of turns of the stack. BM(m) is the time to multiply two boolean matrices.

Another general method for speeding up recursive and stack manipulating programs is the dynamic simulation. It consists in simulating a given program (automaton) in such a way that additional information recorded during the simulation allows in some moments to make in one step what the simulated program (automaton) makes in many steps. One of the adventages is that we do not change considerably the structure of the simulated program. The method is especially interested in the case of nondeterministic algorithms e.g. context free grammars and 2npda′s. In the case of 2dpda′s the exact tabulation method [3] is sufficient. Let K be the set of surface configuration of a 2dpda(k) A for an input w. Assume that A is in the normal form [21]. Introduce the functions:
(1) POP is a boolean function, POP(x) holds if x is a pop configuration; (2) P1(x)=y iff y results from x by a push move; (3) P2(x,y)=z iff z results from y by a pop move and top symbols in x and z are the same; (4) T(x) is the terminator of x. Terminators can be computed recursively:
T(x)= _if_ POP(x) _then_ x _else_ T(P2(x,T(P1(x)))).
Introduce the table TAB which tabulates the computed values of T(x) and speeds up the above program. Initially TAB contains only special values 'undefined'.

```
Algorithm3
  function T(x);
  begin if TAB(x)='undefined' then
  TAB(x):=if POP(x) then x else T(P2(x,T(P1(x)))); T:=TAB(x);
  end;
begin  let x0 be the initial configuration;
   if T(x0) is accepting then ACCEPT
end algorithm.
```

The algorithm computes $T(x_o)$ in $O(n^k)$ time if 2dpda(k) A halts, the looping of A can be detected during the computation (the algorithm should be slightly changed). Let r be the number of configurations reachable from x_o during the computation of A.

There are 2dpda(k)'s such that r is much smaller than n^k [19]
Using the special data structure we can prove (see [26])

Theorem 3

Every 2dpda(k) can be simulated in O(r) time and space.

In the case of 2npda's we can define the function analogous to T, however now the values are sets of terminators, see [30].

Theorem 4

Every loop-free 2npda can be simulated in $O(n^3/\log^2 n)$ time.

We refer to [22] for the more general treatment of the dynamic simulation of deterministic recursive and stack manipulating programs. The classical linear time string matching algorithm can be derived by applying the dynamic simulation to a simple stack manipulating program.

3. Parallel algorithms

We investigate the computation of sequential recursive programs by synchronous parallel machines in the framework of path systems. As a model of parallel computer we consider SIMDAG (see [9]) and P-RAM (see [17]). SIMDAG allows to write and read from the same location by many processors, while P-RAM allows only read conflicts. The action of the instruction:
for $a \in S$ **pardo** instruction(a) consists in (1) assigning a processor to every $a \in S$ (the set S can be characterized by a condition); (2) executing each instruction(a) by every processor simultaneously.

The dependancy graph of the computation of a recursive program is a directed acyclic rooted graph (dag, for short), [3]. The computation of the program can be reduced to the computation of the values associated with the nodes of this dag in a similar manner as evaluation of algebraic expressions represented by trees (or dag's). In the case of nondeterministic programs a nondeterministic dag is required, which is essentially the path system. The crucial problem related to parallel computations bases on path system approach is the problem of the efficient height reduction. The height will correspond to the parallel time, while the size to the number of processors.

By the height of the system G we mean the number $h(G)$ such that if G is solvable then there is a derivation tree of the goal node from terminal nodes whose height is bounded by $h(G)$. Tree-size of the dag is the number of paths from the root. Let tree-size(G) denote the bound on the tree-size of a dag corresponding to the derivation of the goal node. When a dag corresponds to the dependancy graph of a recursive program then tree-size corresponds to the sequential time of such program. In order to get good speed up using parallel machines we need path systems having the height logarithmic with respect to their tree-size. For a path system $G=(N,T,S,\otimes)$ we define the system $bush(G)=(N',T',S',\otimes)$ such that:

$N' = \{(x,y): x,y \in N\} \cup \{(x,\emptyset): x \in N\}$ ($\emptyset$ is a special element)

$T' = \{(x,x): x \in N\} \cup \{(x,\emptyset): x \in T\}$; $S' = (S,\emptyset)$ and the operation

$\otimes$ in bush(G) is defined by the rules

(1) $(y,y) \otimes (z,\emptyset) = \{(x,y): x \in y \otimes z \cup z \otimes y\}$,

(2) $(x,z) \otimes (z,y) = \{(x,y)\}$,

(3) $(x,y) \otimes (y,\emptyset) = \{(x,\emptyset)\}$.

It can be proved that (x,y) is admissible in bush(G) iff $x \in closure(T \cup \{y\})$, and $(x,\emptyset)$ is admissible iff $x \in closure(T)$. The rules (1-3) correspond to implications:

(1′) $x \in y \otimes z$ and $z \in closure(T)$ then $x \in closure(T \cup \{y\})$

(2′) $x \in closure(T \cup \{z\})$ and $z \in closure(T \cup \{y\})$ then $x \in closure(T \cup \{y\})$

(3′) $x \in closure(T \cup \{y\})$ and $y \in closure(T)$ then $x \in closure(T)$

The main property of bush(G) is expressed by the following:

<u>Fact</u>

If G is solvable then $h(bush(G)) = O(\log(\text{tree-size}(G)))$.

Sketch of the proof.

Let D be a derivation tree of the goal node S from terminal nodes of G, such that its size is tree-size(G). D can be treated as a path system and $h(bush(G)) \leq h(bush(D))$. Hence we can assume further that our system is tree-like. Let $G'=bush(D)$. Let size(x) denote the size of the subtree of D rooted in x whose leaves are also leaves of D, and let size1(x,y) denotes the size of the subtree of D rooted in x whose all leaves are also leaves of D but (may be) one leaf which is y.

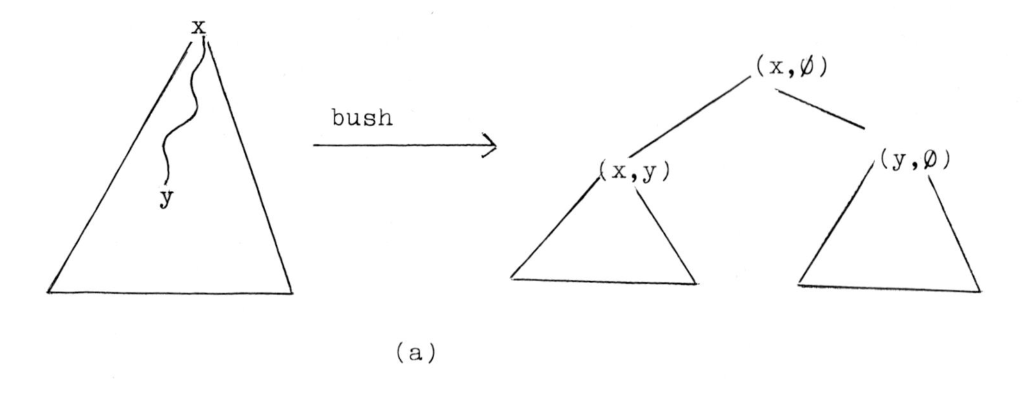

(a)

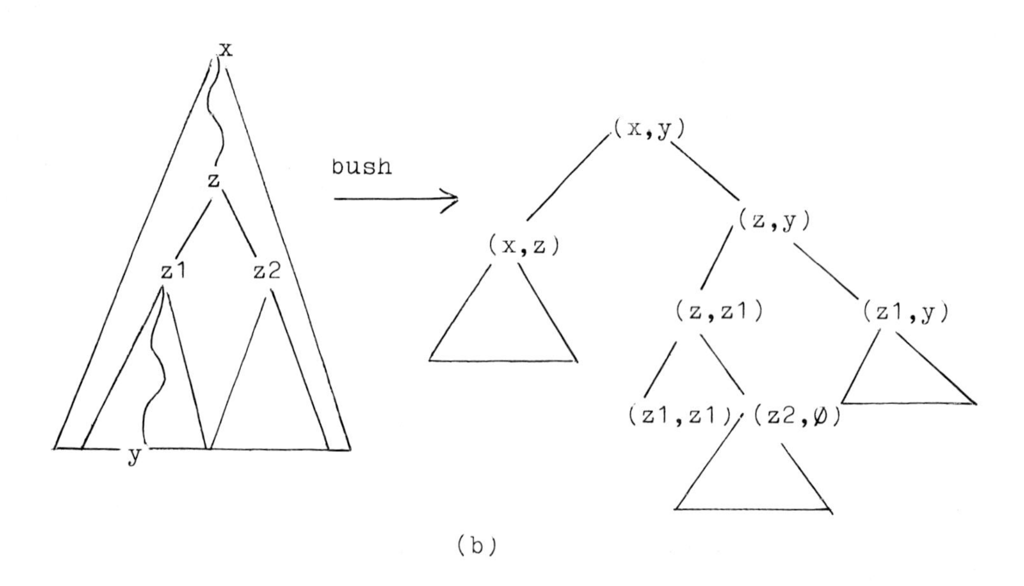

(b)

Fig.1. Constructions used in the proof.

Denote by d(t) the maximal height of the tree needed to derive $(x,\emptyset)$ from the terminal nodes of G' for x satisfying size(x) $\leq$ t and denote by d1(t) the maximal height of the tree needed to derive (x,y) for x,y satisfying size1(x,y) $\leq$ t.

($\times$) $d(t) \leq \max(d1(2\lceil t/3\rceil), d(2\lceil t/3\rceil))+1$.

For the proof see fig.1(a), where y is such that the resulting subtrees have sizes not greater than $2\lceil t/3\rceil$.

($\times\times$) $d1(t) \leq \max(d(t), d1(\lceil t/2\rceil))+3$.

For the proof see fig.1(b). take z1 to be the highest node on the path from y to x such that the subtree rooted in z1 has size not greater than t/2.

The thesis follows easily from ($\times$) and ($\times\times$).

We consider recursive programs given by function-type procedures R. The vector of variables (without input) is denoted by x and X denotes the set of possible values of x for a given input (we know X in advance, the more we know about X the more we know about the computation before we start it). For example recursive programs recognizing context-free languages usually use some integer variables and for a given input data (a word) of the size n we know that these variables have values in the range [0..n]. Let Y denotes the set of potentially possible outcomes of R. $X \times Y$ is the data set of R. Let $d_R(n)$ denote the maximal size of the data set for inputs of the size n, and $time_R(n)$ the maximal execution time of R. In the case of nondeterministic programs the operation choice can be used, whose value is a nondeterministically chosen element from a given set (which is a parameter of the operation). Denote by $\hat{=}$ the input-output relation computed by the nondeterministic program. $R(x_0) \hat{=} y$ holds iff y is one of the possible outcomes (results) of R for x_0.

Theorem 5

If $\log(d_R(n))=O(\log time_R(n))$ then the input-output relation of the nondeterministic recursive program R can be computed on SIMDAG in $O(\log time_R(n))$ time using $O(d_R(n)^3)$ processors.

Proof.

It can be proved that each recursive program R can be reduced to the program f of the following form (*)

$$f(x) = \underset{i=1,2}{\text{choice}} \ (\text{if } p_i(x) \text{ then } c_i(x) \text{ else } f(a_i(x, f(b_i(x))))),$$

where p_i, c_i, a_i, b_i are functions computable in constant time. We want to determine, for given $(x_o, y_o) \in X \times Y$, whether $f(x_o) \hat{=} y_o$. We construct the path system $G=(N,T,S,\otimes)$ where: $N=X\times Y$, $T=\{(x,c_i(x)): p_i(x)\}$, $S=(x_o,y_o)$, $(x,y) \in (x1,y1) \otimes (x2,y2)$ iff $x1=b_i(x), x2=a_i(x,y1), y=y2$ for some $i=1,2$. Now tree-size(G) corresponds to the execution time of f.
The size of G is $O(d_R(n))$. Instead of checking if G is solvable we check if bush(G) is solvable. It will cost log(tree-size(G)) time and $O(d_R(n)^3)$ processors, because the height of bush(G) is sufficiently small. This completes the proof.

Our construction gives directly $O(\log^2 n)$ parallel time recognition of cfl's using $O(n^6)$ processors, the same result can be obtained for poly-time 2npda's. Observe here the power of nondeterminism. There is not known a polynomial time recognition of cfl's on 2dpda's, while the nondeterminism gives linear sequential time on 1npda's.
The method from [18] of constructing bushy versions of path systems can be applied, however this could give m^4 nodes, if the original system has m nodes, while our construction gives m^2 nodes. This simplifies also the recognition of cfl's on stack bounded 2dpda's. The following theorem can be viewed as an application of parallel algorithms to the design of space efficient sequential algorithms.

Theorem 6

Every polynomial-time 2npda can be simulated by multihead 2dpda with stack bounded by $\log^2 n$. Every polynomial-time 2dpda can be simulated by a multihead 2dpda with stack bounded by $\log^2 n$ and working in polynomial time.

Proof.

The first result follows by applying the operation bush. The second result follows from Reif's parallel recognition of dcfl's, where the tables are replaced by recursive functions

Now we show how to solve some optimization problems expressed in recursive form introducing nondeterminism (replacing max,min by nondeterministic choice).

Let dist(v,w) denote the number of positions on which the strings v, w differ. dist(v,w)= ∞ if the strings have different lengths. If L is the language then dist(w.L)= min(dist(v, w): $v \in L$).

Corollary

If L is a given cfl then the function f(w)=dist(w,L) can be computed on SIMDAG in O(log n) time using n^9 processors.

Proof.

Let L be a reduced context free grammar in Chomsky normal form generating L. Let $w=a_1...a_n$ be an input word. Define the function f1(A,i,j)=dist($a_{i+1}..a_j$, L_A), where L_A is the set of terminal words derivable from the nonterminal symbol A. We have:
f1(A,i,i+1)= if $A \rightarrow a_{i+1}$ is a production then 0 else 1;
f1(A,i,j)= min(f1(B,i,k)+f1(C,k,j): $k \in (i..j), A \rightarrow BC$).
Now we can replace the operation min by the operation choice and obtain a nondeterministic recursive program f2. We use the algorithm from Theorem 5 and compute the relation f2(A,i,j)=k. The elements of X are here triples (A,i,j) and elements of Y are integers k, now we can compute in log n parallel time f(w)= f1(S,O,n)=min(k: f2(S,O,n) $\hat{=}$ k), where S is here the axiom of the grammar. This completes the proof.

A second example is the editing problem. We have two texts v, w of the sizes n,m respectively, where $n \geq m$. We are to computethe minimum number of operations insert, change needed to make v=w (the operation insert can insert any symbol and change can change any symbol in the text). Denote such number by edit(v,w).Let $v=a_1...a_n$, $w=b_1...b_m$ and let f (i,j)=0 for i=j =0, f(i,j)=∞ if one of i,j is negative and in other cases f(i,j)=min(f(i-1,j-1)+1,f(i-1,j)+1,f(i,j-1)+1,if $a_i=b_j$ then f(i-1,j-1)). We have f(i,j)=edit($a_1..a_i$,$b_1..b_j$). We can transform f into a nondteterministic recursive program analogously as in the case of dist. Here $d_R(n)=O(n^3)$. This proves:

Corollary

The function edit can be computed on SIMDAG in O(log n) time using n^9 processors.

In a similar manner Theorem 5 can be applied to some other dynamic programming problems,e.g. the cost of the optimal order of multiplying matrices or the cost of the optimal binary

search trees. It is an open problem whether SIMDAG can be replaced by P-RAM in Theorem 5. It was proved ingeniously that it can be made if the programs are deterministic [17]. We show now another method of parallel simulation of deterministic programs on P-RAM based on the idea similar to the construction of bush(G). We construct a new parallel pebble game (also interested in its own) needed to pebble a dag in O(log tree-size) parallel steps. The operations in this game correspond to the rules (1-3) in the construction of bush(G). We reduce (comparing with [17]) the number of processors needed to compute deterministic recursive programs of the form:
(* *) f(x)=if p(x) then c(x) else a(x,f(b1(x)),f(b2(x))), where p, c, a, b1, b2 are computable in constant time.
We could consider more than two recursive calls in this definition, however such generalization is inessentiøal.
Let G be a dag with the set V of nodes , define the following deterministic game on G using one type of pebbles. Associate with each node v a node cond(v). Initially cond is the identity function. The interpretation of cond is: if cond(v) is known then v is also known in the sense of some information associated with the nodes. Nodes of G will correspond to the values of x. The node v is said to be active iff cond(v)≠v. One move in the game is named PEBBLE(G) and it consists in executing: begin activate;square;pebble end.
The operations activate, square, pebble are defined as follows:
activate:

  <u>for each</u> nonleaf node v <u>pardo</u> <u>if</u> v is not active <u>then</u>
  <u>begin</u> <u>if</u> pebbled(leftson(v)) <u>then</u> cond(v):=rightson(v)
  <u>else</u> <u>if</u> pebbled (rightson(v)) <u>then</u> cond(v):=leftson(v)
  <u>end</u> activate.

square: <u>for each</u> node v <u>pardo</u> cond(v):=cond(cond(v)).

pebble: <u>for each</u> v <u>pardo</u> <u>if</u> pebbled(cond(v)) <u>then</u> pebble v.

Assume that each nonleaf node has two sons. The goal of the game is to pebble the root starting with all leaves pebbled. Using the properties of bush(G) the following can be proved:

<u>Fact</u> G can be pebbled using O(log tree-size(G)) moves PEBBLE.

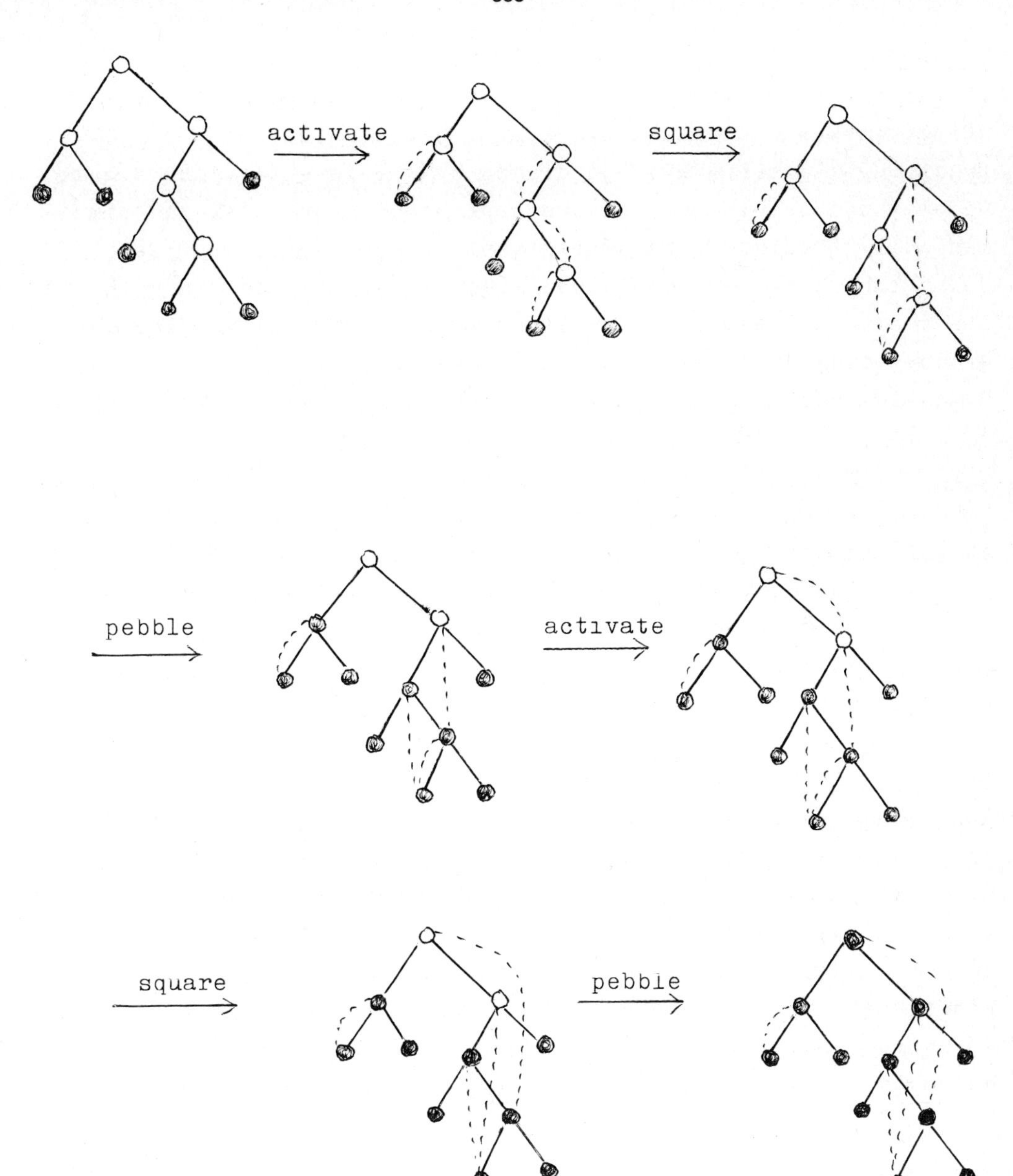

Fig.2. The parallel pebbling of the tree. The black nodes are pebbled. If the node v is active then the dashed line goes from v downto cond(v).

Theorem 7

If f is of the form (**) then it can be computed on P-RAM in $O(\log(time_f(n)))$ time using $O(d_f(n))$ processors.

Proof. Our algorithm will be driven by the invented pebble game. We play the pebble game on the dependency graph of f and simultaneously tabulate f using the table F. The auxiliary table CONDF tabulates the conditional functions , its domain is X and its range is the set of functions from Y into Y. Take the dag G whose nodes are elements of X. The node x is a leaf if p(x) holds and each nonleaf node x has two sons b1(x), b2(x). The main idea is to preserve the invariant:

F(x)= if pebbled(x) then f(x) else ∅,

CONDF(x)(y1)=y iff f(cond(x))=y1 implies f(x)=y.

At the end of the pebble game the root is pebbled and if the invariant holds then $F(x_o)=f(x_o)$. Initially leaves are pebbled and cond is the identity, hence CONDF is also identity function and F(x)=c(x) for each leaf, F(x)=∅ for other nodes.

After executing the operation activate we update:

if cond(x) is set to rightson(x) then CONDF(x):=a(x,F(b1(x)), *) else CONDF(x):=a(x, * ,F(b2(x)).

a(x,y1,*) is the function such that a(x,y1,*)(y)=a(x,y1,y).

After executing square we update:

CONDF(x):= CONDF(cond1(x)) ∘ CONDF(x), where ∘ is the composition of functions (from left to right) and cond1(x) is the value of cond(x) before executing square.

After executing each pebble: F(x):=CONDF(x)(F(cond(x)).

It can be proved that the invariant holds after executing each activate, square, pebble. This completes the proof.

We say that a cfl is input-driven iff there is a 1npda A recognizing it such that the input symbols alone determines whether the move is push or a pop. Such languages can be recognized in O(log n) space on Turing machine, this implies a log n parallel time recognition. However our construction reduces the number of processors and makes parallel algorithms directely. Alternating multihead automata give another example of the recursion of the form (**).

Literature

[1] A.Aho,J.Hopcroft,J.Ullman.Time and tape complexity of push-down automaton languages. Inf.and Contr. 13 (1968)

[2] R.Bird. Improving programs by the introduction of recursion CACM 20:11 (1977)

[3] R.Bird. Tabulation Techniques for recursive programs. ACM Comp.Surveys 12:4 (1980)

[4] S.Cook. Linear time simulation of deterministic two way pushdown automata. IFIP Congress (1971)

[5] S.Cook. Characterizations of pushdown machines in terms of time bounded computers. JACM 18 (1971)

[6] S.Cook. Path systems and language recognition. STOC (1970)

[7] Z.Galil. Some open problems in the theory of computations as questions about two way pushdown automata. Math.Syst.Th. 10(1977)

[8] Z.Galil. Two fast simulations which imply some fast string matching and palindrom recognition algorithms. I.P.L. 4(1976)

[9] L.M.Goldschlager. A universal interconnection pattern for parallel computers. JACM (1982)

[10] J.Gray,M.Harrison,O.Ibarra. Two.way pushdown automata. Inf.and Control 11 (1967)

[11] S.Greibach. The hardest context free language. SIAM J.C. (1973)

[12] E.Gurari,O.Ibarra. Path systems: constructions, solutions, applications. SIAM J.C. 9 (1980)

[13] M.Harrison,O.Ibarra. Multihead and multitape pushdown automata. Inf.and Contr. 13 (1968)

[14] O.Ibarra. Characterizations of some time and tape complexities in terms of multihead and stack automata. JCSS (1971)

[15] D.Knuth, J.Morris,V.Pratt. Fast pattern matching in strings SIAM J.C. 6 (1977)

[16] K.Mehlhorn. Pebbling mountain ranges and its application to dcfl recognition. ICALP (1980)

[17] J.Reif. Parallel time O(log n) acceptance of deterministic dcfl's. FOCS (1982)

[18] W.Ruzzo. Tree-size bounded alternation. JACM 21 (1980)

[19] W.Rytter. An effective simulation of deterministic pushdown automata with many two way and one way heads. Inf.Proc.

Letters 12 (1981)

[20] W.Rytter. Time complexity of unambiguous path systems. Inf. Proc. Lett. 15:3 (1982)

[21] W.Rytter. A simulation result for two way pushdown automata. Inf.Proc.Let. 16 (1983)

[22] W.Rytter. The dynamic simulation of recursive and stack manipulating programs. I.P .L. 13:2 (1981)

[23] W.Rytter. A note on two way nondeterministic pushdown automata. I.P.L. 15:1 (1982)

[24] W.Rytter. Time complexity of loop free two way pushdown automata. I.P.L. 16 (1983)

[25] W.Rytter. A hardest language recognized by two way nondeterministic pushdown automata. I.P.L. 13:5 (1981)

[26] W.Rytter. Remarks on the pyramidal structure. CAAP (1983) Lect.Notes in Comp.Sc. 159, Springer-Verlag

[27] W.Rytter. Some properties of trace languages. Fundamenta Informaticae (1984)

[28] W.Rytter. Techniques for reducing space and time complexity of nondeterministic recursive algorithms. Techn.Report 209, IIMAS, Autonomic Nat. University of Mexico (1979)

[29] W.Rytter. A note on linear cfl's and one way multihead automata. I.P.L., to appear.

[30] W.Rytter. Fast recognition of pushdown automaton and context free languages. MFCS (1984), Lect.Notes in Comp.Sc. Springer Verlag-

[31] L.Valiant,S.Skyum. Fast parallel computations of polynomials using few processors. MFCS (1981)

Context-Free Grammars and Random Number Generation*

Andrew C. Yao
Computer Science Department
Stanford University
Stanford, California 94305

1. Introduction.

In Monte Carlo calculations, one often needs to generate a random quantity X that satisfies certain (cumulative) *distribution* function $F(x)$, i.e. $Pr\{X \leq x\} = F(x)$. Numerous methods have been proposed for this purpose (see Ahrens and Dieter [1], Knuth [4]). An interesting question is: for a given $F(x)$, how difficult is it to generate this distribution?

A complexity theory for generating such nonuniform random numbers was developed in Knuth and Yao [5], in which an algorithm takes an infinite stream of unbiased random bits and outputs the binary representation of X in a bitwise fashion. Two complexity measures were considered: the average number of input bits needed to output k bits of X, and the complexity of program necessary for this purpose. Most of the results [5] were directed at the former complexity measure, with the surprising conclusion that, for any distribution function $F(x)$, there exists an algorithm that will generate X using an average less than $k+4$ input bits for k output bits; this means that, according to this measure, every distribution is easy to generate. It is thus especially interesting to consider the complexity question using the latter complexity measure. A first step was taken in this direction in [5]. The purpose of this paper is to further explore this area.

For simplicity, we restrict ourselves to distribution functions with $F(0) = 0$ and $F(1) = 1$, i.e. X takes on only values in $[0, 1]$. In Knuth and Yao [5], the question "What distribution functions can be generated by algorithms using bounded amount of memories?" was studied. It was shown that, although many distributions can be generated this way, the only such *analytic* distributions are polynomial with rational coefficients. Two problems were left open: (1) What is the class of polynomials that can be generated by such algorithms? It was observed that polynomials $F(x)$ with irrational roots in $F'(x) = 0$ seem to be hard to generate this way. (2) Can more variety of distributions be generated, if a *stack* is made available? We will give partial solutions to these questions.

Most proofs shall be omitted. We give an outline of proof for one of the theorems; complete proofs can be found in a forthcoming technical report [7].

2. Terminology.

Definition 1. A *finite-state generator (fsg)* G consists a finite set of *states* $S = \{S_0, S_1, \ldots, S_m\}$, among them the *initial state* S_0, and a *transition function* $\delta : \{H,T\} \times S \mapsto \{0,1\} * \times S$.

Given any infinte input sequence of unbiased bits $\sigma = \sigma_0\sigma_1\sigma_2 \ldots \in \{H,T\}^\infty$, G starts

* This research was supported in part by National Science Foundation under grant MCS-83-08109.

NATO ASI Series, Vol. F12
Combinatorial Algorithms on Words
Edited by A. Apostolico and Z. Galil

in S_0, successively goes through states $S_{i_0}, S_{i_1}, S_{i_2}, \ldots$ and outputs a sequence of strings $\alpha_1, \alpha_2, \ldots$ according to the transition function; precisely, $S_{i_0} = S_0$ and $(\alpha_{j+1}, S_{i_{j+1}}) = \delta(\sigma_j, S_{i_j})$ for $j \geq 0$. Note that some of the α_i may be the empty string λ. Let $X_G(\sigma)$ denote the real number $0.\alpha_1\alpha_2\alpha_3\cdots$ if the sequence of α's contains infinitely many nonempty α_n; otherwise $X_G(\sigma)$ is undefined.

Definition 2. An fsg G is said to *generate a distribution function* F if, for a random unbiased sequence $\sigma \in \{H, T\}^\infty$, the probability that $X_G(\sigma)$ is undefined is equal to zero and $Pr\{X_G(\sigma) \leq x\} = F(x)$ for all x.

A simple example is given by the fsg with $S = \{S_0\}$ and $\delta(H, S_0) = (0, S_0), \delta(T, S_0) = (1, S_0)$, which generates the uniform distribution $F(x) = x$. In [5], it was demonstrated how a large class of polynomial distribution functions $F(x)$ can be generated by fsg's. On the other hand, it was also proved that not much more can be generated if we are interested in piecewise analytic $F(x)$. Recall that a function is *analytic* on an interval $[a, b]$ if for every point $x \in [a, b]$ the function has a convergent Taylor's series expansion valid in a sufficiently small neighborhood of x.

Theorem 0. (Knuth and Yao [5]). Let $F(x)$ be a distribution function that is analytic on some interval $[a, b] \subseteq [0, 1]$. If $F(x)$ can be generated by some fsg, then $F(x) = Q(x)$ for all $x \in [a, b]$ for some polynomial $Q(x)$ with only rational coefficients.

We will now define the concept of a *pushown-stack generator*. It is a slight modification of a standard pushdown automaton [3], in the same way that an fsg is a variant of a finite automaton. We will only describe it informally here.

Definition 3. A *pushdown-stack generator (psg)* P is a deterministic pushdown automaton with input symbols $\{H, T\}$. At each transition, depending on the current state, the current input symbol and the current stack symbol, P will replace the current stack symbol with a finite (may be empty) string of stack symbols and outputs a string $\alpha \in \{0, 1\}*$ (may be the empty string). The psg P starts in some fixed initial state, with the stack containing a special stack symbol B which will always remain in the bottom of the stack and will never appear on other positions in the stack throughout the computation.

As before, we will use $X_P(\sigma)$ to denote the output real number in $[0, 1]$ corresponding to an input sequence $\sigma \in \{H, T\}^\infty$. The psg P is said to *generate* $F(x)$ if, for all x, $Pr\{X_P(\sigma) \leq x\} = F(x)$ for a random $\sigma \in \{H, T\}^\infty$.

3. Results.

Theorem 1. There are polynomial distribution functions $F(x)$ that can be generated by a psg, but not by any fsg.

For example, the following function can be shown to satisfy the theorem: $F(x) = ax + (1 - a)x^2$, where $a = 4 - 2\sqrt{3}$. Thus, pushdown-stack generators can generate some polynomials with algebraic irrational coefficients. The next result shows that, however, these are about all the analytic distributions that they can generate.

Theorem 2. Let $F(x)$ be a distribution function that is analytic on some interval $[a, b] \subseteq [0, 1]$. If $F(x)$ can be generated by some pushdown-stack generator, then $F(x) = Q(x)$ for all $x \in [a, b]$ for some polynomial $Q(x)$ with only algebraic coefficients.

The third theorem gives further constraints on the type of probability distributions that can be generated by finite-state generators. As a corallory, polynomial distributions $F(x)$ with irrational roots in their probability density functions (e.g., when $F'(x) = \text{constant} \times (2x^2 - 1)^2$) cannot be generated by any fsg.

Let $h(x)$ be any function on $[0,1]$. The *kernel* $Z(h)$ of h is defined to be the set of zeroes for $h(x) = 0$. For any function $f(x)$ on $[0,1]$, and any integers $k \geq 0$ and $0 \leq j < 2^k$, let $f_{k,j}(x) = f((j+x)/2^k)$ for $x \in [0,1]$. Informally, $f_{k,j}$ is the rescaled function f when restricted to the interval $[j/2^k, (j+1)/2^k]$.

Definition 4. For any function $f(x)$ on $[0,1]$, let $\gamma(f)$ be the number of distinct kernels $Z(f_{k,j})$ for all k, j; $\gamma(f)$ may be ∞.

For example, for any $0 < r < 1$, define a function $\ell_r(x) = 1/r$ for $x \in [0, r)$ and $\ell_r(x) = 0$ for all other x. Write $r = 0.a_1a_2a_3\cdots$ and $r_k = 0.a_ka_{k+1}a_{k+2}\cdots$. It is easy to see that the family of kernels $Z(f_{k,j})$ consists of the sets ϕ, $[0,1]$, $[r_1, 1]$, $[r_2, 1]$, $[r_3, 1]$, ..., from which it follows that $\gamma(\ell_r)$ is finite if and only if r is rational.

Theorem 3. Let $F(x)$ be a differentiable distribution function on $[0,1]$. If $F(x)$ can be generated by some fsg, then $\gamma(F') < \infty$.

Corallory. If $F(x)$ is a polynomial distribution, and $F'(x) = 0$ has some irrational root in $[0,1]$, then $F(x)$ cannot be generated by any fsg.

4. A Sketch Of Proof For Theorem 2.

To simplify the discussion, we will assume that $F(x)$ is analytic on $[0,1]$, and that the the Taylor's series expansion of $F(x)$ at $x = 0$ has a radius of convergence greater than 1. We will discuss some auxiliary results in Section 4.1 and 4.2, and apply them to complete the proof in Section 4.3.

4.1. Discrete Stochastic Grammar.

A *discrete stochastic grammar* H is a stochastic context-free grammar (see e.g. [2]) with one terminal symbol 0. Instead of giving a formal definition, we will illustrate it with an example. Consider the following system with nonterminals A, C, with *start symbol* A:

$$A \rightarrow \frac{1}{9}0AAC + \frac{2}{5}00 + \frac{22}{45}\lambda$$

$$C \rightarrow \frac{1}{12}CC + \frac{1}{12}0A + \frac{1}{2}\lambda + \frac{1}{3}0,$$

where λ denotes the empty string. If we substitute these production formulas into the right-hand side of themselves, and use distributive laws, we obtain a system of the same general form, i.e. $A \rightarrow \cdots, C \rightarrow \cdots$. Iterate this process again and again, this system converges to a system of the form

$$A \rightarrow p_0 + p_10 + p_200 + p_3000 + \cdots + p_n0^n + \cdots$$

$$C \rightarrow q_0 + q_10 + q_200 + q_3000 + \cdots + q_n0^n + \cdots.$$

We will say that this grammar generates n 0's *with probability* p_n. Let us call the sequence $p_0, p_1, \ldots$ the *associated sequence of* H, and define the quantity $V_H(x) = \sum_{n\geq 0} p_n x^n$.

Lemma 1. For any discrete stochastic grammar H, the function $V_H(x)$ is an algebraic function.

4.2. Discrete Pushdown-Stack Generator.

A psg P is called a *discrete psg* if (1) the output at each transition is a string of 0's (may be empty), and (2) for a random input sequence $\sigma \in \{H,T\}^\infty$, P enters eventually with probability 1 a designated state, called *dead state*, after which P stays in this state and outputs only empty strings. It is convenient to think of P as generating a discrete probability distribution $p_0, p_1, p_2, \ldots$, where p_n is the probability that P will output n 0's before it enters into the dead state. Let $Q_P(x) = \sum_{n\geq 0} p_n x^n$ be the generating function of this probability distribution.

Lemma 2. Let P be any discrete psg. Then there exists a discrete stochastic grammar H whose associated sequence $p_0, p_1, p_2, \ldots$ is identical to the discrete probability distribution generated by P.

The following lemma follows immediately from Lemmas 1 and 2.

Lemma 3. Let P be any discrete psg. Then $Q_P(x)$ is an algebraic function.

4.3. Completing The Proof.

Let P be a psg that generates $F(x)$. We modify P to yield a discrete psg D as follows: Create a new state R to be the dead state; for any input sequence $\sigma \in \{H,T\}^\infty$, D will operate exactly as P until the first time a string containg a 1, $\alpha = 0^s 1 \beta$, is to be output, at which time D will instead output 0^s and enter the dead state R. It is easy to verify that D is indeed a discrete psg, using the fact that the probability for P to output the all 0 infinite string is zero.

Let $p_0, p_1, p_2, \ldots$ be the discrete probabilty distribution generated by D. Clearly, $p_n = F(1/2^n) - F(1/2^{n+1})$. Let $F(z) = \sum_{i\geq 0} a_i z^i$ be the Taylor's series for $F(z)$ at $z = 0$. Recall that we have assumed that the radius of convergence is greater than 1. Then we have for all $x < 1/2$,

$$\begin{aligned} Q_D(x) &= \sum_{n\geq 0} p_n x^n \\ &= \sum_{n\geq 0} [F(1/2^n) - F(1/2^{n+1})] x^n \\ &= \sum_{n,i} a_i (1 - \frac{1}{2^i})(\frac{x}{2^i})^n \\ &= \sum_i a_i \frac{2^i - 1}{2^i - x}. \end{aligned}$$

By analytic continuation, $Q_D(x)$ has poles at $x = 2^i$ for all nonvanishing a_i. Since

$Q_D(x)$ is an algebraic function by Lemma 3, it has only a finite number of poles, and hence there can be no more than a finite number of nonvanishing a_i's. This proves that $F(x)$ is a polynomial.

5. Remarks.

In this paper we have studied the problem of generating nonuniform random numbers in a formal model. Technically, it is a study of the analytic properties of some formal language systems, and requires methods drawn from both classical analysis and automata theory. We have found the power-series approach to formal language (see e.g. Salomaa and Soittola [6]) useful in some of the proofs. It is likely that other apparatus developed for language theory will also find use in the further study of this subject.

References.

[1] Ahrens and Dieter, *Non-Uniform Random Numbers*, Wiley, New York, to appear.

[2] R. C. Gonzalez and M. G. Thomason, *Syntactic Pattern Recognition*, Addison-Wesley, Reading, Massachusetts, 1978.

[3] J. E. Hopcroft and J. D. Ullman, *Introductin to Automata Theory, Languages, and Computation*, Addison-Wesley, Reading, Massachusetts, 1979.

[4] D. E. Knuth, *The Art of Computer Programming, Vol.2*, Addison-Wesley, Reading, Massachusetts, Second Edition, 1981.

[5] D. E. Knuth and A. C. Yao, "The complexity of nonuniform random number generation," in *Algorithms and Complexity: New Directions and Recent Results*, edited by J.F.Traub, Academic Press, New York, 1976, pp. 357-428.

[6] Arto Salomaa and Matti Soittola, *Automata-Theoretic Aspects of Formal Power Series*, Springer-Verlag, New York, 1978.

[7] A. C. Yao, *On generating nonuniform random numbers with a pushdown stack*, Stanford Computer Science Department Technical Report, to appear, 1985.

NATO ASI Series F

Vol. 1: Issues in Acoustic Signal – Image Processing and Recognition. Edited by C. H. Chen. VIII, 333 pages. 1983.

Vol. 2: Image Sequence Processing and Dynamic Scene Analyisi. Edited by T. S. Huang. IX, 749 pages. 1983.

Vol. 3: Electronic Systems Effectiveness and Life Cycle Costing. Edited by J. K. Skwirzynski. XVII, 732 pages. 1983.

Vol. 4: Pictorial Data Analysis. Edited by R. M. Haralick. VIII, 468 pages. 1983.

Vol. 5: International Calibration Study of Traffic Conflict Techniques. Edited by E. Asmussen VII, 229 pages. 1984.

Vol. 6: Information Technology and the Computer Network. Edited by K. G. Beauchamp. VIII, 271 pages. 1984.

Vol. 7: High-Speed Computation. Edited by J. S. Kowalik. IX, 441 pages. 1984.

Vol. 8: Program Transformation and Programming Environments. Report on an Workshop directed by F. L. Bauer and H. Remus. Edited by P. Pepper. XIV, 378 pages. 1984.

Vol. 9: Computer Aided Analysis and Optimization of Mechanical System Dynamics. Edited by E. J. Haug. XXII, 700 pages. 1984.

Vol. 10: Simulation and Model-Based Methodologies: An Integrative View. Edited by T. I. Ören, B. P. Zeigler, M. S. Elzas. XIII, 651 pages. 1984.

Vol. 11: Robotics and Artificial Intelligence. Edited by M. Brady, L. A. Gerhardt, H. F. Davidson. XVII, 693 pages. 1984.

Vol. 12: Combinatorial Algorithms on Words. Edited by A. Apostolico, Z. Galil. VIII, 361 pages. 1985.